Society of Earth Scientists Series

Series Editor

Satish C. Tripathi, Lucknow, India

The Society of Earth Scientists Series aims to publish selected conference proceedings, monographs, edited topical books/text books by leading scientists and experts in the field of geophysics, geology, atmospheric and environmental science, meteorology and oceanography as Special Publications of The Society of Earth Scientists. The objective is to highlight recent multidisciplinary scientific research and to strengthen the scientific literature related to Earth Sciences. Quality scientific contributions from all across the Globe are invited for publication under this series. Series Editor: Dr. Satish C. Tripathi

More information about this series at http://www.springer.com/series/8785

Tapas Kumar Biswal · Sumit Kumar Ray ·
Bernhard Grasemann
Editors

Structural Geometry of Mobile Belts of the Indian Subcontinent

The Society of Earth Scientists
2008

Springer

Editors
Tapas Kumar Biswal
Department of Earth Sciences
Indian Institute of Technology Bombay
Mumbai, India

Sumit Kumar Ray
Geological Survey of India (Retd)
Kolkata, India

Bernhard Grasemann
Department für Geodynamik
und Sedimentologie
Universität Wien
Vienna, Austria

ISSN 2194-9204 ISSN 2194-9212 (electronic)
Society of Earth Scientists Series
ISBN 978-3-030-40592-2 ISBN 978-3-030-40593-9 (eBook)
https://doi.org/10.1007/978-3-030-40593-9

This Springer imprint is published by the registered company Springer Nature Switzerland AG
The registered company address is: Gewerbestrasse 11, 6330 Cham, Switzerland

Foreword

The tectonic framework of Indian subcontinent is a result of accretion of continental terranes since mid-Archean, and these crustal blocks are bounded by large-scale mobile belts. These surrounding mobile belts are either high-strain, high-grade metamorphic belts or folded basins. Thus, the relatively rigid cratons are surrounded by more ductile zones. The mobile belts have experienced episodes of tectonic re-activation. The study of structural geometry of these mobile belts suggests the tectonic evolution through geological past. The Aravalli–Delhi Mobile Belt, Central Indian Tectonic Zone, Eastern Ghats Mobile Belt, and Himalayan Mobile Belt invited attention of structural geologists over a century, and there are several new dimensions added to their tectonic evolution. The book addresses several such important aspects of structural geometry of mobile belts.

The mega-event of 36th International Geological Congress 2020 in India opened a new chapter on the geology of India. On such an occasion, Society of Earth Scientists Series by Springer decided to bring out 36th IGC Commemorative Volumes on various recent geological and geophysical studies of India. Several well known geoscientists were requested to prepare comprehensive account as monographs or edited volumes. I am personally thankful to all the editors and authors for timely submission of high-quality manuscripts for inviting interest of global community of geoscientists.

Lucknow, India

Satish C. Tripathi

Preface

We express our sincere thanks to Dr. Satish C. Tripathi, General Secretary, The Society of Earth Scientists, Lucknow, India, for inviting us to edit a book on the occasion of 36th International Geological Congress to be held at New Delhi, between February 3 and 8, 2020. We proposed the book entitled *Structural Geometry of Mobile Belts of the Indian Subcontinent* with a vision to bring together important contributions on deformational structures of the mobile belts of Peninsula as well as Extra-Peninsula and their implication on the geodynamic evolution of orogens.

The contributions include papers from the NW to the NE part of the Himalaya, Bangladesh, and Andaman and Nicobar Islands: Ikram et al. showed, by means of structural modeling, a strong shortening of the Kohat fold and thrust belt and emphasized the important role of mechanical heterogeneity to the formations of the fold thrust belt. Banerjee and Srivastava documented NNE-SSW-directed progressive shortening across Main Central Thrust of Garhwal Himalaya by investigation of pre-, syn-, and post-thrusting mesoscopic and microscopic structures. The geological complexities of thrust sheets, tectonic windows, and intermontane basins in the Nepal Himalaya is outlined by Dhital and Adhikari. The tectonic development of the Bengal basin has been investigated by Hossain et al., showing the complex interplay among the Himalayan orogen to the north, the Indo-Burma orogen to the east, and stable Indian Craton to the west. Pal provides a broad view on the tectonic evolution of the Himalaya studying the structural complexities of ophiolite belts of Indo-Burma rage and Andaman and Nicobar Islands.

The Precambrian mobile belts of Peninsular India developed during different periods of global orogeny. Investigations by Singh et al. of multiple stages of folding, westerly verging thrust belts, and the timing of syn-tectonic granite intrusion better constrain the age of the South Delhi orogeny. Gregor et al. studied the tectonometamorphic evolution of the Aravalli Supergroup of the Aravalli–Delhi mobile belt and concluded that the pattern of bivergent thrusts with the Neoproterozoic active continental margin along the western side of the Delhi Fold Belt and the lack of any magmatic signature along the eastern proximity can be interpreted as a retro-wedge setting. Anisotropy of magnetic susceptibility studies on quartzites of the Champaner Group, Gujarat, Western India, by Joshi and Limaye suggest a prolong emplacement of Godhra Granites after syn-tectonic pulse and regional deformation. The Central India Tectonic Zone, that sutures the Bundelkhand and Dharwar protocontinent, bears the signature of multiple orogenic cycles outlined by the structural studies of by Mohanty. Juxtaposition of Eastern Ghats Mobile Belt with Singhbhum Craton along a dextral strike-slip fault was studied by Gupta et al. by mesoscopic and microstructural investigations. The Southern Granulite terrane is marked by several shear zones and most importantly the pseudotachylite-bearing Gangavalli fault, which has been studied by Behera et al. using geochemistry, shear resistance, and melt pressure as a proxy for earthquake source properties.

We are extremely thankful to all our reviewers who spent their valuable time with careful reviews to maintain higher standard of publication. These include S. K. Acharya, A. Carter, A. B. Roy, S. Sengupa, B. Kundu, N. Dasgupta, V. Prasannakumar, Bernhard Grasemann, Tapas Kumar Biswal, Sumit Kumar Ray, S. J. Sangode, D. Mukherjee, O. Bhagrav, R. Carosi, J. K. Nanda, and S. Nielsen.

The edited volume has included all recent works on the mobile belts of the Indian Subcontinent which may find use with the young researchers for further progress in research on Structural Geology and Tectonics. We are extremely thankful to Springer for bringing out the volume on this mega-event of IGC 2020.

Mumbai, India — Tapas Kumar Biswal
Kolkata, India — Sumit Kumar Ray
Vienna, Austria — Bernhard Grasemann

Contents

About the Editors

Tapas Kumar Biswal obtained his M.Sc. from IIT, Kharagpur, India, in 1979 and Ph.D. from Rajasthan University, Jaipur, India, in 1989. He served as a geologist in Geological Survey of India from 1981 to 1994. In 1994, he joined as an assistant professor and later promoted to professor at the Department of Earth Sciences, Indian Institute of Technology Bombay, Mumbai, India. He is specialized in structural geology and tectonics. His current research interest focuses on the analysis of meso- and microscopic strain fabrics of fold belts, ductile shear zones, and faults. He works in the Eastern Ghats Mobile Belt, the Aravalli–Delhi Mobile Belt, and the Southern Granulite Terrane of Peninsular India. He has a great interest in training the students in the field. He received several times Excellence in Teaching award by IIT Bombay. All his class lectures are available on YouTube.

Sumit Kumar Ray studied geology at the Jadavpur University, Kolkata, from 1960 to 1965, and then obtained Ph.D. from IIT, Kharagpur, in 1970. He joined the Geological Survey of India in 1968, and served the department for about 37 years. His knowledge of rock deformation, structure, and ore geology is based on extensive fieldwork carried out in different sectors of the Himalayas, and the Precambrian terrain of Western India. He is currently pursuing research on thrust tectonics, co-seismic ground movement in the Andaman region, and microscopic scale rock deformation at frictional slip interfaces.

Bernhard Grasemann received his Ph.D. in geology from the University of Vienna. He carried out research under several international projects on the geodynamic evolution of orogens like the Himalayas, Zagros, or Hellenides. Since 2007, he has been Professor in geodynamics and tectonics at the University of Vienna. His scientific interests cover the evolution of mountain belts, metamorphic core complex formation, and the physical quantification of structural processes at various scales. Since 2012, Bernhard is the corresponding member of the Austrian Academy of Sciences and has been elected as President of the Austrian National Committee of Geosciences in 2019.

Timing of South Delhi Orogeny: Interpretation from Structural Fabric and Granite Geochronology, Beawar-Rupnagar-Babra Area, Rajasthan, NW India

Subhash Singh, Anjali Shukla, B. H. Umasankar and Tapas Kumar Biswal

Abstract The South Delhi orogeny is constrained by a correlative study between the deformational fabric and geochronology of the metarhyolite and granite of the South Delhi terrane (*SDT*) around Beawar- Rupnagar-Babra, Rajasthan, NW India. The *SDT* contains metaconglomerate (Bar conglomerate), calcareous schist, mica schist, amphibolite, metarhyolite (G_1) and three phases of granites ($G_{2\text{-}4}$), which underwent three stages of deformations ($D_{1\text{-}3}$). The D_1 produced F_1 reclined/recumbent folds and S_1 fabric in a greenschist facies metamorphic conditions. The D_2 produced F_2 folds which were coaxial with F_1 along NE-SW axis, S_2 crenulations and ductile shear zones. The D_3 produced NW-SE trending F_3 folds. The S_1 fabric in the metarhyolite (G_1) and, G_2 and G_3 granites differs from that of the G_4 granite. It shows low temperature of formation in the former ($G_{1\text{-}3}$), characterized by recrystallization of quartz by bulging and development of biotite, muscovite, epidote. In the later (G_4), it is characterized by high temperature of formation as indicated by dynamic recrystallization of the plagioclase by subgrain rotation and grain boundary migration. Based on these differences, we interpreted that the intrusion of the G_4 is syntectonic with the D_1 deformation while the metaryolite (G_1), G_2 and G_3 granites were pretectonic to the D_1. The G_4 (Sewariya granite) was dated previously at 860 Ma that constrains the age D_1 deformation and South Delhi orogeny. The G_1, G_2 and G_3 (Sendra granite equivalent) were earlier dated at ca. 970 Ma which probably indicates age of rifting.

Keywords Deformation fabric · Metarhyolite · Granite · South Delhi terrane · 860 Ma South Delhi orogeny

---

S. Singh (✉) · A. Shukla · B. H. Umasankar · T. K. Biswal
Department of Earth Sciences, Indian Institute of Technology Bombay, Powai, Mumbai, Maharashtra 400076, India

T. K. Biswal
e-mail: tkbiswal@iitb.ac.in

T. K. Biswal et al. (eds.), *Structural Geometry of Mobile Belts of the Indian Subcontinent*, Society of Earth Scientists Series,
https://doi.org/10.1007/978-3-030-40593-9_1

1 Introduction

Like Phanerozoic orogens, the Precambrian mobile belts/orogens, are characterized by multiple phases of folding, shearing, metamorphism and intrusion of magmatic rocks. Therefore, the Precambrian tectonics were visualized in terms of Phanerozoic—Wilson cycle comprising of rifting, formation of passive margins, closing of the ocean by subduction/collision and creation of mountain belt. Plate tectonics operated since late Neoarchean when lithosphere acquired appreciable thickness (ca. 100 km). Several major global orogenies have been identified that were responsible for creation of supercontinents. The orogenies are mainly Nuna orogeny (ca. 2.1–1.8 Ga, Rogers and Santosh 2002; Zhao et al. 2002), Grenvillian orogeny (ca. 1.3–1.0 Ga; Valentine and Moores 1970; McMenamin and McMenamin 1990; Meert and Torsvik 2003; Cawood 2005; Li et al. 2008) and Pan-African orogeny (Stern 1994; Kröner and Stern 2005; Fritz et al. 2013; Oriolo et al. 2017). The Nuna orogeny created Columbia, the Grenvillian orogeny created Rodinia and the Pan-African orogeny created Gondwanaland Supercontinents. Each orogeny consists of several phases of subduction and arc accretion that vary temporally and spatially at different parts of the Supercontinent. As for example the Pan-African orogeny consists of an earlier phase marked by subduction and arc-accretion (870–630 Ma) which was present in Arabian-Nubian shield (Kröner and Stern 2005) and a late phase collision (ca. 650–500 Ma) which was present in Madagascar, Tanzania and other parts of Africa (Meert and Lieberman 2008; Lehmann et al. 2016).

A number of geochronometers are used to date orogenic events that include folding, shearing, metamorphism and magmatic intrusion. Granites are dated by zircon geochronology because of higher closure temperature of zircon to U-Th-Pb system (>900 °C; Dahl 1997; Cherniak and Watson 2001). An orogen consists of several phases of granite intrusions namely rift related granites that intrude during sedimentation in a basin, granite produced by melting or metasomatism of meta-sediments at amphibolite to granulite facies metamorphism, granite intruded during late phase extensional tectonics that may be related to exhumation and others. The age of the granite broadly constrain the major tectonic events. For further resolution of the major events, it is essential to correlate the rock fabric with growth of monazite, because monazite crystallizes in response to fluid activity at different period of shearing (Williams and Jercinovic 2002; Pyle and Spear 2003; Foster et al. 2004; Mahan et al. 2006). ^{40}Ar-^{39}Ar dating constrains the age of cooling or crystallization events of low temperature minerals (Harrison et al. 1985). Thus multiple geochronometers are applied to constrain the tectonic evolution of an orogen.

In this paper one of Precambrian orogens in the Peninsular India has been studied namely the Archean-Neoproterozoic Aravalli-Delhi mobile belt (*ADMB*) of NW India (Fig. 1a). Structural mapping and microscopic studies have been carried out to establish the time-relationship between magmatism and deformation fabric of the northern part of the South Delhi terrene (*SDT*) of the above mobile belt to constrain the South Delhi orogeny.

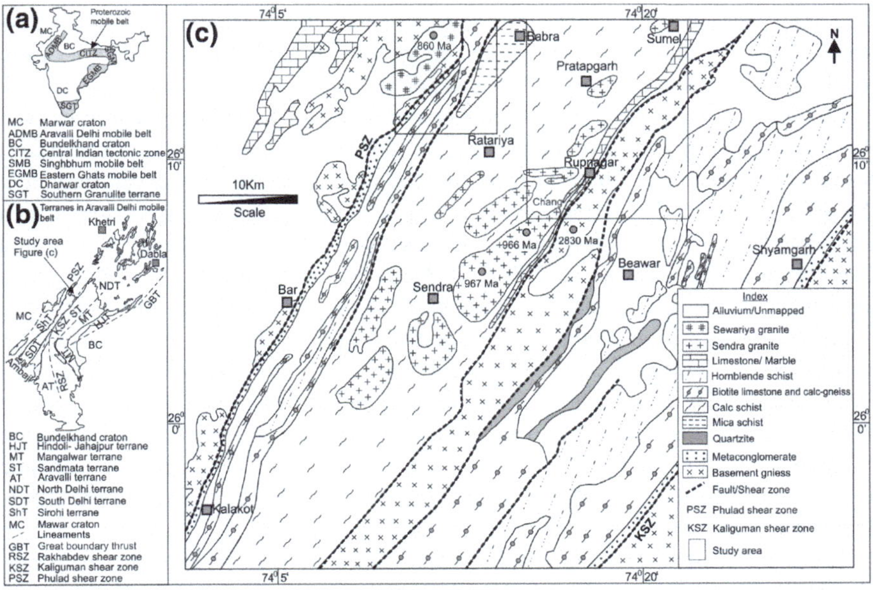

Fig. 1 **a** Precambrian mobile belt and craton in Peninsular India, **b** the *ADMB* with different terranes and shear zones (Singh et al. 2010), **c** geological map modified after Heron (1953) and Gupta et al. (1995). The ages of Sendra and Sewariya granites are after Tobisch et al. (1994), Pandit et al. (2003), Sivasubramaniam et al. (2019)

2 Geological Setting

Our knowledge on geology of the *ADMB* owes a lot to Heron's work (Heron 1953) that envisaged the three fold stratigraphic classification namely Banded Gneissic Complex, Aravalli system and Delhi system. Subsequently, extensive input of geochronological data helped in revising the stratigraphy and tectonic settings. The present understanding is that the *ADMB* consists of collage of terranes juxtaposed along shear zones and marked by contrasting tectonic history (Fig. 1b, Gupta et al. 1995; Roy and Jakhar 2002; Singh et al. 2010). The Archean rocks in the *ADMB* are represented by the Mangalwar-Sandmata-Hindoli-Jahazpur terranes which were grouped under Bhilwara Supergroup (Gupta et al. 1995). The Mangalwar terrane and Sandmata terrane, earlier referred to as Banded Gneissic Complex (BGC) (Heron 1953), consist of tonalite-trondjhemite-granodiorite gneisses of 3.3–2.8 Ga age (Kaur et al. 2019 and reference therein). The Hindoli-Jahazpur terrane represents Archean greenstone belts (Gupta et al. 1995; Sinha-Roy et al. 1998). The Archean rocks went through Bhilwara orogenic cycle that ended with intrusion of Berach granite at 2.6 Ga.

The Paleoproterozoic rocks of the *ADMB* occur in the Aravalli and North Delhi terranes. The Aravalli terrane comprises shallow water stromatolites bearing rocks in the east, and carbonate and pelitic rocks in the west, deposited over 2.5 Ga old continental to oceanic crust, with conglomerate and palaeosol at the base (Pandit et al.

2008; De Wall et al. 2012). The terrane bears syn-tectonic granite intrusion at 2.0–1.8 Ga. The Aravalli rocks were deformed by E-W folds that make it distinct from North Delhi and South Delhi rocks (Naha et al. 1984). The northern margin of the Aravalli terrane was migmatised and appears similar with the Mangalwar terrane. The North Delhi terrane consists of quartzite, calc-schist and mica-schist deposited in fault bounded sub-basins at Khetri, Alwar and Lalsot-Bayana, over Archean basement (Mehdi et al. 2015). The terrane contains several rift related granite intrusions which were dated at ca. 1711–1854 Ma. Another set of granites indicate arc-setting and dated at 1820 Ma (Kaur et al. 2006, 2013). The rocks in the North Delhi terrane were metamorphosed at 952 Ma (Pant et al. 2008). The Aravalli orogenic cycle closed with subduction along Rakhabdev shear zone at 1.7 Ga, where ophiolites were emplaced (Deb et al. 1989; Verma and Greiling 1995). The Aravalli orogeny is further marked by the intrusion of several granulite and charnockite plutons within Sandmata terrane at ca. 1.7 Ga, which were later metamorphosed at high pressure granulite facies during 0.9 Ga (Bhowmik et al. 2010); and rhyolite and volcanic tuffs intrusion in the Hindoli-Jahazpur terrane at 1.8 Ga (Deb and Thorpe 2001) representing arc related magmatism (Raza and Siddiqui 2012).

The South Delhi (*SDT*) and Sirohi terranes contain Meso-Neoproterozoic rocks. The *SDT* shows diversity in lithological association from north to south. In the north, it is represented by low grade metamorphic rocks resting unconformably over gneissic-granulitic basement rocks of Magalwar and Sandmata terranes (Heron 1953). There occurs tectonic slices of basement rocks at Beawar and Anasagar (2.8 Ga, Tobisch et al. 1994). There are exotic granulitic outcrops at Pilwa-Chinwali which produces an age of 1.7–1.5 Ga to 1.0 Ga (Fareeduddin et al. 1994; Bhowmik et al. 2018). The *SDT*- metasediments were intruded by several granite plutons namely Sendra and Sewariya granites. The central part of the *SDT* is dominated by low grade metasediments and metavolcanics; the metavolcanics were interpreted as ophiolites (Gupta et al. 1995). The metasediments and metavolcanics were intruded by diorite (ca. 1.0 Ga) and granites (840 Ma) (Volpe and MacDougall 1990). The southern part of the *SDT* comprises quartzite and calcareous rocks intruded by several phases of Ambaji granite (G_0, ca. 960 Ma, during sedimentation; ca. 860 Ma, G_1, during peak metamorphism and folding; ca. 840 Ma, G_2, during shearing and F_2 folding; ca. 760 Ma, G_3, during brittle fracturing, Singh et al. 2010). Granulites are exposed near Ambaji (Biswal et al. 1998a) which were constrained at 860 Ma by Singh et al. (2010) through zircon geochronology and at 875–857 Ma by Tiwari and Biswal (2019a) through monazite geochronology. The Sirohi terrane occurs on the western trans-Aravalli plain and is represented by isolated pockets of low grade metasediments and large scale granite intrusion. The metasediments produced a zenocrystic zircon age of ca. 992 Ma (Purohit et al. 2012) and the granites vary in age from ca. 1.0 Ga to ca. 765 Ma (Sarkar et al. 1992; Meert and Lieberman 2008; Just et al. 2011; Dharma Rao et al. 2013; Meert et al. 2013; Ashwal et al. 2013). The *SDT* and Sirohi terrane together evolved through South Delhi orogeny that probably happened by westward subduction along Kaliguman shear zone (Sugden et al. 1990; Biswal et al. 1998b; Singh et al. 2010). Abundance of granitic intrusion towards west and formation of extensional basin over the granitic basement at Sindreth and

Punagarh support a westward subduction. The deformation in the *SDT* is characterized by NE-SW fold (Naha et al. 1984, 1987; Biswal 1988). The Sirohi terrane has undergone folding along NE-SW axis, thrusting, extensional tectonics (Roy and Sharma 1999; Sharma 2005; De Wall et al. 2014; Bhardwaj and Biswal 2019).

In the present study an attempt has been made for detailed structural analysis of metasediments and granitic intrusions in the northern part of the *SDT*, west of Beawar (Fig. 1b; N74° 19′ 8.91″ E26°5′ 60″). Previously published geochronological ages of the granites have been used to constrain the deformation in the *SDT*. The geological map of the area by Heron (1953) indicates exposure of the basement rocks with prominent erosional unconformity on either flank of the *SDT*, at Bar and close to Shyamgarh (Fig. 1c). Low grade metasediments comprising calcareous schist, mica schist, quartzite, biotite marble and amphibolite are exposed in two synclines (NW, SE synclines) which are separated by a slice of basement rocks immediately west of Beawar (age of the gneissic rocks, ca. 2.8 Ga, Tobisch et al. 1994). The Sendra granite (ca. 980 Ma, Tobisch et al. 1994; Pandit et al. 2003) and the Sewariya granite (ca. 860 Ma, Sivasubramaniam et al. 2019) intruded the belt. The Sendra granite occurs in the core of the NW syncline and the Sewariya granite occurs on the western flank close to Babra. There are several shear zones and faults such as Phulad shear zone in the west and Kaliguman shear zone in the east. We have mapped a part of NW syncline where Sendra and Sewariya granite plutons are exposed, having distinct intrusive contact with metasediments.

3 Methodology

The area was mapped on an enlarged scale (1:10,000) in parts of Survey of India toposheets, 45 J/3, 4, 7, 8 between Rupnagar and Babra (Fig. 2a, b). Structural data related to different generations of folding were measured and oriented samples from shear zones were collected. Different generations of folds were identified based on overprinting relationship of strain fabric and interference pattern of folds (Ramsay 1967). The area is divided into number of sectors depending on uniformity in the orientation of structural fabric. Separate stereoplots were prepared for each sub-area (Fig. 2c–i). Structural cross sections were drawn taking into consideration the structural data and rock types (Fig. 3a, b). Four shear zones were mapped namely SZ-I to SZ-IV from west to east; the SZ-I coincides with Phulad shear zone. Thin sections were prepared (XZ section, parallel to stretching lineation and perpendicular to foliation, the sections are mostly vertical) of the oriented samples collected from shear zones.

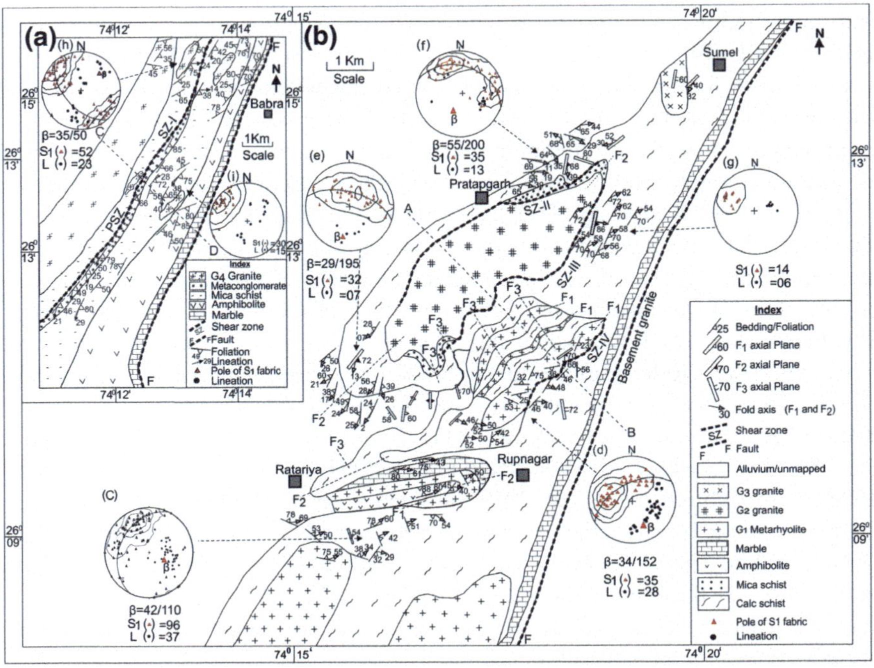

Fig. 2 **a** Structural map of the *SDT* west and southwest of Babra, **b** structural map of the *SDT* around Rupnagar, Pratapgarh and Ratariya, West of Beawar, **c**–**i** stereoplots of S_1 fabric and F_1 and F_2 fold axes (L), **c** β axis 42/110, contour 1-2-3%, **d** β axis 34/152, contour 1-2-3%, **e** β axis 29/195, contour 1-2-3%, **f** β axis 55/200, contour 1-2-3%, **g** contour: 1%, **h** β axis 35/50, contour 1-2-3%, **i** contour 1-2-3%

4 Result

4.1 *Lithology*

The study area is represented by greenschist facies metasedimentary rocks namely metaconglomerate, mica schist, calcareous schist and quartzite. Parallel bands of amphibolite and metarhyolite (G_1) occur within the metasediments. These rocks were intruded by three phases of granite G_2, G_3 and G_4 which are distinct in mineralogical assemblage and strain fabric (Figs. 2a, b and 3a, b). The G_4 is the youngest phase and it cross cut the other granites. The G_1, G_2 and G_3 may be syntectonic as cross cutting relationship amongst them is yet to be established. The metaconglomerate varies in outcrop width from meter to tens of meter and occurs within the shear zone SZ-I in the western part of the area near Babra (Fig. 1d). The metaconglomerate zone consists of several quartzite and granite gneiss pebbles set in mica schist and quartzite matrix (Fig. 4a). The quartzite bed within this zone has been stretched vertically to produce boudins that resemble a sedimentary conglomerate.

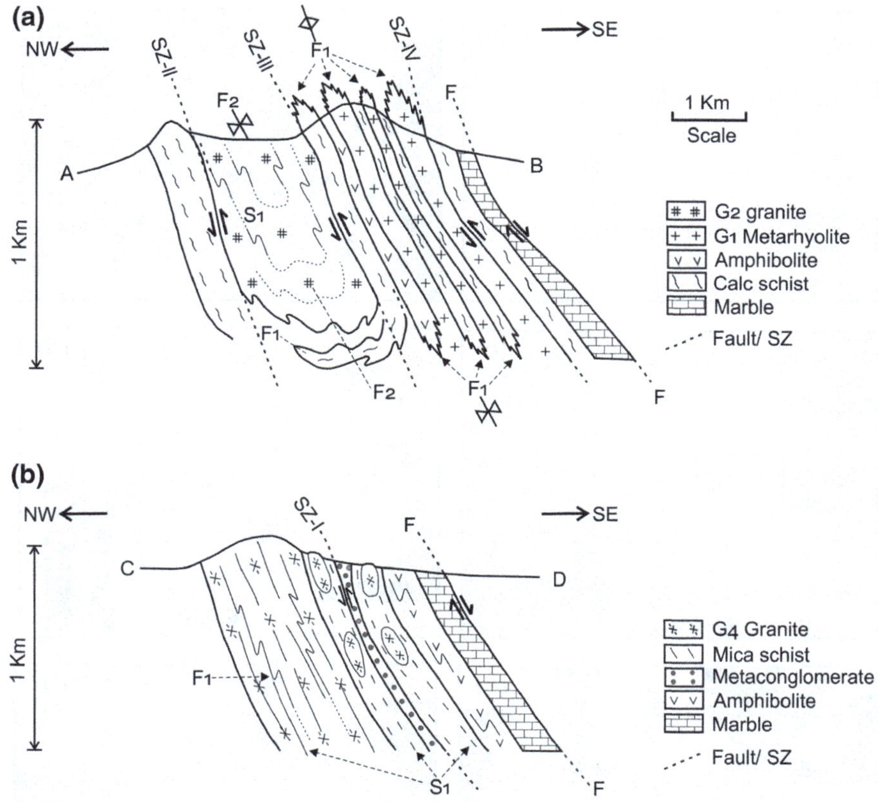

Fig. 3 Structural sections A-B and C-D (refer Fig. 2 for location of the cross sections). **a** A-B depicts alternate sequence of mica schist, calc schist, G_1 metarhyolite and amphibolite which are intruded by G_2 (and G_3, not seen on the section) granite. These rocks were folded by the F_1 and F_2 folds. The G_2 granite occurs in the core of a large scale F_2 fold. G_1 metarhyolite and amphibolite show series of F_1 antiform and synforms. The F_1 fold and S_1 fabric are refolded around the hinge of the F_2 fold. Shear zones SZ-II, SZ-III and SZ-IV (thrusts) and normal fault occur along this section. **b** C–D depicts a sequence of mica schist, amphibolite and marble with the metaconglomerate. The metaconglomerate is emplaced along shear zone SZ-I (thrust). High grade basement rocks are also emplaced along the thrust (see the text). The G_4 granite intrudes the rocks during F_1 folding. Granite and the surrounding host rocks bear the impression of F_1 fold and S_1 fabric

Those boudins have been referred here as tectonic conglomerate. Further, slivers of pelitic-gneissic rocks consisting of garnet-staurolite-sillimanite-tourmaline and quartzofeldspathic minerals occur within the metaconglomerate zone. These slivers show higher grade metamorphism compared to the surrounding. They may represent exhumed basement rocks. The mica schist on either side of the metaconglomerate is constituted of biotite, muscovite and quartz minerals and is marked by close spaced schistosity (S_1) defined by parallel alignment of mica. The schistosity is crenulated (S_2, Fig. 4b). Calcareous schist is a layered rock (Fig. 4c) consisting of alternate tremolite-actinolite-epidote-carbonate rich layer and quartz-plagioclase-microcline

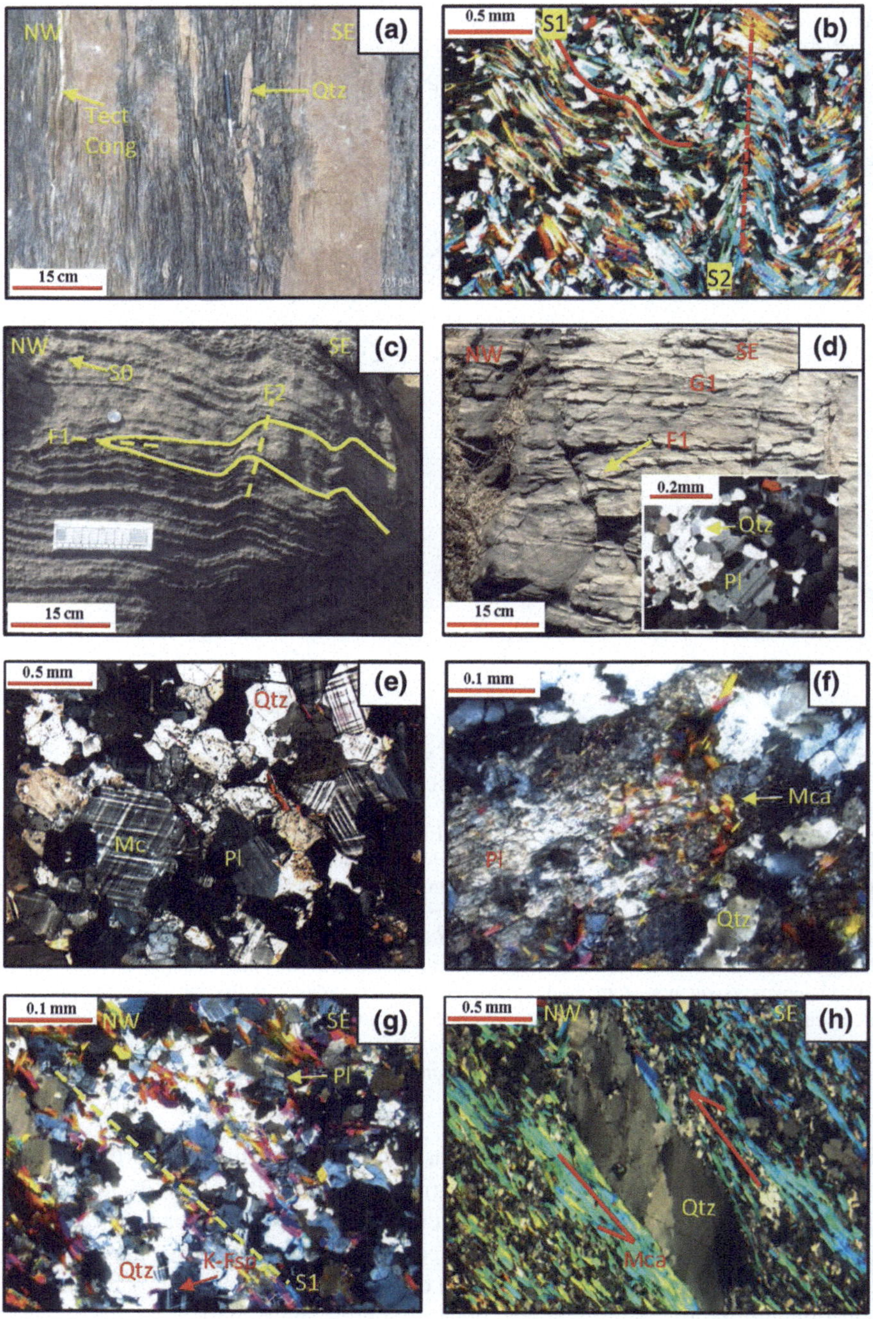
(a)
NW
SE
Tect Cong
Qtz
15 cm
(b)
0.5 mm
S1
S2
(c)
NW
SE
S0
F1
F2
15 cm
(d)
NW
SE
G1
F1
0.2mm
Qtz
Pl
15 cm
(e)
0.5 mm
Qtz
Mc
Pl
(f)
0.1 mm
Mca
Pl
Qtz
(g)
0.1 mm
NW
SE
Pl
Qtz
K-Fsp
S1
(h)
0.5 mm
NW
SE
Qtz
Mca

◂**Fig. 4** **a** Metaconglomerate zone, quartzite(Qtz) and granite gneiss pebbles within mica schist matrix. Quartzite beds are stretched to produce tectonic conglomerate (Tect. Cong.), vertical section, 26° 12′ 16″ N: 74° 11′ 10″ E. **b** Mica schist, the schistosity (S_1) is represented by muscovite and biotite. Quartz, feldspar and tourmaline are aligned parallel to schistosity plane. S_1 is folded by F_2 in form of crenulation (S_2 fabric), vertical section, 26° 09′ 01″ N: 74° 15′ 43″ E. **c** Calc schist, illustrating coaxial folding between F_1 (recumbent) and F_2 (upright). Bedding is denoted by S_0, vertical section, 26° 10′ 29″ N: 74° 17′ 04″ E. **d** G_1 metarhyolite with quartz rich layers standing out as ridges. Those ridges show F_1 isoclinal recumbent folds. Inset shows photomicrograph of the G_1 metarhyolite; plagioclase is euhedral, probably inhertiing magmatic origin, quartz shows bulging recrystallization, these features are indicative of low temperature deformation, 26° 10′ 26″ N: 74° 17′ 40″ E. **e** G_2 granite containing equigranular quartz (Qtz), plagioclase (Pl) and microcline (Mc). Quartz shows deformation. But feldspars show magmatic texture, location 26° 12′ 20″ N: 74° 18′ 03″ E. **f** G_4 granite containing porphyritic plagioclase (Pl) grains with mica (Mca) inclusions. Quartz (Qtz) shows deformation, location 26° 16′ 02″ N: 74° 03′ 10″ E. **g** Deformed G_4 granite. S_1 fabric defined by shape preferred orientation of mica. Plagioclase(Pl) shows dynamic recrystallization. Quartz shows incipient chessboard extinction. These features suggest high temperature deformation, vertical view, location 26° 16′ 04″ N: 74° 12′ 54″ E. **h** Metaconglomerate shows quartzite (Qtz) pebbles in mica schist matrix. The pebbles are aligned parallel to mylonitic foliation (Shear zone SZ-I). But recrystallised quartz grain show oblique arrangement. Top- to- NW thrusting is interpreted, vertical view, location 26° 12′ 16″ N: 74° 11′ 10″ E

rich layers. The rock contains rib structure due to differential weathering of the layers. The tremolite and actinolite minerals show shape preferred orientation (S_1) and are also crenulated (S_2). The plagioclase and epidote porphyroblasts contain inclusions of tremolite and actinolite. Calcareous schist grades into calcareous gneisses which is rich in carbonate minerals, and calcareous quartzite, rich in quartz. At Rupnagar, epidote-garnet bearing skarnoid rocks developed in the carbonate rock which lies in contact with metarhyolites (G_1) and amphibolite.

Amphibolite and metarhyolite (G_1) occur as interlayered bands within calcareous schist and mica schist. Amphibolite is the metamorphic product of basalt. It shows granoblastic to well foliated (S_1) character and contains hornblende- epidote- biotite- quartz- plagioclase mineral assemblages. The metarhyolite (G_1) occurs as thick layers (thickness up to tens of meter, Fig. 4d) to fine layers (few cm thickness) inter-fingered with amphibolite. Metarhyolite preserves vesicles at places and contains deformation fabric (S_1). Mineralogically, the metarhyolite contains quartz (40–50%), plagioclase (30–40%), hornblende and epidote with the grain size varying between 100–120 μm (Fig. 4d inset). Quartz is sub-rounded with bulging. Plagioclase has euhedral shape that suggest preservation of magmatic character. Hornblende is euhedral and alters to epidote along the margin. Hornblende and epidote grains show shape preferred orientation (S_1). Bulging character of the quartz indicates low temperature of dynamic recrystallization during D_1.

The G_2 and G_3 granites occur near Pratapgarh and Sumel respectively. These plutons show a coarse grained texture with discrete deformation fabric (S_1). They have sharp contact with the metasediments. The G_2 and G_3 differ in their mineralogy. The G_2 is dominated by microcline and quartz; the plagioclase occurs in subordinate amount. These minerals form an equigranular mosaic (Fig. 4e; grain size

450–500 μm). Biotite, muscovite and rare garnet, sillimanite and tourmaline occur in the rock. Quartz shows bulging and microcline and plagioclase maintains euhedral character. Mica minerals show shape preferred orientation defining deformation fabric (S_1) in the rock.

The G_3 is an in equigranular coarse grained porphyritic rock containing plagioclase phenocrysts, K–feldspar, quartz, tourmaline and biotite. Plagioclase is dominating, which makes it distinct from G_2 granite. Microscopically, the quartz shows bulging recrystallization, but the plagioclase still preserves euhedral shape. Biotite shows shape preferred orientation defining deformation fabric (S_1) in the rock. Absence of dynamic recrystallization in feldspar and presence of bulging recrystallization of quartz indicate low temperature of deformation (D_1) in G_2 and G_3.

The G_4 granite occurs to the west of Babra (Fig. 2a). Several G_4 granite veins occur within the adjoining mica schist and cross cut the other granites. The G_4 differs from other granites in that it is extremely rich in muscovite and biotite in addition to presence of coarse plagioclase, K-feldspar and quartz. The G_4 is marked by deformational fabric (S_1) defined by shape preferred orientation of biotite and muscovite. The intensity of deformation increases towards the contact with mica schist so much so that the G_4 appears as banded gneiss close to mica schist contact. Probably, these G_4-banded gneisses occurring near Kalakot, Bar and Babra, were interpreted as basement gneisses (Fig. 1c, Heron 1953). The plagioclase and K-feldspar in the core of the G_4 pluton preserves magmatic character. The minerals show poikilitic texture with small biotite and muscovite inclusions (Fig. 4f). The inclusions show random orientation, but two directions are prominent, one parallel and the other perpendicular to feldspar cleavage. In addition to that some of the minerals are dynamically recrystallized. The recrystallized quartz shows undulose extinction, chessboard extinction and grain boundary migration. Further, plagioclase and K-feldspars are dynamically recrystallized by grain boundary migration. Away from the core of the pluton and close to the mica schist contact, the rock is completely recrystallized into medium grained quartz, plagioclase, microcline, biotite and muscovite with prominent gneissosity fabric (Fig. 4g). Dynamic recrystallization of plagioclase indicates higher temperature of deformation (D_1) in the G_4.

4.2 Structures

Three stages of deformation ($D_{1\text{-}3}$) affected the rocks. The D_1 deformation produced NE-SW trending F_1 isoclinal and recumbent to reclined folds on the beddings of the quartzite, mica schist and calcareous schist (Fig. 4c). Alternate layers of amphibolite and G_1 metarhyolite, and the quartz rich layers in the G_1 metarhyolite preserve the F_1 fold (Fig. 4d). These layers represent the primary foliation in the rock. The F_1 folds carry Z and S asymmetric folds on the limbs and boudinage structure on the competent beds. The F_1 axial planar schistosity (S_1) is marked by muscovite, biotite in mica schist (Fig. 4b); hornblende, tremolite, actinolite and epidote in amphibolite and calcareous schist. The S_1 fabric is further, defined by shape preferred orientation of hornblende- epidote in the G_1 metarhyolite, biotite- muscovite in the G_2, G_3 and G_4 granites. The S_1 fabric is further associated with dynamic recrystallization of plagioclase, K-feldspar and quartz in the G_4 granite. Contrarily, the plagioclase and K-feldpsar do not show dynamic recrystallization in G_1 metarhyolite and, G_2 and G_3 granites. The S_1 fabric cross cuts the bedding/primary foliation at the F_1 fold hinge that serves criteria for identifying F_1 folds in different rocks.

The D_2 deformation produced NE-SW trending upright F_2 folds (Fig. 4c) and crenulation cleavage, S_2 (Fig. 4b). The F_1 fold and S_1 fabric were coaxially refolded by the F_2 fold producing type 3 interference pattern of Ramsay (1967). The D_1-muscovite, biotite and tremolite were crenulated to produce S_2 fabric. The D_3 deformation includes NW-SE to N-S trending F_3 folds which refolded the axial plane and axis of the F_1 and F_2 folds producing type-1 and type-2 interference patterns.

Structural map and the NW-SE profile sections of the area depict several large scale F_1, F_2 and F_3 folds (Figs. 2 and 3). The G_1 metarhyolite, marble and amphibolite to the west of Rupnagar, were folded by the F_2 fold, the axis of the fold represented by "β" in the stereoplot (Fig. 2c) plunges 42/110. On the southern limb of the F_2 fold, the G_1 metarhyolite describes an F_1 fold, marked by crosscutting relationship between S_1 fabric and the primary foliation in the G_1. The axial trace of F_1 fold is refolded by F_2 fold, hence a type 3 interference pattern is produced. Further, to the north of Rupnagar, alternate bands of the G_1 metarhyolite, amphibolite and calcareous schist were refolded by the NW-SE trending F_3 folds. The stereoplot of S_1 fabric shows girdle distribution and the "β" axis (F_3 fold axis) plunges towards SE (34/152, Fig. 2d). The G_1 metarhyolite bands pinch out to NE as well as SW. The thicker eastern and westernmost bands show parasitic F_1 folds at the contact with calcareous schist. Furthermore, to the NE of Ratariya, calcareous schist describes an F_2 fold. Stereoplot shows the "β" axis (F_2 fold axis) plunging to S (29/195; Fig. 2e). The G_2 granite occurs at the core of the F_2 fold. Mica schist unit occurring between calcareous schist and the G_2 granite contains parasitic F_2 and F_3 folds, axial planes of these folds lay nearly right angle to each other. Additionally, to the NE of Pratapgarh, the calcareous schist and mica schist are intruded by several meter scale G_4 granite veins that have been folded by F_2 and F_3 folds. The S_1 fabric poles show girdle distribution and "β" axis of the girdle (F_2 fold axis) plunges to SW (55/200; Fig. 2f). Further, the areas far east of Pratapgarh show uniformly SE dipping calcareous schist that represents the limb of an F_2 fold (Fig. 2g).

The area west of Babra shows NE-SW striking G_4 granite, metaconglomerate, mica schist, amphibolite and marble (Fig. 2a). Several centimeter to meter thick G_4 granite veins intruded parallel to S_1 fabric in the mica schist. The S_1 fabric in mica schist occurring to the west of the metaconglomerate shows girdle distribution and the "β" axis (F_2 fold axis) plunges to NE (35/50; Fig. 2h). The mica schist and amphibolite to the SE of the metaconglomerate show uniform orientation, and their poles lie in the NW quadrant, and F_1 and F_2 fold axes are distributed from NE to E-to south direction in the stereoplot (Fig. 2i).

4.2.1 Shear Zones and Faults

Four shear zones, SZ-I, SZ-II, SZ-III and SZ-IV were mapped in the area (Fig. 2, 3). The shear zones are synkinematic with the D_2 deformation. The SZ-I passes through the metaconglomerate host rock, west of Babra (Figs. 2a and 3b) and it coincides with Phulad shear zone (Fig. 1b, c). Sillimanite-staurolite-garnet bearing gneissic- slivers occur within the shear zone. Minor F_3 folds are developed in the mylonites of the shear zone, but the shear zone maintains overall NE-SW trend dipping towards SE, in large scale. The SZ-II passes through contact between mica schist and G_2 granite near Pratapgarh. The SZ-III passes on the G_2 granite protolith. The SZ-IV passes over the G_1 metarhyolite, to the N and NW of Rupnagar. These shear zones are folded by the NW-SE striking open F_3 folds near Rupnagar. The shear zones contain mylonites having biotite-muscovite-quartz minerals assemblage. The mylonitic foliation dips steeply to the SE (Fig. 3a, b) and contain down the dip stretching lineation defined by mica and stretched quartz grains. Lensoidal quartz aggregates, mica fishes and S-C fabric on the XZ section of the mylonites indicate top- to- NW sense of thrust slip kinematics (Fig. 4h). The lensoidal quartz aggregates show dynamic recrystallization of quartz grains by subgrain rotation. Temperature of deformation is around 450 °C (e.g., Stipp et al. 2002). The shear zones suffered excessive flattening; as a result the quartzite beds were stretched vertically and boudinaged to form tectonic conglomerate in the SZ-I (Fig. 4a). The S and C fabric are nearly parallel in SZ-III and SZ-IV. Further, the F_1 and F_2 fold axes were rotated to shear direction and the S_1 fabric was transposed parallel to mylonitic foliation in the shear zones. In addition to the ductile shear zones, two brittle normal faults occur in the area (Figs. 2a, b and 3a, b), the rocks were deformed into a cataclasites along the fault. Probably, these normal faults were produced from reactivation of the shear zones.

5 Discussion

5.1 Deformation of the Rocks of the South Delhi Terrane (SDT)

The present study brings out an initial stage of coaxial folding in NE–SW axis (between F_1 and F_2) during the D_{1-2} deformations. This was followed by NW-SE folding (F_3) during D_3 deformation. The F_1 and F_2 folds are associated with parasitic folds, cleavage refraction and wavelength to thickness characteristics of buckling origin. A NW-SE compression was responsible for the D_{1-2} deformations. The compression was generated due to collision/subduction during South Delhi orogeny. The rocks underwent greenschist facies metamorphism during F_1 folding as indicated by muscovite, biotite, quartz, tremolite, actinolite and epidote minerals defining the S_1 fabric in the rock. The D_3 was due to NE-SW shortening of the *SDT* that probably rose as a result of stress relaxation along NW-SE direction, an explanation similar to what Naha et al. (1987) argued for the development of NW-SE folds in the SE syncline. The ductile shear zones developed synkinematic with D_2 deformation; the shear zones are greenschist facies shear zones and show top- to- the NW sense of thrusting. Thrusting exhumed the rocks. Several thin slices of garnet-staurolite-sillimanite gneissic rocks with basal conglomerate were exhumed due to thrusting along the SZ-I. The mineral assemblage of these gneisses indicates higher grade metamorphism (upper amphibolite to granulite facies) which is incompatible to metamorphic assemblage of the surrounding metasediments (mostly greenschist facies). Therefore, we interprete them to be part of exhumed basement high grade rocks.

The above deformation pattern is common to all parts of the *SDT*. The entire 500 km long and 50 km wide *SDT* shows uniformity in major folding pattern (Gangopadhyay and Mukhopadhyay 1984; Naha et al. 1984, 1987; Biswal 1988; Sengupta 1988; Mukhopadhyay 1989; Mukhopadhyay and Matin 1991; Singh et al. 2010; Ruj and Dasgupta 2014; Tiwari and Biswal 2019a, b). The earliest part of folding history is marked by coaxial folding between a recumbent/reclined fold and an upright open to tight fold along NE-SW axis. The entire terrane is studded with pervasive type 3 interference pattern that resulted in repetition of rocks in profile section and running of low plunging fold axes for many kilometers. This was superimposed by NW-SE trending folds, which is at right angle to two early folds. The overprinting of the NW-SE fold on NE-SW folds produced type 1 and type 2 interference pattern, axial culminations and depression and veering of lithological trend at several places. In addition to folds, the *SDT* is marked by longitudinal ductile shear zones at Phulad (Ghosh et al. 1999; 2003; Sengupta and Ghosh 2004, 2007), Ambaji (Tiwari and Biswal 2019a, b), and Bar (Dasgupta et al. 2012). The Phulad shear zone which coincides with the SZ-I, shows thrust kinematics. Further, it is transpressive in nature and has rotated the earlier fabric. The shear zones at Ambaji were synkinematic with the D_2 deformation and had an earlier history of high temperature transpressive top-to-NW sense of thrusting that brought the lower-middle crustal rocks to the upper crust. This was overprinted by sinistral retrograde shearing. Shear zone at Bar shows dextral shear. The *SDT* has been deformed by many faults, mostly normal faults, produced from NW-SE extension (Tiwari and Biswal 2019b). The faults host 760 Ma granite intrusion near Ambaji (Singh et al. 2010). The high grade assemblages along the SZ-I in the study area has similarity with granulitic assemblages of the Pilwa-Chinwali (Fareeduddin et al. 1994; Bhowmik et al. 2018) and Srinagar areas (Bose et al. 2017). These belong to the high grade assemblage of the basement rocks underlying the *SDT* was opined by Fareeduddin et al. (1994) and Tiwari and Biswal (2019a).

In contrast to the *SDT*, the older terranes like Aravalli and Mangalwar terranes differ in their folding history in that these were marked by recumbent/reclined folding along E-W axis which was overprinted by NE-SW trending recumbent/reclined and upright folds characteristic of the *SDT* (Naha and Halyburton 1977a, b; Naha et al. 1984).

5.2 *Deformation Fabric and Previous Geochronology of Granite, and Timing of the South Delhi Orogeny*

Petrographic analysis of the metarhyolite and granites of the study area has brought out a marked contrast in mineralogy and deformational fabric. The G_1 is a metarhyolite, the G_2 is K-rich granite, the G_3 is a plagioclase rich granite and G_4 is a mica rich granite. The G_1 occurs as volcanic flow interbanded with amphibolite. The flow layers which appears as quartz rich mesoscopic ridges are folded into F_1 folds. The amphiobolite and G_1 describe series of large scale F_1 folds in the area (Fig. 3a cross section A–B). We interpreted that the extrusion of the G_1 (metarhyolite) and basic volcanics (now amphibolite) represent rift related volcanic rocks. The G_1, G_2, G_3 and G_4 carry S_1 deformational fabric, however, the temperature of deformation varies. The S_1 fabric in G_1, G_2, and G_3 was developed by low temperature solid state deformation. This is indicated by plagioclases and K-feldspars minerals which still maintain euhedral shape and are randomly oriented indicating magmatic character (Fig. 2d inset, e). Further, quartz shows bulging recrystallization and hornblende was retrograded to epidote; these corroborate to low temperature deformation. We therefore interpreted that $G_{1\text{-}3}$ were pretectonic to the D_1 deformation and probably emplaced during rifting of the South Delhi basin. They were deformed and metamorphosed along with the sediments in greenschist facies conditions during the D_1 deformation.

Contrastingly, the G_4 granite shows high temperature solid state deformation. The plagioclase was dynamically recrystallized into smaller equant grains, that indicates the temperature of deformation would be at ca. 700 °C (e.g., Pryer 1993; Kruhl 1996; Rosenberg and Stünitz 2003; Passchier and Trouw 2005). Close to the contact with metasediments, the G_4 pluton was completely recrystallized dynamically into a gneissic rock which was previously referred to as basement gneiss by Heron (1953). The S_1 fabric in the G_4 granite gneiss and the host mica schist are coplanar. It was interpreted that the G_4 granite at the margin was solidified fast due to its contact with cold mica schist. As the temperature of the pluton was high, the G_4 underwent high temperature dynamic recrystallization during the D_1 deformation. Away from the margin in the core of the G_4 pluton, magmatic fabric is preserved, as the pluton was still in molten state. The K-feldspar and plagioclase still maintain poikilitic texture and are aligned parallel to S_1 fabric. In the core also, around the plagioclase and K-feldspar minerals, smaller mineral grains are dynamically recrystallized by grain boundary migration. High temperature solid state deformation in the margin and presence of magmatic fabric at the core lead to suggest that the intrusion of the G_4 granite was syntectonic with the D_1 deformation. A transition from magmatic to high temperature solid state deformation fabric and coupling between pluton and host rock fabrics are the best evidence of synkinematic emplacement (Vernon et al. 1989; Bouchez et al. 1990; Miller and Paterson 1994; Paterson et al. 1998; Biswal et al. 2007).

We are in the process of dating the G_1 metarhyolite, and G_2, G_3 and G_4 granites. As per the published literature, the Sewariya granite (same as G_4) was dated at ca.

860 Ma (Rb-Sr magmatic cooling age, Sivasubramaniam et al. 2019). Since the intrusion of the G_4 was syntectonic with D_1 deformation, we interpret that the D_1 was at ca. 860 Ma. Heron (1953) and Gupta et al. (1995) included the G_1, G_2 and G_3 under broad Erinpura and Sendra Granites which were dated at ca. 970 Ma ($^{87}Sr/^{86}Sr$, Tobisch et al. 1994, U-Pb TIMS Pandit et al. 2003). As the G_1 was syn-rifting, we constrain the age of rifting and sedimentation in the South Delhi basin at ca. 970 Ma. Similar ages were obtained from southern part of the *SDT*. The D_1 deformation was at ca. 860 Ma and the metarhyolite extruded during sedimentation was ca. 960 Ma (Deb et al. 2001; Singh et al. 2010; Tiwari and Biswal 2019a). From the above age correlation, we can argue that the south Delhi orogeny is <860 Ma. Rifting, formation of basin and sedimentation were around ca. 1.0 Ga.

The granulites at Pilwa-Chinwali areas that lies on the western flank of the *SDT*, in continuity of the Phulad shear zone, produced ages at ca. 1.7, 1.5 and 1.0 Ga (Fareeduddin et al. 1994; Bhowmik et al. 2018). Suggestion that the Delhi orogeny is as old as 1.0 Ga based on the ages of Pilwa-Chinwali granulites (Bhowmik et al. 2018), does not fit to our study on correlation between deformation fabric and geochronology. The above mentioned granulites belong to older sequence, likely to be part of Sandmata Terrane that produces equivalent ages. Similarly, the Phulad shear zone in the south shows metamorphism age of 970 Ma (Chatterjee et al. 2017), that probably represents the age of xenocrystic monazite derived from Sendra granite.

5.3 South Delhi Orogeny in Relation to Global Orogeny

The South Delhi orogeny resulted from subduction/collision between the Marwar and Bundelkhand cratons in the northwestern part of Peninsular India during Proterozoic period (Synchanthavong and Desai 1977; Sinha-Roy 1988; Volpe and Macdougall 1990; Biswal et al. 1998a, b; Deb et al. 2001; Khan et al. 2005; Singh et al. 2010; Dharma Rao et al. 2013). The Marwar craton was extensively intruded by Neoproterozoic granite and hence scarcley preserved. However, its equivalent blocks exist in the Arabian-Nubian shield (Affif-Abas terrane), central Madagascar (Azania block), Tanzania and Dharwar craton (Collins and Pisarevsky 2005). Interpretations varies regarding timing of the South Delhi orogeny; one school interpreted that it was synchronous with globally occurring Grenvillian orogeny (Roy 2001; Bhowmik and Dasgupta 2012; Meert et al. 2013; Dharma Rao et al. 2013; Chatterjee et al. 2017 and references therein). The Grenvillian orogeny was responsible for assembly of the Rodinia supercontinent between 1.3 and 1.0 Ga ago (Valentine and Moores 1970; McMenamin and McMenamin 1990; Meert and Torsvik 2003; Cawood 2005; Li et al. 2008). The other school interpreted that it was coeval with Pan-African orogeny (Singh et al. 2010; De Wall et al. 2014; Tiwari and Biswal 2019a). The Pan-African orogeny led to Gondwanaland assembly by juxtaposition of several landmasses including Africa, Madagascar and India in multiple orogenic phases spanning over ca. 900–500 Ma. Neoproterozoic belts, namely Seychelles, Madagascar, Arabian-Nubian shield, Eastern Granulite form continuous tectonic belts in

Gondwanaland, that came together in discrete phases during Pan-African orogeny (Collins and Windley 2002; Sommer et al. 2003; Biswal et al. 2007; Singh et al. 2010). Pan-African orogeny has been divided into two parts, namely ca. 900–630 Ma which is characterized by terrane accretion as seen in Arabian-Nubian shield, and 630–540 Ma that shows collision and escape tectonics in many parts of East African orogen (Stern 1994; Kröner and Stern 2005; Fritz et al. 2013). Another interpretation was that the Pan-African orogeny belongs to <650 Ma (Oriolo et al. 2017). Signature of ca. < 650 Ma old orogenic phase (comparable with Kuunga or Malagasy orogeny; e.g., Meert 2003) is not extensive in the *SDT*. However, extensional tectonics of similar ages are observed in the southern part of the *SDT* (Tiwari and Biswal 2019a) and in the Sirohi terane (Bhardwaj and Biswal 2019). In addition to those, isolated incidences of thermal resetting and fluid activity at ca. 500 Ma were also observed (Crawford 1975; Ashwal et al. 2013; Pandey et al. 2013; Sen et al. 2013). In these perspectives, the South Delhi orogeny could be either part of early part of Pan-African orogeny or it represents a transition between Grenvillian and Pan-African orogeny.

The Aravalli- Delhi mobile belt (*ADMB*) evolved through three distinct orogenic phases. The Bhilwara orogeny culminated with intrusion of Berach granite at ca. 2.8 Ga. The Aravalli orogeny left impression in the Aravalli terrane, Mangalwar-Sandmata terrane and North Delhi terrane and culminated at 1.7 Ga with intrusion of Rakhabdev ophiolites, intrusion of granulite and nepheline syenite in Sandmata terrane (Roy et al. 2016), and arc related granite in the North Delhi terrane (Kaur et al. 2013). The Aravalli orogeny may be related to Nuna orogeny. The Grenvillian orogeny was recorded in some parts of Sandmata terrane. The South Delhi orogeny was ca. 0.8 Ga that was marked by multiple phases of folding, granulite to greenschist facies metamorphism at different parts, tectonic emplacement of basement rocks along ductile shear zones and intrusion of multiple phases of granitic rocks.

6 Conclusions

(i) Present study suggests that the *SDT* around Beawar- Rupnagar- Babra area, Rajasthan, NW India had undergone an early stage of deformation $D_{1\text{-}2}$ that produced coaxial F_1 and F_2 folds along NE-SW axis. These were superimposed by the NW-SE trending F_3 folds during the D_3 deformation. Rocks underwent greenschist facies of metamorphism during D_1 deformation. The D_3 deformations have no impact on metamorphism of rocks except that the S_1 fabric has been crenulated. During D_2 deformation greenschist grade ductile shear zones were developed axial planar to F_2 fold (four number SZ-I to SZ-IV). The shear zones show top- to- the NW thrusting. High grade basement rocks with basal conglomerates were thrusted up along the SZ-I.

(ii) The G_1-metarhyolite alternates with amphibolite and metasediments, suggesting that this was erupted as rhyolite flow in a rift setting. The G_1-metarhyolite and G_2, G_3 and G_4 granites bear the imprint of S_1 fabric. The G_1, G_2 and

G_3 preserve euhedral shape of the feldspar and plagioclase, inherited from magmatic stage. The quartz grains were dynamically recrystallized by bulging suggesting low temperature deformations. We interpreted the G_1, G_2 and G_3 were pre-kinematic with the D_1 deformation. The G_4 granite was deformed at higher temperature as indicated by the dynamic recrystallization of the plagioclase and K-feldspar minerals. It has preserved relicts of magmatic fabric at the core of the pluton, marked by parallel alignment of plagioclase and feldspar grains. Textural study implies that the G_4 intruded syntectonic with D_1 deformation.

(iii) We used the previous geochronology data pertaining to the granites of the study area. Based on the age of Sewariya granite which is same as G_4 granite the D_1 deformation is constrained at ca. 860 Ma. Sendra granite which was probably syntectonic with G_1, G_2 and G_3 suggest the rifting of the basin was probably at ca. 970 Ma. We interpret that the South Delhi orogeny is ca. 860 Ma old, which is much younger than Grenvillian orogeny and older than Pan-African orogeny. It may be a transitional phase between the above two.

(iv) The Aravalli Delhi mobile belt (*ADMB*) underwent three orogenies, the Bhilwara orogeny was at ca. 2.8 Ga, the Aravalli orogeny at ca. 1.8 Ga and the South Delhi orogeny at ca. 0.8 Ga with an interval of one billion year. The Aravalli orogeny may be related to Nuna orogeny. Imprint of Grenvillian orogeny was recorded in some parts of Sandmata Terrane. Thus the period of orogeny is characteristic features of an individual terranes, it may coincide with major global orogenic phases.

Acknowledgements We thank Department of Earth Sciences, IIT Bombay for financial support during field work. Review by anonymous reviewers greatly improved the MS.

References

Ashwal LD, Solanki AM, Pandit MK, Corfu F, Hendriks BW, Burke K, Torswik TH (2013) Geochronology and geochemistry of the Neoproterozoic Mt, Abu granitoids, NW India: regional correlation and implications for Rodinia paleogeography. Precambr Res 236:265–281

Bhardwaj A, Biswal TK (2019) Deformation and tectonic history of Punagarh Basin in the Trans-Aravalli Terrane of North-Western India. In: Geological evolution of the Precambrian Indian shield. Springer, pp 159–178

Bhowmik SK, Bernhardt HJ, Dasgupta S (2010) Grenvillian age high-pressure upper amphibolite–granulite metamorphism in the Aravalli-Delhi mobile belt, northwestern India: New evidence from monazite chemical age and its implication. Precambr Res 178:168–184

Bhowmik SK, Dasgupta S (2012) Tectonothermal evolution of the Banded Gneissic Complex in central Rajasthan, NW India: present status and correlation. J Asian Earth Sci 49:339–348

Bhowmik SK, Dasgupta S, Baruah S, Kalita D (2018) Thermal history of a Late Mesoproterozoic paired metamorphic belt during Rodinia assembly: new insight from medium-pressure granulites from the Aravalli-Delhi Mobile Belt, Northwestern India. Geosci Front 9(2):335–354

Biswal TK (1988) Polyphase deformation in Delhi rocks, south-east Amir-garh, Banaskantha district, Gujarat, in Precambrian of the Aravalli Mountain, Rajasthan, India. Memoir Geol Soc India 7:267–277

Biswal TK, Gyani KC, Parthasarathy R, Pant DR (1998a) Implications of the geochemistry of the Pelitic Granulites of the Delhi Supergroup, Aravalli Mountain Belt, Northwestern India. Precambr Res 87:75–85

Biswal TK, Gyani KC, Parthasarathy R, Pant DR (1998b) Tectonic implication of geochemistry of gabbro-norite-basic granulite suite in the Proterozoic Delhi Supergroup, Rajasthan, India. J Geol Soc India 52:721–732

Biswal TK, De Waele B, Ahuja H (2007) Timing and dynamics of the juxtapositionof the Eastern Ghats Mobile Belt against the Bhandara Craton, India: a structural and zircon U–Pb SHRIMP study of the fold–thrust belt and associated nepheline syenite. Tectonics 26:TC4006. https://doi.org/10.1029/2006tc002005

Bose S, Seth P, Dasgupta N (2017) Meso-Neoproterozoic mid-crustal metamorphic record from the Ajmer-Srinagar section, Rajasthan, India and its implication to the assembly of the greater Indian Landmass during the Grenvillian-age orogenesis. In: Pant NC, Dasgupta S (eds) Crusstal evolution of India and Antarctica, The supercontinent Connection, vol 457, Geological Society London, pp 291–318

Bouchez JL, Gleizes G, Djouadi T, Rochette P (1990) Microstructure and magnetic susceptibility applied to emplacement kinematics of granites: the example of the Foix pluton (French Pyrenees). Tectonophysics 184(2):157–171

Cawood A (2005) Terra Australis Orogen. Rodinia breakup and development of the Pacific and Iapetus margins of Gondwana during the Neoproterozoic and Paleozoic. Earth Sci Rev 69:249–279

Chatterjee SM, Roy Choudhury M, Das S, Roy A (2017) Significance and dynamics of the Neoproterozoic (810 Ma) Phulad Shear Zone, Rajasthan, NW India. Tectonics 36(8):1432–1454

Cherniak DJ, Watson EB (2001) Pb diffusion in zircon. Chem Geol 172:5–24

Collins AS, Windley BF (2002) The tectonic evolution of central and northern Madagascar and its place in the final assembly of Gondwana. J Geol 110(3):325–339

Collins AS, Pisarevsky SA (2005) Amalgamating eastern Gondwana: the evolution of the Circum-Indian Orogens. Earth Sci Rev 71:229–270

Crawford AR (1975) Rb-Sr age determination for the Mount Abu Granite and related rocks of Gujarat. J Geol Soc India 16:20–28

Dahl PS (1997) A crystal-chemical basis for Pb retention and fission track annealing systematics in U-bearing minerals with implications for geochronology. Earth Planet Sci Lett 150:277–290

Dasgupta N, Mukhopadhyay D, Bhattacharyya T (2012) Analysis of superposed strain: a case study from Barr Conglomerate in the South Delhi Fold Belt, Rajasthan, India. J Struct Geol 34:30–42

De Wall H, Pandit MK, Chauhan NK (2012) Paleosol at the Archean–Proterozoic contact in Udaipur. Precambr Res 216:120–131

De Wall H, Pandit MK, Sharma KK, Schobel S, Just J (2014) Deformation, granite intrusion in the Sirohi area, SW Rajasthan—constraints on cryogenian to Pan African crustal dynamics of NW India. Precambr Res 254:1–18

Deb M, Thorpe RI, Cumming GL, Wagner A (1989) Age, source and stratigraphic implication of Pb isotope data for conformable, sediment-hosted, base metal deposits in the Proterozoic Aravalli-Delhi Orogenic Belt, Northwestern India. Precambr Res 43:1–22

Deb M, Thorpe RI, Krstic D, Corfu F, Davis DW (2001) Zircon U-Pb and galena Pb isotope evidence an approximate 1.0 Ga terrane constituting the western margin of the Aravalli-Delhi orogenic belt, northwestern India. Precambr Res 108:195–213

Deb M, Thorpe RI (2001) Geochronological constraints in the Precambrian Geology of Northwestern India and their Metallogenic Implication. In: Deb M, Goodfellow WD (eds) Sediment-hosted Lead-Zinc sulfide deposit in the Northwestern Indian Shield. Proceedings of an international workshop, Delhi-Udaipur, India, pp 137–152

Dharma Rao CV, Santosh M, Kim SW, Li S (2013) Arc magmatism in the Delhi Fold Belt. SHRIMP U-Pb zircon ages of granitoids and implications for Neoproterozoic convergent margin tectonics in NW India. J Asian Earth Sci 78:83–99

Fareeduddin SM, Basavalingu B, Janardhan A (1994) PT conditions of pelitic granulites and associated charnockites of Chinwali area, west of Delhi fold belt. Rajasthan. J Geol Soc India 43(2):169–178

Foster G, Parrish RR, Horstwood MSA, Cheney S, Pyle J, Gibson HD (2004) The generation of prograde P-T-t points and paths; a textural, compositional, and chronological study of metamorphic monazite. Earth Planet Sci Lett 228:125–142

Fritz H, Abdelsalam M, Ali KA, Bingen B, Collins AS, Fowler AR, Ghebreab W, Hauzenberger CA, Johnson R, Kusky TM, Macey P (2013) Orogen styles in the East African Orogen: a review of the Neoproterozoic to Cambrian tectonic evolution. J Afr Earth Sci 86:65–106

Gangopadhyay A, Mukhopadhyay D (1984) Structural geometry of the Delhi supergroup near Sendra. In: Saha AK (ed) Geological evolution of peninsular India, Recent Researches in Geology, pp 45–60

Ghosh SK, Hazra S, Sengupta S (1999) Planar, non-planar and refolded sheath folds in the Phulad Shear Zone, Rajasthan, India. J Struct Geol 21(12):1715–1729

Ghosh SK, Sen G, Sengupta S (2003) Rotation of long tectonic clasts in transpressional shear zones. J Struct Geol 25(7):1083–1096

Gupta SN, Arora YK, Mathur RK, Iqbaluddin BP, Prasad B, Sahai TN, Sharma SB (1995) Geological map of the Precambrians of the Aravalli region, southern Rajasthan and northwestern Gujarat, India (4 sheets, scale 1:250,000). Geological Survey of India, Western Region, Jaipur, India

Harrison TM, Duncan I, McDougall I (1985) Diffusion of 40Ar in biotite: temperature, pressure, and compositional effects. Geochim Cosmochim Acta 49:2461–2468

Heron AM (1953) The geology of central Rajputana. Mem Geol Sur India 79:389

Just J, Schulz B, de Wall H, Jourdan F, Pandit MK (2011) Monazite CHIME/EPMA dating of Erinpura granitoid deformation: Implications for Neoproterozoic tectono-thermal evolution of NW India. Gondwana Res 19:402–412

Kaur P, Chaudhri N, Okrusch M, Koepke J (2006) Palaeoproterozoic A-type felsic magmatism in the Khetri Copper Belt, Rajasthan, northwestern India: petrologic and tectonic implications. Mineral Petrol 87(1–2):81–122

Kaur P, Zeh A, Chaudhri N, Gerdes A, Okrusch M (2013) Nature of magmatism and sedimentation at a Columbia active margin: Insights from combined U-Pb and Lu–Hf isotope data of detrital zircons from NW India. Gondwana Res 23:1040–1052

Kaur P, Zeh A, Chaudhri N (2019) Archean crustal evolution of the Aravalli Banded Gneissic Complex, NW India: constraints from zircon U-Pb ages, Lu-Hf isotope systematics, and whole-rock geochemistry of granitoids. Precambr Res 327:81–102

Khan MS, Smith TE, Raza M, Huang J (2005) Geology, geochemistry and tectonic significance of mafic–ultramafic rocks of Mesoproterozic Phulad Ophiolite Suite of South Delhi Fold Belt, NW Indian shield. Gondwana Res 8:553–566

Kröner A, Stern RJ (2005) Pan-African orogeny. In: Selley RC, Cooks LRM, Plimer IR (eds) Encyclopedia of geology, vol 1. Elsevier, Amsterdam, pp 1–12

Kruhl JH (1996) Prism-and basal-plane parallel subgrain boundaries in quartz: A microstructural geothermobarometer. J Metamorph Geol 14(5):581–589

Lehmann J, Saalmann K, Naydenov KV, Milani L, Belyanin GA, Zwingmann H, Charlesworth G, Kinnaird JA (2016) Structural and geochronological constraints on the Pan-African tectonic evolution of the northern Damara Belt, Namibia. Tectonics 35:103–135

Li ZX, Bogdanova SV, Collins AS, Davidson A, De Waele B, Ernst RE, Fitzsimons ICW, Fuck RA, Gladkochub DP, Jacobs J, Karlstrom KE (2008) Assembly, configuration, and break-up history of Rodinia: a synthesis. Precambr Res 160:179–210

Mahan KH, Goncalves P, Williams ML, Jercinovic MJ (2006) Dating metamorphic reactions and fluid flow: application to exhumation of high-P granulites in a crustal-scale shear zone, western Canadian Shield. J Metamorph Geol 24:193–217

McMenamin MAS, McMenamin DLS (1990) The emergence of animals: the cambrian breakthrough, p 217

Mehdi M, Kumar S, Pant NC (2015) Low grade metamorphism in the Lalsot-Bayana Sub-basin of the North Delhi Fold Belt and its tectonic implication. J Geol Soc India 85(4):397–410

Meert JG (2003) A synopsis of events related to the assembly of eastern Gondwana. Tectonophysics 362:1–40

Meert JG, Torsvik TH (2003) The making and unmaking of a supercontinent: Rodinia revisited. Tectonophysics 375:261–288

Meert JG, Lieberman BS (2008) The Neoproterozoic assembly of Gondwana and its relationship to the Ediacaran–Cambrian radiation. Gondwana Res 14:5–21

Meert JG, Pandit MK, Kamenov GD (2013) Further geochronological and paleomagnetic constraints on Malani (and pre-Malani) magmatism in NW India. Tectonophysics 608:1254–1267

Miller RB, Paterson SR (1994) The transition from magmatic to high-temperature solid-state deformation: implications from the Mount Stuart batholith, Washington. J Struct Geol 16(6):853–865

Mukhopadhyay D (1989) Structural history of the central section of the Delhi orogenic belt. In: Proceedings of 28th international geological congress, Abstract 2, pp 479–480

Mukhopadhyay D, Matin A (1991) Early major folds in the Delhi Supergroup around Hatankhera, Ajmer district, Rajasthan. Indian J Geol 63:67–74

Naha K, Halyburton RV (1977a) Structural pattern and strain history of a superposed fold system in the Precambrian of Central Rajasthan, India. Structural pattern in the 'main Raialo syncline', Central Rajasthan. Precambr Res 4(1):39–84

Naha K, Halyburton RV (1977b) Structural pattern and strain history of a superposed fold system in the Precambrian of Central Rajasthan, India. Strain history. Precambr Res 4(1):85–111

Naha K, Mukhopadhyay DK, Mohanty R, Mitra SK, Biswal TK (1984) Significance of contrast in the early stages of the structural history of the Delhi and the Pre-Delhi rock groups in the Proterozoic of Rajasthan, western India. Tectonophysics 105:193–206

Naha K, Mitra SK, Biswal TK (1987) Structural history of the rocks of the Delhi Group around Todgarh, Central Rajasthan. Indian J Geol 59:126–156

Oriolo S, Oyhantçabal P, Wemmer K, Siegesmund S (2017) Contemporaneous assembly of Western Gondwana and final Rodinia break-up: implications for the supercontinent cycle. Geosci Front 8(6):1431–1445

Pandey M, Pant NC, Santosh K (2013) Criteria to distinguish between regional and contact zone monazite. A case study from Proterozoic North Delhi Fold Belt (NDFB), India. Episodes 36:275–289

Pandit MK, Carter LM, Ashwal LD, Tucker RD, Torsvik TH, Jamtveit B, Bhushan SK (2003) Age, petrogenesis and significance of 1 Ga granitoids and related rocks from the Sendra area, Aravalli Craton, NW India. J Asian Earth Sci 22(4):363–381

Pandit MK, de Wall H, Chauhan NK (2008) Paleosol at the Archean-Proterozoic contact in NW India revisited: evidence for oxidizing conditions during paleoweathering? J Earth Syst Sci 117:201–209

Pant NC, Kundu A, Joshi S (2008) Age of metamorphism of Delhi Supergroup rocks-electron microprobe ages from Mahendragarh district, Haryana. J Geol Soc India 72(3):365–372

Passchier CW, Trouw RA (2005) Microtectonics. Springer Science and Business Media

Paterson SR, Fowler TK Jr, Schmidt KL, Yoshinobu AS, Yuan ES, Miller RB (1998) Interpreting magmatic fabric patterns in plutons. Lithos 44:53–82

Pryer LL (1993) Microstructures in feldspars from a major crustal thrust zone: the Grenville Front, Ontario, Canada. J Struct Geol 15(1):21–36

Purohit R, Papineau D, Kröner A, Sharma KK, Roy AB (2012) Carbon isotope geochemistry and geochronological constraints of the Neoproterozoic Sirohi Group from northwest India. Precambr Res 220:80–90

Pyle JM, Spear FS (2003) Four generations of accessory-phase growth in low pressure migmatites from SW New Hampshire. Am Miner 88:338–351

Ramsay JG (1967) Folding and fracturing of rocks. McGraw-Hill Companies, p 568

Raza M, Siddiqui MZ (2012) Geochemistry and tectonic significance of mafic volcanic rocks of the Hindoli belt, southeastern Rajasthan: Implications for continent assembly. J Geol Soc India 80(4):553–562

Rogers JJ, Santosh M (2002) Configuration of Columbia, a Mesoproterozoic supercontinent. Gondwana Res 5(1):5–22

Rosenberg CL, Stünitz H (2003) Deformation and recrystallization of plagioclase along a temperature gradient: an example from the Bergell tonalite. J Struct Geol 25(3):389–408

Roy AB, Sharma KK (1999) Geology of the region around Sirohi town, western Rajasthan–story of Neoproterozoic evolution of the Trans-Aravalli crust. Geological Evolution of Western Rajasthan, pp 19–33

Roy AB (2001) Neoproterozoic crustal evolution of northwestern Indian shield: implications on break-up and assembly of supercontinents. Gondwana Res 4:289–306

Roy AB, Jakhar SR (2002) Geology of Rajasthan (NW India) Precambrian to recent. Scientific Publishers (India) Jodhpur, p 387

Roy AB, Dutta K, Rathore S (2016) Development of ductile shear zones during diapiric magmatism of nepheline syenite and exhumation of granulites—examples from central Rajasthan, India. Curr Sci 110(6):1094–1101

Ruj T, Dasgupta N (2014) Tectonic imprints within a granite exposed near Srinagar, Rajasthan, India. J Earth Syst Sci 123(6):1361–1374

Sarkar G, Bishui PK, Chattopadhyay B, Chowdhury S, Chowdhury I, Saha KC, Kumar A (1992) Geochronology of granites and felsic volcanic rocks of Delhi Fold Belt. Geol Surv India 125(2):21–23

Sen A, Pande K, Sheth HC, Sharma KK, Sarkar S, Dayal AM, Mistry H (2013) An Ediacaran-Cambrian thermal imprint in Rajasthan, western India: evidence from 40 Ar–39 Ar geochronology of the Sindreth volcanics. J Earth Syst Sci 122(6):1477–1493

Sengupta S (1988) Development of successive sets of structures in a process of continuous deformation: a case study from the Delhi metasediments near Kharwa, Ajmer District, Rajasthan, India. J Earth Sci 15:116–131

Sengupta S, Ghosh SK (2004) Analysis of transpressional deformation from geometrical evolution of mesoscopic structures from Phulad shear zone, Rajasthan, India. J Struct Geol 26:1961–1976

Sengupta S, Ghosh SK (2007) Origin of striping lineation and transposition of linear structures in shear zones. J Struct Geol 29(2):273–287

Sharma KK (2005) Malani magmatism: an extensional lithospheric tectonic origin. Geological Society of America Special Papers 388, pp 463–476

Singh YK, De Waele B, Karmarkar S, Sarkar S, Biswal TK (2010) Tectonic setting of the Balaream–Kui–Surpagla–Kengora granulites of the South Delhi Terrane of the Aravalli Mobile Belt, NW India and its implication on correlation with the East African Orogen in the Gondwana assembly. Precambr Res 183:669–688

Sinha-Roy S (1988) Proterozoic Wilson cycles in Rajasthan. In: Roy AB (ed) Precambrian of the Aravalli Mountain, Rajasthan, India. Geological Society of India, Memoirs vol 7, pp 95–108

Sinha-Roy S, Malhotra G, Mohanty M (1998) Geology of Rajasthan. Geological Society of India Memoire, p 278

Sivasubramaniam R, Anand SV, Pandian MS, Balakrishnan S (2019) Geological, geochemical and Rb–Sr isotopic studies on tungsten mineralised Sewariya-Govindgarh granites of Delhi Fold Belt, Rajasthan, NW India. J Earth Syst Sci 128(1):19

Sommer H, Kröner A, Hauzenberger C, Muhongo S, Wingate MTD (2003) Metamorphic petrology and zircon geochronology of high-grade rocks from the central Mozambique Belt of Tanzania. crustal recycling of Archean and Palaeoproterozoic material during the Pan-African orogeny. J Metamorph Geol 21:915–934

Stern RJ (1994) Arc assembly and continental collision in the Neoproterozoic East African Orogen. Implications for the consolidation of Gondwanaland. Annu Rev Earth Planet Sci 22:319–351

Stipp M, Stunitz H, Heilbronner R, Schmid SM (2002) The eastern Tonale fault zone: a 728 'natural laboratory' for crystal plastic deformation of quartz over a temperature range from 250 to 700 °C. J Struct Geol 24:1861–1884
Sugden TJ, Deb M, Windley BF (1990) The tectonic setting of mineralisation in the Proterozoic Aravalli-Delhi orogenic belt, NW India. In: Naqvi SM (ed) Precambrian continental crust and its economic resources. Elsevier, New York, pp 367–390
Synchanthavong SP, Desai SD (1977) Protoplate tectonics controlling the Precambrian deformation and metallogenic epochs in NW India. Mineral Sci Eng 1:218–236
Tobisch OT, Collerson KD, Bhattacharya T, Mukhopadhyay D (1994) Structural relationship and Sm-Nd isotope systematics of polymetamorphic granitic gneisses and granitic rocks from central Rajasthan, India-Implications for the evolution of the Aravalli craton. Precambr Res 65:319–339
Tiwari SK, Biswal TK (2019a) Dynamics, EPMA Th-U-total Pb monazite geochronology and tectonic implications of deformational fabric in the lower-middle crustal rocks, a case study of Ambaji granulite, NW India. Tectonics 38(7):2232–2254. https://doi.org/10.1029/2017TC004891
Tiwari SK, Biswal TK (2019b) Paleostress and magma pressure measurement of granite veins in the Neoproterozoic Ambaji granulite, South Delhi terrane, Aravalli-Delhi mobile belt, NW India: implication towards extension driven exhumation of middle-lower crustal rocks. J Earth Syst Sci 128(6):150. https://doi.org/10.1007/s12040-019-1187-5
Valentine JW, Moores EM (1970) Plate-tectonic regulation of animal diversity and sea level: a model. Nature 228:657–659
Verma K, Greiling RO (1995) Tectonic evolution of the Aravalli orogen (NW India): an inverted proterozoic rift basin? Geol Rundsch 84:683–686
Vernon RH, Paterson SR, Geary EE (1989) Evidence for syntectonic intrusion of plutons in the Bear Mountains fault zone, California. Geology 17:723–726
Volpe AM, Macdougall JD (1990) Geochemistry and isotopic characteristics of mafic (Phulad ophiolite) and related rocks in the Delhi Supergroup, Rajasthan, India: Implications for rifting in the Proterozoic. Precambr Res 48:167–191
Williams ML, Jercinovic MJ (2002) Microprobe monazite geochronology; putting absolute time into microstructural analysis. J Struct Geol 24:1013–1028
Zhao G, Cawood PA, Wilde SA, Sun M (2002) Review of global 2.1–1.8 Ga orogens: implications for a pre-Rodinia supercontinent. Earth Sci Rev 59:125–162

Deformation in the Aravalli Supergroup, Aravalli-Delhi Mobile Belt, NW India and Tectonic Significance

Gregor Hahn, Georg Kodl, Helga de Wall, Bernhard Schulz, Michel Bestmann and Narendra Kumar Chauhan

Abstract Sediments of the Aravalli Supergroup have acquired a complex superposed fold pattern during Neoproterozoic amalgamation (Grenvillian Orogeny) of north Indian blocks (Marwar block, Bundelkhand craton, North Delhi crustal block) and collision with the Central Indian Craton. The Upper Aravalli metasediments reveal a prograde mono-metamorphic history with continuous increase in metamorphic conditions towards the Delhi Supergroup contact. The transition from phyllites to garnet-bearing mica schist is accompanied by strain increase seen in tightening of F_2 folds. Microstructures and deformation mechanism in quartz document an increasing temperature gradient from ~280 to 500 °C towards the Aravalli-Delhi contact. Garnets, grown prior to the contact-parallel, pervasive second cleavage indicate crystallisation along a prograde P-T-path from 400 °C/4 kbar to 500 °C/7 kbar. Older metamorphic records in the northern Aravalli-Delhi mobile belt sector (e.g. Sandmata Complex, Mangalwar Complex, and North Delhi Terrane) cannot be related with deformation and metamorphism in the Aravalli Supergroup. Considering the pattern of bivergent thrusts with the Neoproterozoic active continental margin along the western side of the Delhi Fold Belt, the lack of any magmatic signature and the inferred contractional fold and thrust deformation along the eastern proximity can be interpreted as a retro-wedge setting.

G. Hahn (✉) · G. Kodl · H. de Wall · M. Bestmann
GeoZentrum Nordbayern, Friedrich-Alexander-Universität Erlangen-Nürnberg, Erlangen, Germany
e-mail: gh156@leicester.ac.uk

B. Schulz
Institut für Mineralogie der TU Bergakademie, Freiberg, Germany

N. K. Chauhan
Department of Geology, M.L.S. University, Udaipur 313001, India

G. Hahn
University of Leicester, Leicester, UK

G. Kodl
Helmholtz-Zentrum, Deutsches GeoForschungsZentrum, Potsdam, Germany

T. K. Biswal et al. (eds.), *Structural Geometry of Mobile Belts of the Indian Subcontinent*, Society of Earth Scientists Series,
https://doi.org/10.1007/978-3-030-40593-9_2

Keywords Aravalli-Delhi contact · Aravalli supergroup · Multiple phases of folding · Quartz microstructure · Garnet thermobarometry · NW India

1 Introduction

The Aravalli-Delhi Mobile Belt (*ADMB*) is a major lineament in NW India and can be traced over a distance of ~700 km. In its southern part this NE-SW trending structure bends into an E-W trend and probably continues in the Central Indian Tectonic Zone (*CITZ*). There is evidence for a long history of this *ADMB-CITZ* belt from the Paleoproterozoic to the Neoproterozoic (Acharyya and Roy 2000; Roy and Prasad 2003). The *ADMB* marks the suture between Marwar block and Bundelkhand Craton (e.g. Meert et al. 2010) and the amalgamation with the South Indian block (Dharwar–Bastar–Singhbhum craton) as proposed by Bhowmik et al. (2010, 2012).

The *ADMB* is commonly subdivided into a northern sector (North Delhi Terrane, NDT) and a southern sector (South Delhi Terrane, SDT) on account of differences in the geological history (Fig. 1). The tectono- metamorphic and magmatic history of the NDT with ages of 1.7–1.4 Ga has been interpreted in the geodynamic context of the Columbia supercontinent (Bhowmik and Dasgupta 2012 and references therein, Kaur et al. 2013). In contrast, the evolution of the SDT gives evidence for younger sedimentation, metamorphism and magmatism related to the Delhi orogeny at 1.20–0.85 Ga (Deb et al. 2001; Pandit et al. 2003; Singh et al. 2010; McKenzie et al. 2013).

In context of the South Delhi orogeny, the western margin of the *SDT* is identified as a Neoproterozoic active convergent margin with volcanic arc activity, clastic deposits and calc-alkaline intrusions (Pandit et al. 2003; Dharma Rao et al. 2013). The Middle- to Paleoproterozoic Aravalli Terrane (*AT*) occurs to the east of the *SDT*. So far, there is no working model for welding of the *SDT* with *AT*. Interpretations range from an unconformity of Delhi sedimentation on Aravalli basement (Heron 1953; Sharma and Rahman 2000) to models of subducted oceanic crust emplaced during the closure of the Aravalli basin. In this context lineaments containing ultramafic remnants (Rakhabdev Lineament, Kaliguman Lineament; Fig. 2b) are regarded as suture zones (Sen 1981; Deb and Sarkar 1990; Sugden et al. 1990; Sinha-Roy et al. 1998; Singh et al. 2010; Bhowmik and Dasgupta 2012; Wang et al. 2018). We have attempted a structural study of Paleo- to Mesoproterozoic Aravalli Supergroup, exposed in a key section in the Udaipur Valley and its continuation towards the contact to the Meso- to Neoproterozoic Delhi units. This section is furthermore prominent for the superb record of superimposed folding as compiled in very detailed structural maps (Gupta et al. 1997; Roy and Jakhar 2002). We present petrological, microstructural and field observations along an E-W transect through the Jharol Formation (Upper Aravalli) towards the contact with the higher grade (amphibolite facies) sediments of the Gogunda Group as part of the Delhi Supergroup. We relate

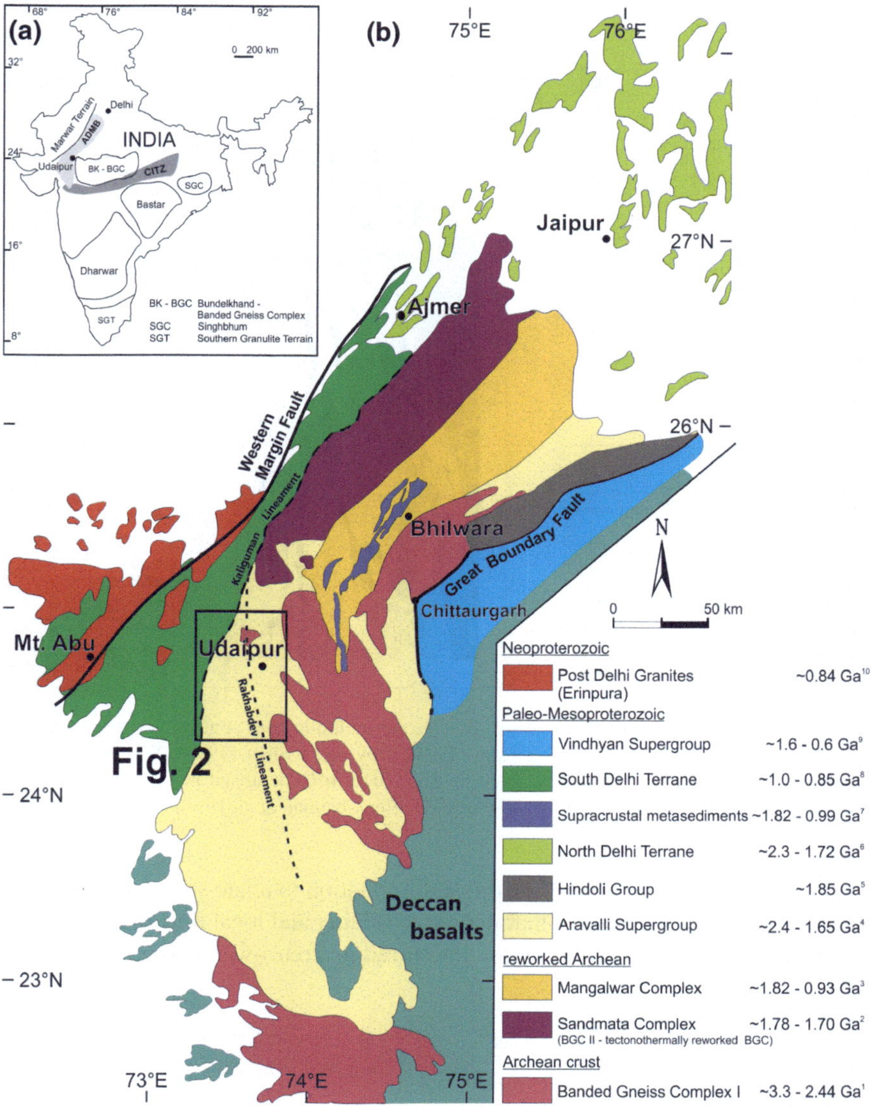

Fig. 1 **a** Simplified map showing major Archean and Early Proterozoic cratons in India. **b** Geological map of distribution of Precambrian Aravalli-Delhi Supergroups in NW India with magmatic emplacement ages (modified after Heron 1953; Gupta et al. 1997; Roy and Jakhar 2002). Age data sources: (1) Roy and Kröner (1996), Wiedenbeck and Goswami (1994), Wiedenbeck et al. (1996); (2) Sarkar et al. (1989), Buick et al. (2006), Bhowmik et al. (2010); (3) Hazarika et al. (2017); (4) Deb (1999), McKenzie et al. (2013), Wang et al. (2018); (5) Deb et al. (2002); (6) Biju-Sekhar et al. (2003), Kaur et al. (2009, 2011, 2017), McKenzie et al. (2013), Wang et al. (2017); (7) Hazarika et al. (2017), Ozha et al. (2016); (8) Volpe and Macdaugall (1990), Deb et al. (2001), Pandit et al. (2003), Wang et al. (2017); (9) Venkatachala et al. (1996), Sarangi et al. (2004), McKenzie et al. (2011); (10) Deb et al. (2001), Just et al. (2011)

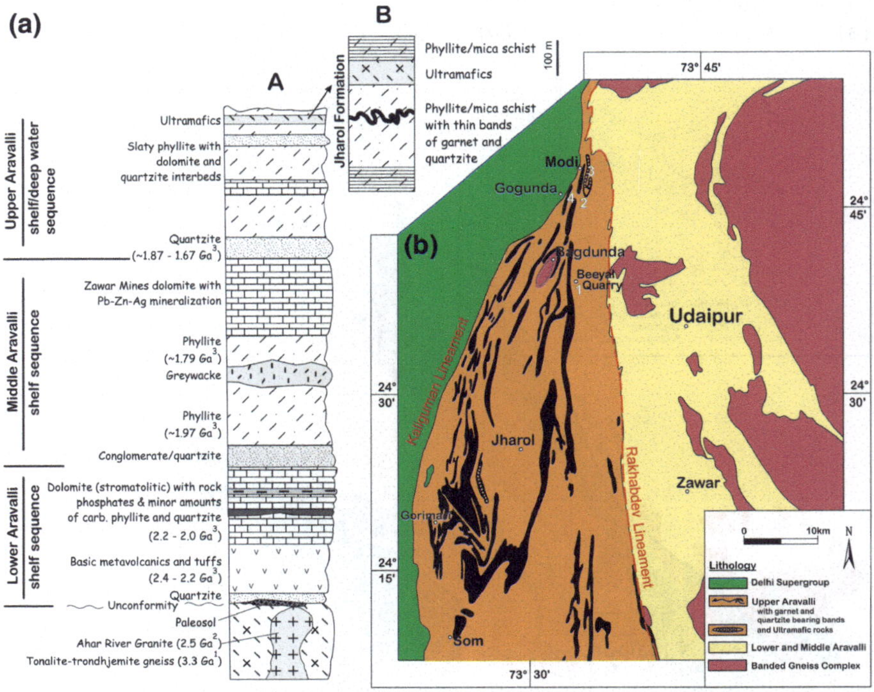

Fig. 2 **a** Lithostratigraphy of the Aravalli Supergroup including age constraints from following data sources: (1) Wiedenbeck and Goswami (1994), Roy and Kröner (1996), Wiedenbeck et al. (1996); (2) Gopalan et al. (1990); (3) Wang et al. (2018). **b** Generalised geological map of the Udaipur region (based on Heron 1953 and modified after Chauhan et al. 1996)

crustal convergence and associated thrust deformation to a late stage of Grenvillian orogenic events. It is proposed that pervasive folding and local thrusting during D_2 deformation in the Aravalli Supergroup occurred in a retro-wedge setting in respect to the Delhi orogenic belt.

2 Geological Setting

The Archean cratonic nucleus, named Banded Gneiss Complex (BGC), forms the basement of the Aravalli Supergroup sediments. Oldest record in this metamorphic complex are 3.3 Ga old zircons and a final stabilization occurred at ~2.5 Ga (Gopalan et al. 1990; Wiedenbeck and Goswami 1994; Roy and Kröner 1996; Wiedenbeck et al. 1996; Mondal et al. 2002). New tectonic models derived from geochemical and isotopic studies propose a Paleo- to Mesoproterozoic evolution (2.4–1.6 Ga) during transition from passive to an active continental margin (Absar and Sreenivas 2015; Wang et al. 2018). Considerable uplift and denudation exposed the craton to atmospheric weathering (Pandit et al. 2008; de Wall et al. 2012) forming a mature and thermally equilibrated basement with locally preserved paleosol on top for deposition of the Aravalli Supergroup (2.1–1.6 Ga; McKenzie et al. 2013; Wang et al. 2018). Sedimentation and associated mafic magmatism of the Aravalli Supergroup have been interpreted as a continuous event representing a progressive intracratonic rift basin as depositional environment (Verma and Greiling 1995; Roy and Kröner 1996). However, geochemical data and new age constraints on detrital zircons give indication for a deposition in a more complex and diverse geodynamic setting (McKenzie et al. 2013; Absar and Sreenivas 2015), and a transition from passive to active continental margin setting during the middle Proterozoic (2.4–1.67 Ga). In the central part of the *ADMB*, the Archean *BGC* is tectono-thermally reworked during an igneous and magmatic event at 1.72–1.70 Ga, (Sandmata and Mangalwar complexes, Fareeduddin and Kröner 1998; Roy et al. 2005, 2012; Buick et al. 2006). The migmatites and biotite schists are overlain by supracrustal metasediments, now preserved in folded curvilinear outcrop pattern (Gupta et al. 1997; Sinha-Roy et al. 1998). The Great Boundary Fault (*GBF*) marks the eastern limit of the Aravalli orogen and constitutes a structural boundary to the Meso- to Neoproterozoic platform sediments of the Vindhyan Supergroup (Fig. 1). Opening of the Delhi rift basin followed this scenario and there is now evidence that deposition of Delhi sediments proceeds until the end of the Mesoproterozoic, as zircon ages of as young as 1047 ± 42 Ma and 1091 ± 42 Ma are found in the Kumbalgarh Group and the Gogunda Group, respectively (Wang et al. 2018).

These newly available information on age and depositional setting of the Aravalli sediments (McKenzie et al. 2013; Absar and Sreenivas 2015; Wang et al. 2018) and on timing of metamorphic and structural imprint in higher grade metamorphic rocks of the adjoining Sandmata and Mangalwar Complexes (Ozha et al. 2016; Hazarika et al. 2017; Bhowmik et al. 2018) require a revision or refinement of pre-existing tectonic models.

A generalized lithostratigraphy for the Aravalli Supergroup is presented in Fig. 2a. Oldest members are mafic volcanics and tuffs which are intercalated with 500 m–3 km thick basal siliciclastic sediments (Aravalli quartzite). These are overlain by the Jhamarkotra Formation (Lower Aravalli), a mixed succession of dolomite, stromatolite bearing rocks, carbonaceous phyllite and quartzite. A deposition age between 2.4 and 1.9 Ga for this Lower Aravalli unit is inferred from detrital zircon dating (Wang et al. 2018). These Lower Aravalli Units are unconformably overlain by the monotonous Middle Aravalli siliciclastic-dominated units of quartzite, conglomerate and phyllite as well as carbonatic rocks. Provenance and geochemical signatures of greywackes from the Middle Aravalli have been interpreted to reflect a continental arc setting, active at about 1.6 Ga (Absar and Sreenivas 2015).

The boundary between the Middle and Upper Aravalli is marked by the Rakhabdev Lineament (Figs. 1b and 2b). Along this prominent structure, ultramafic rocks are exposed (Purohit et al. 2015 and references therein) and the significance of this lineament as long-lasting crustal scale structure is underlined by activity and reactivation during the Proterozoic and Cenozoic (Bakliwal and Ramasamy 1987; Roy and Jakhar 2002). The Upper Aravalli comprises a succession of consistently merging shelf deposits of conglomerate, quartzite and phyllite with pelitic deep-water deposits of phyllite, mica schist and fine grained quartzite/chert (Roy et al. 1993). The succession starts with sandstones (Debari Formation) and continues into slaty phyllites with dolomite intercalations (Lakhawali Formation). Youngest detrital zircons in these units are dated at ~1.6 Ga (Wang et al. 2018). The upper part of the Upper Aravalli, termed Jharol Formation, comprises phyllites with intensely folded garnet-bearing quartzite bands and ultramafics (serpentinites, talc schists), intercalated within metapelites. The ultramafic assemblage is associated with mafic rocks named Gopir Volcanics and Gopir Dykes or combined Gopir Mafics (*GM*) (Abu-Hamatteh et al. 1994). The appearance of mafic intercalations within the predominately pelitic succession gives an outline of the general regional structural grain in the Upper Aravalli (Fig. 2b).

The structural history of the Aravalli Supergroup of southern Rajasthan is marked by multi-stage deformation (D_1–D_3), which is visible in outcrop scale (Fig. 3a). Two major pervasive deformation phases constitute a complex superposition fold pattern

Fig. 3 **a** Undulating hilly landscape with erosion underlining superposition of F_3 over F_1/F_2 structures (region between Modi and Gogunda, view towards east, 24° 47′ 23.52″ N 73° 34′ 05.85″ E). Higher ridges in the background belong to quarzitic members (Gogunda Group) of the Delhi Supergroup. **b** Slaty cleavage in the Jharol Formation related to upright F_2 folding (view towards north, 24° 37′ 17.7″ N 73° 34′ 00.1″ E). **c** Isoclinal upright fold. S_1 almost parallel to S_0 (view towards north, near Gogunda). **d** Refolded isoclinal F_1 folds in hornblende schists at river NE of Bagdunda (view towards east, 24° 43′ 05.6″ N 73° 33′ 58.1″ E)

on various scales. The first folding event (F_1) is characterized by tight to isoclinal folds with varying axes and shallow dipping axial planar traces. Bedding-parallel cleavage S_0/S_1 in quartzite and quartz schist indicates the isoclinal nature of F_1 (Fig. 3b) and transposition of bedding and first cleavage. F_1 folds are superposed (Figs. 3c and 4a, b) by N-S to NNE-SSW trending, open to isoclinal, upright F_2 folds (Chauhan et al. 1996). This second deformation F_2 is predominant, controls the regional outcrop pattern (Chauhan et al. 1996; Gupta et al. 1997) and has imposed a pervasive upright cleavage in the Upper Aravalli metapelites (Fig. 3b).

A third phase of folding with fold axes perpendicular to strongly oblique in respect to F_2 axial traces has created open folds with large wavelength, low amplitudes and sub-vertical axial planes. This is best recorded in the Bagdunda domal structure (Fig. 2b), there Lower-Aravalli units occur in a tectonic window within the Upper Aravalli Jharol Formation. In the southern sector of the Upper Aravalli (Lunavada Sector, e.g. Roy and Jakhar 2002) this folding over E-W to NE-SW axes becomes more pronounced (Gupta et al. 1997; Mamtani et al. 1999).

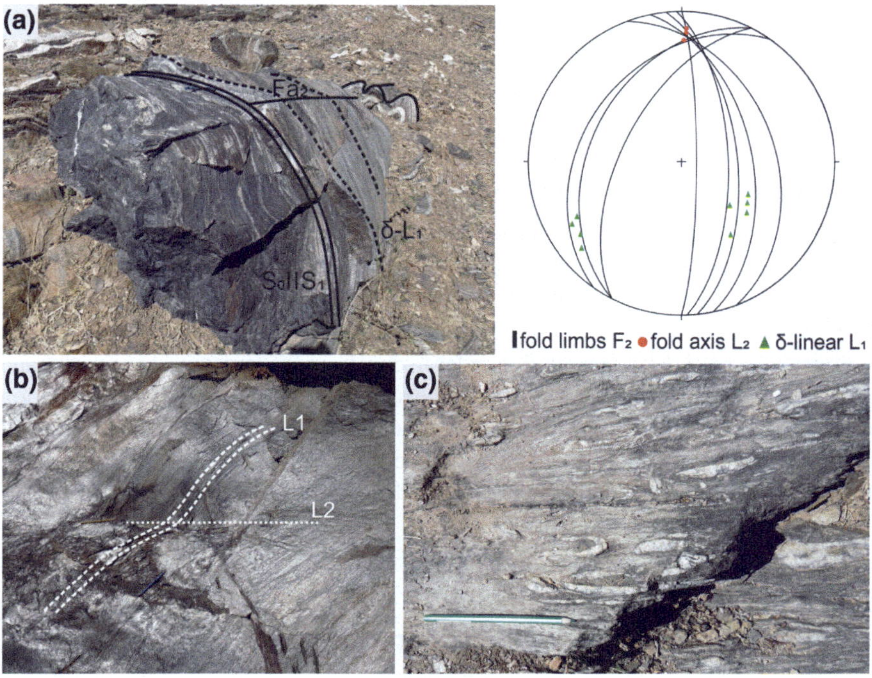

Fig. 4 Structural features in quartzites and mica schists of the M27 road section. **a** 3D fold exposure in quartzites (24° 44′ 14.24″ N 73° 33′ 18.47″ E) documenting the multiple deformation history and the corresponding stereonet; S_0 and S_1 are parallel forming macroscopically visible 'layering' with δ-L_1 on planes of F_2 fold limbs. Superposition of F_3 results in a curvature of F_2 fold axes. **b** Features of superimposed folding: L_1 got folded by S_2 and L_2 crosscuts L_1. **c** Folded multilayer of quartzite within phyllite (24° 44′ 15.78″ N 73° 33′ 39.58″ E). In the fold limbs, late stretching has led to the formation of well-aligned lens shaped boudins

Regional structural discordances along lineaments in the Delhi-Aravalli transition have been mapped on base of Landsat images and interpreted as thrusts (Din and Iqbaluddin 2000). The thrust zone marking the tectonic contact between the Delhi Supergroup and the Archean Sandmata Complex in the north has been referred to as Kaliguman Lineament (Gupta et al. 1997), which in its southern extremity splits up into a fan of thrust sheets (Din and Iqbaluddin 2000), one of which (local name Som thrust) marks the boundary between Delhi Units (Gogunda Group) and Aravalli Units (Jharol Group) in our study area (Fig. 2b). Basement-cover synclines and anticlines of Archean gneisses and Paleo- to Mesoproterozoic units form the larger structural grain in the here studied area (Fig. 1).

Increasing metamorphism conditions towards the contact zone of the Aravalli and Delhi Supergroups has been observed in earlier studies (Roy et al. 1993; Roy and Jakhar 2002 and references therein), however detailed descriptions are rare. The co-folding of basement and cover units can be explained by a simplified model as shown in Fig. 12b. Contractional deformation in the supra-crustal Aravalli Supergroup results in folding as well as thrusting along localized high strain regions. At higher depth, strain is accommodated by listric intra-basement shear zones.

3 Analytical Methods

3.1 Sample Collection and Sample Preparation

For structural and petrographical analysis 24 orientated samples were collected from key localities within the Jharol Formation west of Udaipur around Gogunda and Bagdunda. A progressive increase in metamorphic grade within the Upper Aravalli phyllites towards *DFB* is unveiled in order to constrain the structural kinematics and petrological P-T conditions to understand the relative timing of the major deformation events. Samples were cut in respect of structural elements of finite and incremental strain and subsequently mechanically polished. Samples were analysed by optical microscopy (transmitted light) under plane and cross polarized light and an additional gypsum plate.

3.2 Microstructural Analyses of Quartz Fabrics

Quartz microfabrics can provide information on temperature conditions during deformation. Quartz fabrics are generally very sensitive to record imposed strains and stress. We are aware that the quartz deformation microstructures and the concluded mechanism can give only estimated and evaluated temperatures and cannot be taken as absolute indication since their dependence on strain rates, stress, water content and other parameters (Handy et al. 2007 and references therein). However, change in prevalence of quartz microstructures in progressively metamorphosed rock sequences in the Italian Alps (Stipp et al. 2002) has shown a gradual change in recrystallization mechanism with increasing temperature inferred from mineral reaction isogrades. Virtually monomineralic quartz layers and veins in the progressively metamorphosed Aravalli metasediments are therefore examined to gain information on deformation mechanism in here studied section.

3.3 SEM-Based Automated Mineralogy (MLA) and EPMA

Thin sections of garnet mica schist were analysed by automated mineralogical methods (Fandrich et al. 2007), based on a Scanning Electron Microscope (SEM) Quanta 650-FEG-MLA by FEI Company, equipped with Bruker Dual X-Flash energy dispersive spectrometers for EDX analyses. Electron beam conditions were set at 25 kV acceleration voltage and a beam current of 10 nA when focused. The software package for mineral liberation analysis (MLA version 2.9.0.7 by FEI Company) was used for the automated steerage of the electron beam for EDX identification of the mineral grains and the collection of numerous EDX spectra. The GXMAP routine produces a narrow grid of ~1600 single EDX-ray spectra per mm^2. For the classification of mineral phases and compositions, a list of identified reference EDX-ray spectra was established by collecting spectra from matrix phases and from defined parts of several garnet porphyroblasts (core–mid–rim). Garnet reference spectra are characterized by EDX-ray single spot elemental analyses, which reveal strong variations in Fe, Mg, Mn and Ca from the porphyroblasts. The reference spectra were then labelled with the corresponding garnet Fe–Mg-Mn–Ca compositions in a generic way (Schulz 2017; Oriolo et al. 2018). The GXMAP measurements were classified against the reference EDX-ray spectra list with a high degree of probability of match and with an algorithm marking the non-matching spectra as unknowns. Therefore the classified spectra represent semi-quantitative compositional data. When the labelled spectra are arranged on a color scale, semi-quantitative garnet zoning maps are established.

The GXMAP measurements allowed the selection of zonation profiles in garnet porphyroblasts for quantitative WDS analysis with the electron probe microanalyser (EPMA). These mineral-chemical analyses were performed using a JEOL JXA-8900-RL instrument at beam conditions of 15 kV, 20 nA, and 2 μm, and with the corresponding ZAF correction procedures (Table 1).

Based on the same assumptions of thermodynamic equilibrium, there are two principal methods for estimating pressure-temperature conditions recorded in metamorphic rocks. The classical thermobarometry involves continuous cation-exchange and net-transfer solid-solid reactions (Spear 1993; Powell and Holland 1994, 2008). Alternatively, pseudosection calculations define phase assemblage stability fields for given bulk compositions and allow the contouring of divariant fields (Holland and Powell 1998, 2011). However, apart from the basic pre-requisites and common uncertainties, the pseudosection calculations also involve the additional assumption regarding the effective local chemical bulk composition in a rock domain of unknown

Table 1 Electron microprobe analyses of garnet (Grt), biotite (Bt), muscovite (Ms) and plagioclase (Pl) in mica schist. Garnet in almandine (Alm), grossular (Grs), pyrope (Prp) and spessartine (Sps) components in %. Garnet and plagioclase analyses are from core (c), intermediate parts (m) and rim (r) of porphyroblasts. Combinations of analyses for geothermobarometry are: Grt105-Bt37-Ms40-Pl51; Grt96-Bt56-Ms40-Pl75

Grt	105c	100 m	111 m	96r	Mica	Bt37	Bt56	Ms40	Ms63	Pl	Plc74	Plc51	Plr75	Plr45
SiO_2	36.88	36.59	35.96	36.35	SiO_2	35.65	35.34	45.22	45.79	SiO_2	64.28	64.85	65.63	65.10
Al_2O_3	20.71	20.91	21.98	20.97	Al_2O_3	18.18	18.69	34.35	34.69	Al_2O_3	22.60	22.76	21.35	21.15
FeO	25.77	28.52	30.44	33.87	TiO_2	1.27	1.54	0.21	0.24	CaO	4.11	3.92	2.91	2.73
MnO	9.93	7.20	5.44	3.08	FeO	22.09	23.60	1.76	2.10	Na_2O	9.56	8.66	10.05	10.80
MgO	0.73	0.79	0.89	1.16	MnO	0.06	0.10	0.02	0.00	K_2O	0.07	0.08	0.05	0.07
CaO	5.19	5.51	4.29	3.63	MgO	6.49	7.21	0.51	0.49		100.62	100.27	99.99	99.84
	99.21	99.52	99.00	99.05	CaO	0.13	0.18	0.02	0.05					
					Na_2O	0.17	0.14	0.69	1.25	Si	2.821	2.841	2.886	2.873
Si	3.007	2.979	2.938	2.980	K_2O	8.08	8.41	10.61	9.75	Al	1.169	1.175	1.107	1.100
Al	1.989	2.006	2.116	2.027		92.12	95.23	93.40	94.35	Ca	0.193	0.184	0.137	0.129
Fe	1.757	1.941	2.079	2.322						Na	0.813	0.735	0.857	0.924
Mn	0.686	0.497	0.376	0.214	Si	2.821	2.731	3.085	3.085	K	0.004	0.004	0.003	0.004
Mg	0.089	0.096	0.108	0.141	Al^{IV}	1.179	1.269	0.915	0.915		5.000	4.939	4.989	5.030
Ca	0.453	0.480	0.375	0.319	Al^{VI}	0.516	0.434	1.847	1.840					
	7.980	7.999	7.992	8.002	Ti	0.076	0.090	0.011	0.012	*X*An	0.19	0.20	0.14	0.12
					Fe	1.461	1.525	0.101	0.119					
Alm	58.9	64.4	70.7	77.5	Mn	0.004	0.007	0.001	0.000					
Sps	23.0	16.5	12.8	7.1	Mg	0.766	0.830	0.052	0.049					
Prp	3.0	3.2	3.7	4.7	Ca	0.011	0.015	0.002	0.004					
Grs	15.2	15.9	12.8	10.6	Na	0.026	0.022	0.092	0.163					

(continued)

Table 1 (continued)

Grt	105c	100 m	111 m	96r	Mica	Bt37	Bt56	Ms40	Ms63	Pl	Plc74	Plc51	Plr75	Plr45
*X*Mg	0.030	0.032	0.037	0.047	K	0.815	0.829	0.924	0.838					
*X*Ca	0.152	0.159	0.128	0.106		7.674	7.751	7.029	7.024					
					*X*Mg	0.34	0.35	0.34	0.29					

size. Especially for the assemblage with the garnet cores the uncertainty upon the reactant bulk compositions appears to be critical.

The classical thermobarometry has been chosen as the preferred method for P-T estimations. For the P-T estimations for garnet core crystallisation, analyses of garnet 1 cores were combined to enclosed mica and cores of plagioclase in the matrix. For the later stage of the P-T evolution, analyses of garnet rims were related to mica in pressure shadows, in the main foliation, and the rims of zoned matrix plagioclases (Table 1). Temperatures were estimated by garnet-biotite geothermometers (Perchuk and Lavrenteva 1983; Bhattacharya et al. 1992; Holdaway 2001). The temperatures were combined with the garnet-biotite-muscovite-plagioclase (GBMP) geobarometers. These calibrations are based on the internally consistent thermodynamic mineral data set (Holland and Powell 1998), with the activity models for garnet and plagioclase taken from Ganguly et al. (1996), Powell and Holland (1993) and the updated calibration by Wu (2015). The calculations from the analyses of each mineral combination resulted in a P-T box in which the temperatures are limited by the garnet-biotite thermometers (Powell and Holland 1994). The lower and upper pressures of the P-T box are given by the isopleths of the GBMP-barometers. All geothermobarometric estimates include a minimum error of ±50 °C and ±1 kbar (Fig. 7d). The calculated P-T path is mainly dependent on the garnet *X*Ca and *X*Mg evolution, as the corresponding mica and plagioclase have only slight compositional variations. The shapes or the relative $\Delta P/\Delta T$ trends of the P-T paths appear to be mainly preserved when uncertainties about the corresponding plagioclase and biotite are considered (Spear 1993).

4 Results

4.1 Structural and Lithological Characteristics

Superb macroscopic deformation features of superimposed fold geometries are exposed along the recently constructed M27 road section from Udaipur to Pindwara, an example is documented in Fig. 4a. The S_0 bedding is subparallel to the S_1 schistosity and forms intersection δ-lineation which is folded during the superposition of F_2 over F_1 and visible at the fold limbs of F_2. S_0 and S_1 are being refolded by F_2 open upright folds which are shallowly northwards plunging and undulating due to F_3. Intercalated quartz veins are aligned in the phyllite to the formative F_2 cleavage planes exposing sheared boudinage structures and residual fold hinges of F_1 (Fig. 4c).

Restricted to the westernmost part of the Jharol Formation near the contact to the DFB there occur garnet-rich quartz bands. This prominent intensely folded, thin and competent marker horizon can be traced for ~150 km between Gogunda and Jharol (Roy and Jakhar 2002) and impressively documents the structural and metamorphic imprint along the Aravalli-Delhi contact. Thin garnet-quartzite bands (mm to cm) within mica schists show ptygmatic folding indicating strong viscosity contrasts of

Fig. 5 **a** Orthorhombic folds in a garnet quartzite layer within mica schist in a F_2 fold hinge. **b** Refolded (during D_2) isoclinal F_1 fold in garnet-quartzite band at Modi. **c** Broadly cylindrical N-S trending F_2 fold hinges of garnet quartzite layer seen in road cuts along the M27 (viewed towards north; 24° 44′ 18.76″ N 73° 33′ 28.23″ E)

layers (Fig. 5a, b). Superposition of F_2 folding over isoclinal F_1 folds is evident (Fig. 5c). Thickened hinge zones, compression in the inner arc and dilation in the outer arc are typical, and buckling can be inferred as dominant process of folding. At places a very regular parallel pattern of N-S trending F_2 fold traces gives a spectacular impression of this ptygmatic folding (Fig. 5c).

In general, a transition from lower greenschist to lower amphibolite facies is recorded in the metapelites (Fig. 6). The lower greenschist facies mineral assemblage is composed of quartz + chlorite + magnetite + rutile (Fig. 6a, b). Zircon is observed as an accessory mineral. Cleavage and microlithons are both well defined by mica, converted entirely to chlorite, proceeding as thin layers (80–140 μm) and single mineral ribbons between the quartz layers. Towards the west, a coarsening of the major constituents is evident within the phyllites where chlorite and quartz become consistently less abundant. At ~2 km distance to the Aravalli-Delhi contact, growth of white mica and garnet starts to occur (Fig. 6c). Increasing content in white mica along with the porphyroblastic growth of garnet indicates the transition from lower greenschist facies phyllites to higher grade garnet mica schist. The mica schist is host to the prominent north-south striking garnet-bearing quartzite bands in the westernmost part of the Jharol Formation near the Delhi-contact zone (Fig. 5). In

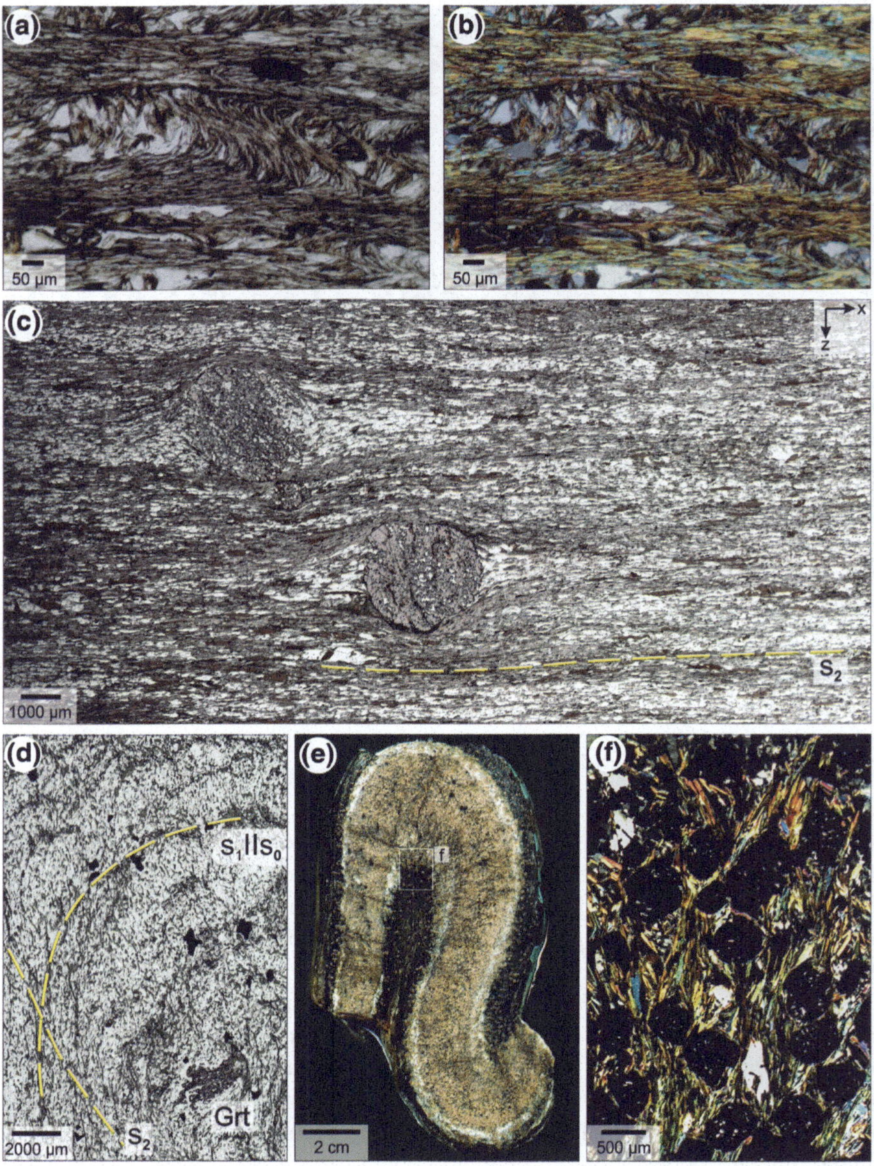

Fig. 6 **a** Well defined crenulation cleavage with microfolds of F_2 refolding F_1 marked by chlorite and white mica (plane polars). **b** Under cross polars. **c** Garnet porphyroblasts with internal fabric (S_1 cleavage) oblique to the pervasive second cleavage (S_2), (plane polars). **d** Subparallel layered bedding planes in S_0–S_1 and mineral phases aligned in S_2. **e** F_2 fold of quartz-garnet band with thickened hinges. **f** Microphotograph (cross polars) documenting the high abundance of garnet porphyroblasts, rotated and arranged within the parallel cleavage (S_2)

such bands fine grained (70–350 μm) idiomorphic to xenomorphic garnet crystals constitute up to 45% of the rock fabric (Fig. 6e, f).

During microscopic survey, special attention was given to the detection of newly formed monazite in order to obtain time constraints on metamorphism and deformation in the Upper Aravalli. Monazites in metapelites of the here studied section are generally too small (<20 μm) to allow any reliable dating.

4.2 Garnet Mineral Chemistry and Geothermobarometry

Utilizing the GXMAP mode of SEM automated mineralogy, zoned garnet porphyroblasts with diameters up to 3 mm were recognised (Fig. 7a). The large garnet porphyroblasts typically have Mn-rich cores at ~23% spessartine, ~15% grossular and ~3.0% pyrope. Toward the rims the spessartine (7%) and grossular (10%) decrease, whereas almandine (28%) contents increase (Fig. 7c). In the grossular-pyrope-spessartine ternary diagram (Fig. 7b), all analysed porphyroblasts display the same evolution trend. The garnets crystallised in a mineral assemblage with biotite, muscovite, plagioclase and quartz. Microstructures show that the micas coexisted with garnets, as both occur as inclusions, in pressure shadows, and in the main foliation of S_1. Also plagioclase, which occurs in porphyroblasts in the matrix, coexisted throughout the garnet crystallisation. Cores of biotites and muscovites enclosed in garnet porphyroblasts, in pressure shadows of such, and aligned along the main foliation were analysed (Table 1). In all samples the XMg of biotites are at ~0.35 and at Al^{IV} between 0.18 and 0.27 (per formula unit, p.f.u). The Si^{4+} in muscovites are low with 3.08 at Al^{IV} of 0.92 and Na of ~0.1 (p.f.u). Plagioclase in the matrix is zoned with An_{20} in the core to An_{14} in the rim.

The garnets recorded a crystallisation along a prograde P-T path from 400 °C/4 kbar to 500 °C/7 kbar (Fig. 7d). These are conditions which are within the stability field of xenotime as outlined by Spear (2010). Correspondingly, xenotime is observed along the foliated matrix. Furthermore, the P-T conditions of garnet crystallisation are throughout at lower temperatures as are indicated for monazite crystallisation, even at low Ca and high Al bulk compositions (Janots et al. 2007; Spear 2010; Spear and Pyle 2010). This may explain that various hydrous small LREE-bearing phosphate and silicate grains, but no monazite, are observed in the sample. There are also no indications that pre-existent monazite has been decomposed during a retrogressive metamorphism, as described by Budzyń et al. (2011) and Goswami-Banerjee and Robyr (2015).

4.3 Quartz Microstructural Fabrics

In the following microstructural characteristics are presented for four samples located in decreasing distance to the Aravalli-Delhi contact. Sample locations are indicated

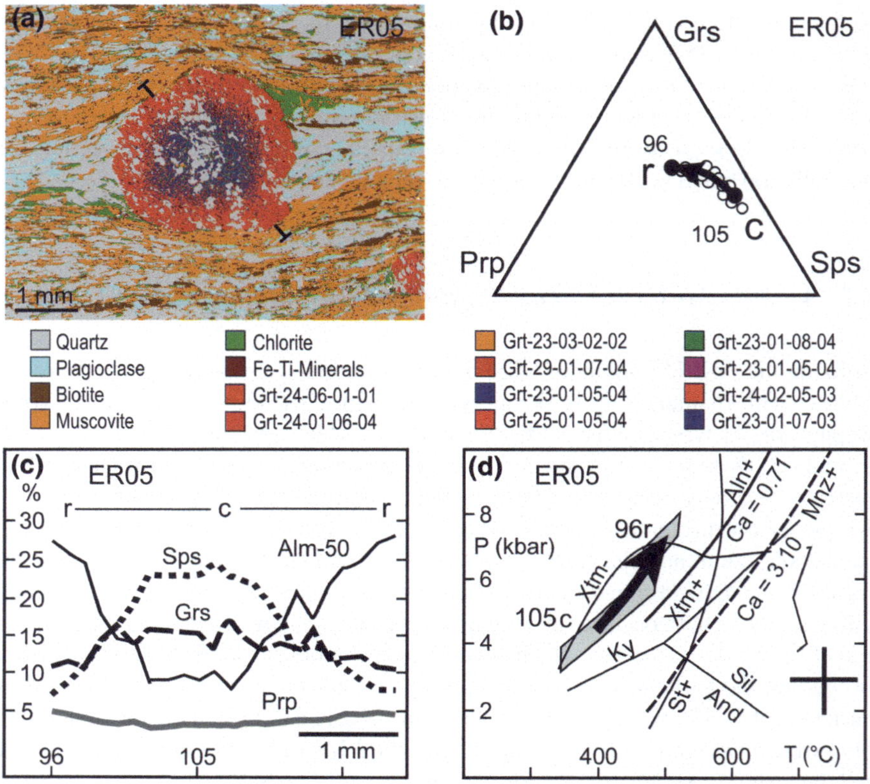

Fig. 7 **a** Map of classified and color-coded EDX spectra (GXMAP measurement mode of automated SEM) from sample GH-15-05 (ER05). EDX spectra are related to distinct minerals. EDX spectra from chemically zoned garnet (Grt) are generically labelled according to their Fe–Mg-Mn–Ca contents, in complete analyses normalised to 100 wt%. **b** Garnet core-to-rim (c, r) zonation profile plotted in grossular-pyrope-spessartine (Grs-Prp-Sps) ternary diagram. **c** Garnet core-to-rim (c, r) zonation profile in spessartine (Sps), almandine (Alm-50%, due to scale), grossular (Grs) and pyrope (Prp) components, in mole%. Numbers are selected garnet analyses, used for geothermobarometry in Table 1). **d** Prograde metamorphic P-T evolution during garnet crystallization, with error bar to the left. Lines indicate stability fields of andalusite (And), kyanite (Ky), sillimanite (Sil), staurolite (St) and xenotime (Xtm), after Spear (1993, 2010). Stability fields of monazite (Mnz) and allanite (Aln) at different bulk rock contents as a function of Ca wt%, according to Janots et al. (2007) and Spear (2010)

in Fig. 2b and are from E to W: (1) veins in phyllites exposed in the Beeyal quarry (GK-15-09: 24° 39′ 18.65″ N 73° 34′ 34.06″ E), (2) folded quartz bands within quartz-rich mica schists exposed near the road cut of the highway No. 27 (GK-16-01: 24° 44′ 14.24″ N 73° 33′ 18.47″ E), (3) metasediments with strongly folded quartz-veins near Modi (GK-16-02: 24° 48′ 12.86″ N 73° 34′ 26.24″ E), (4) quartz-rich metapelites with planar fabrics in contact with ultramafic units, exposed along the highway 27 (GK-15-8: 24° 44′ 51.23″ N 73° 33′ 10.70″ E) in 500 m distance to the Aravalli-Delhi contact.

Observed microstructures are presented in Figs. 8, 9, 10 and 11, following the numbering scheme 1 to 4 of outcrops as shown in Fig. 2b, with decreasing distance towards the Aravalli-Delhi contact. Inferred deformation temperatures are also discussed.

Typical low-temperature intracrystalline deformation microstructures (Groshong 1988; Derez et al. 2015 and references within) are seen in the Beeyal quartz (GK-15-09, locality 1): Observed deformation microstructure comprises crack-assisted lattice distortion (Fig. 8a–d) which is typical for low temperatures indicating the transition from brittle to crystal-plastic deformation. Under such conditions mobility of dislocation in the quartz lattice is reduced. This type of deformation fabric is typical for condition between 250 and 280 °C (Stipp et al. 2002 and references within). At

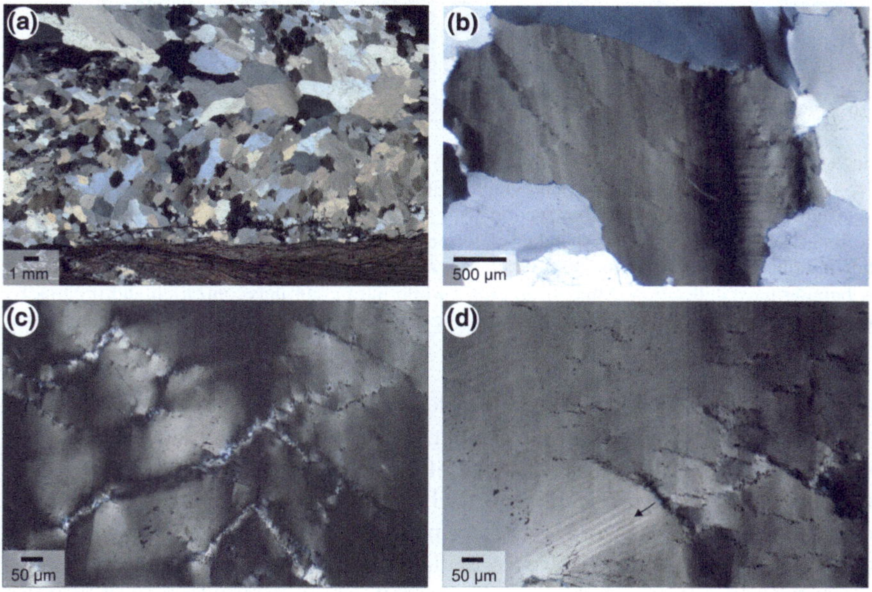

Fig. 8 Quartz microstructures in a folded vein (F_2 fold) within low grade phyllites (Beeyal quarry, sample GK-15-09); all photographs with cross polars. **a** Scan of thin section documents the heterogenous quartz fabric locally with curved mineral shapes. Grain boundaries of the coarse quartz grains are locally irregular which could be related to diffusion along grain boundaries (Kruhl and Nega 1996) or fluid-mediated dissolution–precipitation processes (Spruzeniece et al. 2017). The microphotographs (**b**–**d**) give an insight in several deformation microstructures related to low-temperature deformation of quartz (Derez et al. 2015). **b** Characteristic are strong undulose extinction and deformation lamellae in combination with deformation bands at a high angle to the deformation lamellae indicating a state of high stress (Vernooij and Langenhorst 2005). **c** Patchy extinction in combination with localized extinction bands (bLEB, sLEB, gLEB; following the terminology of Derez et al. 2015). **d** Lamellar interfingering of repetitive orientation domains (black arrow) parallel to deformation lamellae may reflect crystal plastic recovery resulting in increasing misorientation along deformation lamellae after the main high stress event

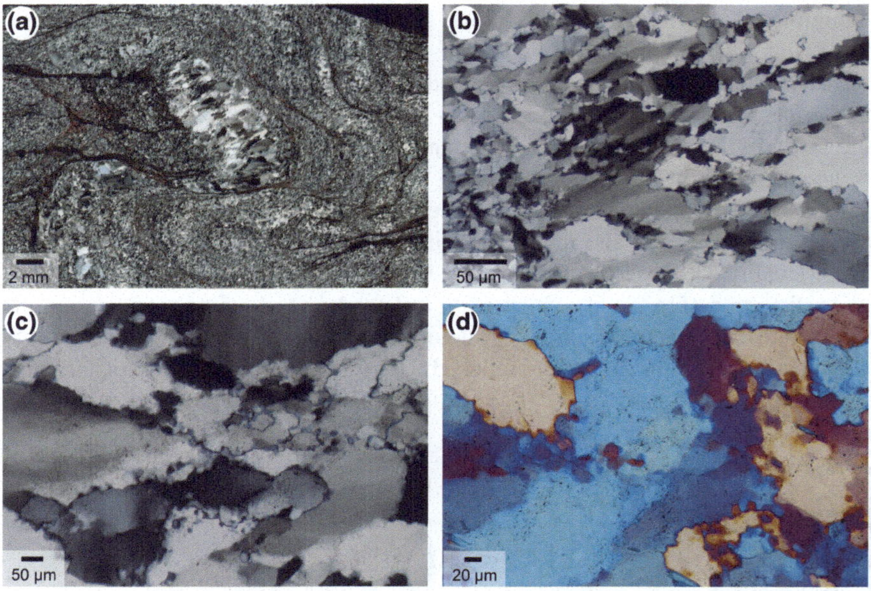

Fig. 9 Quartz microstructures in a folded vein (F_2 fold) within mica schists (M27 road cut, sample GK-16-01, see also Fig. 4a); **a–c** photographs with cross polars and **d** additional overlayed with a gypsum plate. **a** Scan of the thin section with trace of F_2 fold axial plane oriented parallel to the long axes of the image. **b** Abundant deformation bands in large quartz grains and domains showing quartz recrystallization. **c, d** Grain size reduction along grain boundaries mainly occurs by bulging recyrstallization (Stipp et al. 2002) while intra-grain polygonization and progressive subgrain rotation is also abundant. **d** Note the fine grain size (c. 10 μm) of recrystallized quartz

locality 2 (GK-16-01, Fig. 9a–d) higher deformation temperatures are indicated by crystal plastic behavior of quartz. Prevailing microstructures are bulging and subgrain polygonization indicating the occurrence of bulging recrystallization (BLG) together with subgrain rotation recrystallization (SGR). This indicates activity of temperature-related dislocation climb processes and points to temperatures between 280 and 350 °C (Stipp et al. 2002). However, the latter becomes more dominant and larger areas of quartz grains are subjected to recrystallization and can be related to higher deformation temperatures towards 400 °C (Stipp et al. 2002).

In both samples taken near the Aravalli-Delhi contact (localities 3 and 4, Figs. 10 and 11), beside subgrain rotation, clear evidence for grain boundary migration (GBM) is seen. Mobility of grain boundaries is typically related to a higher temperature regime (>500 °C, Stipp et al. 2002) triggered by intergranular fluids while water-deficient conditions hinder grain boundary mobility (Mancktelow and Pennacchioni 2004). Distinct differences in quartz fabrics of both samples are evident (Figs. 10 and 11). The sample from Modi (GK-16-2, locality 3) shows a very heterogeneous fabric (Fig. 10) with individual grains growth most likely as consequence of differences in dislocation density between neighboring grains resulting in prominent GBM microstructures. All in all, for both samples with different fabrics (planar,

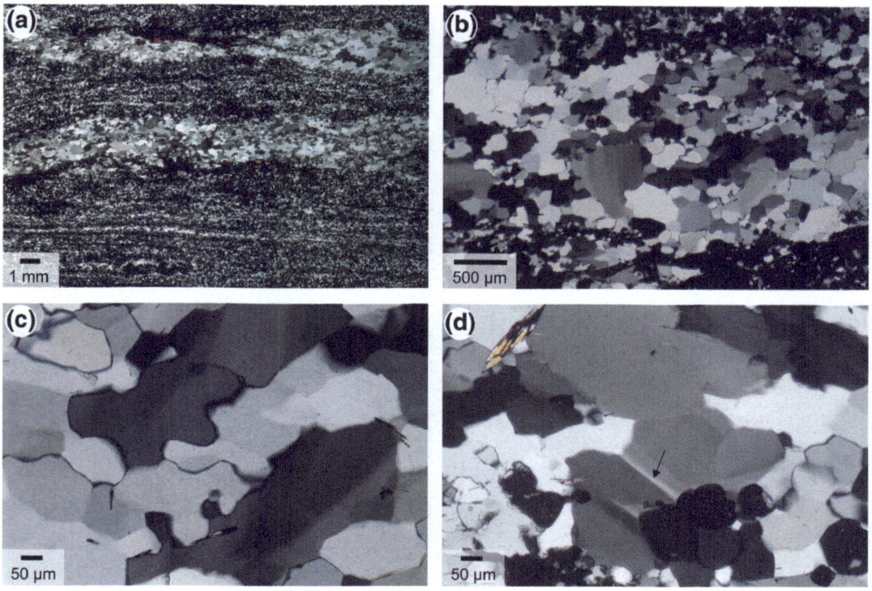

Fig. 10 Quartz microstructures in a folded quartz vein within mica schists near Modi (GK-16-02); all photographs with cross polars. **a** Scan of the thin section F_2 cleavage planes parallel to the long axes of the image. **b** Very heterogeneous quartz fabric with variation in grainsize and grain orientation. **c** A closer look shows straight as well as lobate (wavy to bulged) grain boundaries indicating grain boundary migration (GBM); in general the grains show homogeneous extinction except larger grains with subgrain boundaries. **d** Locally, a low-temperature overprint (black arrow) is seen which is mainly related to inter-grain contacts with high-angle lattice misorientation

strongly folded), the coexistence of SGR and GBM quartz microstructures indicates deformation temperatures around 500 °C.

5 Discussion and Interpretation

Considering the recently constrained time frame for deposition of the Aravalli Supergroup (McKenzie et al. 2013; Absar and Sreenivas 2015; Wang et al. 2018) and in context of new thermobarometric and age data for the polymetamorphic history of the adjoining Mangalwar Complex (Dey et al. 2019; Ozha et al. 2016) we discuss the Meso- to Neoproterozoic evolution and compile our findings to give an interpretation for the depositional setting and regional structural evolution.

Petrographic and microscopic analyses show an increase in mafic components and iron oxides towards the top of the Upper Aravalli sequence, culminating in occurrence of ultramafic intercalations in the uppermost part (Jharol Formation). We therefore favor a model of stratiform deposition of the mafic to ultramafic units in the Jharol as has been proposed in earlier studies (Sharma 1988). Occurrence of thin

Fig. 11 Quartz microstructures in a quartz-rich metasediment sample from M 27 road cut near to the Aravalli-Delhi transition (GK-15-08). **a** Scan of the thin section (crossed polars), sample is cut perpendicular to the planar foliation and parallel to the distinct stretching lineation (xz-section); note grain shape preferred orientation (SPO) oblique to the foliation. **b** Quartz microstructure under cross polars with various grains size related to amount of second phase particle. **c** Same area as **b** (plane polars): Actinolithe, white mica, epidote and magnetite are accessory minerals and define parallel foliation planes. **d** Details shown under cross polars: in quartz bands straight to lobate grain boundaries prevail, grain boundary migration (GBM) is evident from pinning structures (Jessel 1987) and overgrowth of foliation parallel accessory minerals by quartz

conformable ultramafic bands within the upper Jharol Formation as reported in Roy and Jakhar (2002) support this contention. Distribution of ultramafic rocks on a larger map scale (Fig. 2b) and the strong variation in thickness and strain between exposures as observed in the field is comparable with the structural pattern of superimposed folding. A comparable pattern is seen in the distribution of quartzite layers along the western proximity of the Aravalli sediments, discussed in a previous study (de Wall et al. 2012).

A model of stratiform deposition based on lithological observations and structural field relationship instead of dismembered ophiolites (Sugden et al. 1990; Singh et al. 2010) is supported by geochemical signatures which infer an extensional continental setting with characteristic continental tholeiites and continental flood basalts (Abu-Hamatteh et al. 1994; Abu-Hamatteh 2005; Ahmad et al. 2008). Continental rifting and associated mafic magmatism, contemporaneous with clastic sedimentation is therefore a likely setting for deposition of the Jharol Group. In contrast, the linear arrangement of ultramafic linear bodies along the Rakhabdev lineament, which are crosscutting the general fold trends (Gupta et al. 1980), are more likely fault-bounded intrusions into the Aravalli rocks, suggested previously by Chattopadhyay and Gangopadhyay (1984).

The Aravalli basement-cover sequence exposed in the Udaipur–Gogunda sector shows Archean crystallization ages in the basement (Banded Gneiss Formation, Fig. 1b) and Mesoproterozoic detrital ages in the cover units (Fig. 2a). Monazite in metapelites of the here studied section are generally too small (<20 μm), and also show hydrous compositions which both allowed no reliable dating of the metamorphic imprint. Also the reconstruction of the prograde P-T path by geothermobarometry of garnet-bearing assemblages revealed metamorphic conditions outside the monazite stability field, even for aluminous and Ca-poor bulk compositions. However, a crystallisation of monazite could be expected in Ca-poor metapelites, when decompression/further heating would have followed the P-T conditions of garnet rim crystallisation at ~500 °C/7 kbar (Fig. 7d). In consequence, such monazites would be related to the post-peak pressures and the retrograde evolution in the Jharol units.

By evaluation of deformation mechanisms in quartz fabrics, ambient condition of deformation in the Upper Aravalli could be deciphered. Temperatures are estimated from optical microstructural analysis of quartz veins sampled in an E to W transect. Prevalence of features of low-temperature plasticity is seen in quartz-veins within lower greenschist-facies phyllites in the eastern sector of the Jharol formation and a successive change in recrystallization mechanism from bulging recrystallization and subgrain rotation to grain boundary migration is observed in the mica schists towards the contact of Upper Aravalli with the Delhi sequences. Considering established temperature estimates (Stipp et al. 2002; Derez et al. 2015), this can be taken as indication for an increasing temperature gradient from ~280 to ~500 °C. Inferred deformation temperatures can only be regarded as an estimate as other parameter such as strain-rate and fluid network can influence the deformation mechanism (Handy et al. 2007). However, the data have been substantiated by a P-T path from 400 °C/4 kbar to 500 °C/7 kbar in metapelites of the Jharol Formation near the Aravalli-Delhi contact. Garnet overgrowing the first cleavage S_1 and rotation of S_1 during subsequent deformation (F_2 folding), relate peak metamorphic conditions definitely after D_1 and sustaining during D_2. As there is no indication for any retrogression in the metamorphic assemblages, we infer a close timely relationship between the deformation events. The superimposed folding, impressively documented in the garnet-quartz veins (Fig. 5), has to be related to a continuous deformation path. High angle between traces of F_1 and F_2 fold axes, e.g. seen in folding of L_1 (intersection of bedding plane and first cleavage) by the subsequent F_2 (Fig. 4) indicate non-coaxial orientation of F_1 and F_2 fold axial traces. The diversity of F_1 fold axes within the Aravalli units has been noted in several studies (for compilation see Roy and Jakhar 2002). We could not examine the complex and variegated interference patterns, documented in regional structural maps (Gupta et al. 1997; Roy and Jakhar 2002 and references therein) and was not within the scope of this study, as such this objective needs further detailed field work. As outlined in many previous field studies and analogue experiments the type of interference pattern, introduced by Ramsay (1967) and extended by Grasemann et al. (2004), can change completely during progressive deformation or superposition of deformation events whereby initial fold geometry is a major parameter for the final type of interference pattern (Ghosh et al. 1992; Grujic 1993).

In contrast to the mono-metamorphic low-temperature Aravalli metasediments, the adjoining higher-grade Sandmata and Mangalwar Complexes and the basement in the granitic terrain in the North Delhi Fold Belt give record of several Paleo- to Mesoproterozoic episodes of crustal reworking (Buick et al. 2006; Kaur et al. 2007; Bhowmik et al. 2010, 2018), assigned to the 1.8–1.5 Ga period for accretion and dispersal of Columbia supercontinent (Zhao et al. 2003). During this orogeny, the Archean basement (Banded Gneiss Complex, BGC) hosting the Sandmata and Mangalwar metasediments has also been reworked (Dharma Rao et al. 2011), previously referred to as BGC II (Heron 1953), in contrast to the unaffected basement underlying the Aravalli Supergroup (BGC I). Age of metamorphism in the Sandmata Complex (1.70–1.73 Ga, Buick et al. 2006; Bhowmik et al. 2010) and the major metamorphic event in the Mangalwar Complex (1.82 Ga, Ozha et al. 2016) give information on timing of this Mesoproterozoic crustal reworking. Volcanic activity along an active plate margin and exogene processes during denudation of this Mesoproterozoic continent (as part of the Columbia supercontinent, e.g. Rogers and Santosh 2002), can account for the accumulation of 1.87–1.65 Ga old zircons in detritus of the Upper Aravalli (Wang et al. 2018) and could constitute the postulated arc setting, identified as provenance for part of the Middle Aravalli successions (Absar and Sreenivas 2015). The structural imprint and the low-grade metamorphic overprint in the Aravalli Supergroup is therefore a younger event and care should be given, not to relate older metamorphic records, seen in the northern sector of the *ADMB* (e.g. Sandmata Complex, Mangalwar Complex, NDT), to constrain the age of rifting, deformation and metamorphism in the Aravalli Supergroup. Stratigraphic correlations including detrital zircon ages by Wang et al. (2017) indicate younger deposition of the sedimentary successions in the *SDT* (lower age limits: 1.31–1.04 Ga for Kumbalgarh and Gogunda Groups) as compared to the *NDT* (youngest zircon ages ~1.75 Ga), underlining the different evolution. New combined structural and geochronological studies on supracrustal metasediments of the northern sector of the *ADMB* indicate a major tectonometamorphic event at ~1.3 Ga (D'Souza et al. 2019), supporting earlier data by Ozha et al. (2016). So this terrain could be suspected as provenance for the relatively young *SDT* successions.

These new constraints give a time frame for crustal convergence, associated folding and finally welding of the Aravalli and Delhi Supergroups. Considering the zircon ages of 1091 $\pm$ 42 Ma for deposition of the Gogunda Group in combination with the monometamorphic nature of the Aravalli Supergroup as inferred in our study, a closure of the South Delhi Basin and timing of the prograde metamorphism of the Jharol Formation has to be younger than 1 Ga. Precise constraints on timing of crustal convergence can be obtained from a compilation of available age data, which attests a bracket of 980–950 Ma for magmatism and metamorphism (Table 2) and supports the Grenvillian age for the Delhi orogeny as suggested by Bhowmik et al. (2010).

Table 2 Compiled age constraints for evolution of the South Delhi Terrane (SDT)

Event	Age	Event	Method	References
Sedimentation				
SDT, granulite terrane	1214 ± 25 Ma	Maximum deposition age		Singh et al. (2010)
SDT, Gogunda Group	1091 ± 43 Ma	Maximum deposition age		Wang et al. (2017)
SDT, Kumbalgargh group	1047 ± 42 Ma	Maximum deposition age		Wang et al. (2017)
Arc magmatism				
Sendra granites	968 ± 1 Ma	Crystallization, Chang intrusion	U-Pb, zircon	Pandit et al. (2003)
SDT, rhyolites	990 ± 6 Ma 987 ± 6.4 Ma	Crystallization, rhyolites	U-Pb	Deb et al. (2001)
SDT, granites	1015 ± 4.4 Ma 966.5 ± 3.5 Ma	Crystallisation	U-Pb	Dharma Rao et al. (2013)
Metamorphism				
SDT	970 ± 9 Ma	Metapelites	EPMA, monazite	Chatterjee et al. (2017)
SDT	980 ± 22 Ma	Staurolite schist	EPMA, monazite	Bose et al. (2017)
Mangalwar Complex	921 ± 43 Ma (Xn) 945 ± 21 Ma (Mz)	Metapelites	Xenotime, Monazte, EPMA	Hazarika et al. (2017)
Pur-Banera Basin	987 ± 7 Ma	Monazite	EPMA,	Ozha et al. (2016)
Rampura–Agucha sulfide deposit	~1010–960 Ma ~1000 Ma	Recrystallization/alteration of existing uraninite monazite	EPMA,	Ozha et al. (2017), Hazarika et al. (2013)
Mangalwar Complex	~970–930 Ma		U-Pb	Buick et al. (2010)
Mangalwar Complex, Sandmata Complex	~980–950 Ma	Metamorphism in metapelites (MC), intrusive granitoids in granulites (SC)	Monazite (U–Th–Pb + REE), EPMA	Bhowmik et al. (2010)
Exhumation and granite intrusion				
Siwaya granite	836 Ma	Zircon	U-Pb	Deb et al. (2001)
Erinpura granite-gneiss	863 ± 23 Ma	Magmatic monazite	EPMA	Just et al. (2011)

The enhanced temperature field during folding, in respect to the generally low-grade metamorphism in the Jharol Formation, is restricted to a very small corridor (<2 km) parallel to the transition of the Aravalli to Delhi Supergroups. Higher thermal gradients during orogeny cannot account for the temperature increase of ~200 °C inferred from the findings in our study. The continuous temperature increase in metasediments of the Upper Jharol Formation also argues against stacking of deeper crustal units. Garnet-quartz bands were traced from Modi in the north to Gorimari in the south of the Jharol Formation (for locations see Fig. 2b) for ~50 km, additionally along strike, detached bands of garnet-rich rocks occur within mica schists. A structural control of the temperature enhancement along a structural lineament is therefore most likely. Under such conditions the thin and more competent garnet-bearing bands within the thick micaceous more ductile strata respond by tight ptygmatic folding. Tectonic thickening of supracrustal units associated with superposition of very tight and vertical crenulation cleavage can occur during contraction in orogenic forelands accommodating subhorizontal shortening, e.g. in the intra-cratonic Proterozoic basins of the Eastern Ghats (Saha and Chakraborty 2003).

Detailed geological maps of the contact between Aravalli and Delhi units in the Gogunda region (Chauhan, unpublished) and in the Modi region (Saini et al. 2006) show the structural discordance and the angular relationship between foliation trends of both the units and support the larger scale interpretation as thrust-bounded units based on interpretation of Landsat images (Din and Iqbaluddin 2000). Field studies (Saini et al. 2006 and references therein) describe the highly tectonized nature of the Delhi-Aravalli contact, however, due to erosion the direct contact is not exposed in the here studied Gogunda section. While a reverse E-directed thrusting of the SDT is more pronounced in the northern continuation of the structure (Ruj and Dasgupta 2014) referred to as Kaliguman lineament (Fig. 1) in the here studied southern continuation (Som thrust) a lateral shift has been proposed on base of interpretation of satellite images (Din and Iqbaluddin 2000). A simplified model with the strain concentration in the supra-crustal cover sequences and along localized high strain regions, alongside thick skinned thrusting with listric intra-basement shear zones accommodate the strain at higher depths is shown in Fig. 12b.

The pre-collision design with discontinuity in mechanical properties between the BGC II in the north (reworked during Columbia orogeny) and the BGC I in the southern Aravalli region could have an impact to the overall structural pattern. The southern sector (BGC I) can be regarded as a rigid block with a strong and competent basement and weaker low-metamorphic multilayer cover. During the initial convergence, thin-skinned thrusting of the sedimentary cover amalgamation with reworked basement (basement-involved thin-skinned tectonics, Pfiffner 2017) could be considered e.g. as observed at the western proximity of the Udaipur syncline (Barodya section of the M27, de Wall et al. 2012). Steepening of thrusts and thick-skinned deformation of basement and cover units might account for the final structural pattern. There is no indication for any magmatic activity accompanying this fold and thrust tectonics. The superimposed complex fold pattern may be related to oblique convergence and shortening in respect to the preexisting craton (Columbia continent, BGC II) or change in the maximum shortening direction during folding and thrusting. A final E–W shortening can be constrained from general trend of thrust sheets in BGC II. Deformation in the western part (Rajasthan sector) of the 1.6 Ga (U-Pb zircon ages of tuff strata, Rasmussen et al. 2002) Lower Vindyan unit could also be a consequence of this late-Grenvillian crustal convergence. N-S trending fold structures are seen in the western sector of the basin and show increasing tightness of folding from E to W and bending of fold axis parallel to the GBF along the outer arc of the basin (Verma 1996). Folding occurred prior to deposition of the undeformed

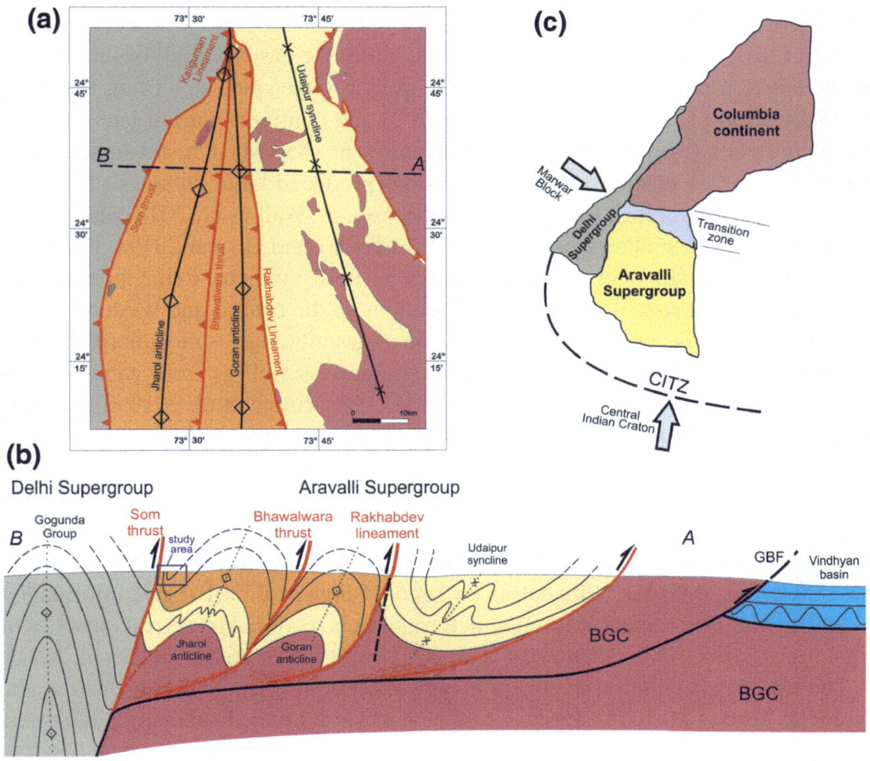

Fig. 12 **a** Simplified map of suspected thrusts and large-scale F_2 folds in the Aravalli Supergroup. **b** Schematic E-W cross section with interpreted structural grain by D_2 thrusting, folding and amalgamation of Delhi and Aravalli Supergroups. **c** Interpretation of tectonometamorphic units constituting the Aravalli-Delhi Mobile Belt

Upper Vindyan sequence, constrained at ~1050 Ma on base of kimberlite intrusion into sediments of the Kaimur Group (e.g. Gregory et al. 2006).

Southern continuation of the Aravalli-Delhi fold Belt into the Central India Tectonic Zone (also known as Satpura Mountain Belt, Mohanty 2012) has been proposed (Mamtani et al. 1999). Amalgamation of the North Indian Blocks (Bundelkhand-Aravalli Block, Marwar Block) (Delhi Orogeny) with the Central Indian Bastar Craton (Sausar orogeny) occurred approximately coeval during late-Grenvillian plate convergence (Roy et al. 2001, 2006; Bhowmik et al. 2012). The Upper Aravalli supracrustal rocks, situated in the triangle between the two orogenic belts (Fig. 12c), were squeezed by multidirectional compressional forces imparting complex the observed complex folding pattern.

The Delhi orogenic belt reveals a pattern of bivergent thrusting with E-vergence in the eastern foreland (this study) and W-vergence along the western flank (de Wall et al. 2014) but with remarkable asymmetry in timing of deformation and associated thrusting. Motor of thrust sheet motion are generally the compressional forces during crustal convergence and all available age data indicate a bracket of 980–920 Ma for deformation and metamorphism along the eastern flank of the orogen. At the western margin thrusting proceeds over a longer period starting at ~1 Ga with indication of plate convergence (Table 2) and continues during late-orogenic granite intrusion (Erinpura Granite emplacement) at ~850 Ma. Crustal anatexis during this late-phase of the orogenic process has produced large volumes of granitic melts (Erinpura Granites, for age compilation see de Wall et al. 2018) and is very pronounced in the southern part of the DFB (Tiwari and Biswal 2019a, b) and its western foreland (Gupta et al. 1997). We assign this asymmetry to the more mature stage and buoyant rise of deeper crustal sections in the central and western part of the orogeny. This is evident from exhumation of granulite facies metamorphic rocks derived from pelitic and calcareous metasediments as well as from gabbro-norite intrusions (≥850 °C, 5.5–6.8 kb; Biswal et al. 1998; Singh et al. 2010, Tiwari and Biswal 2019a). Exhumation of these high grade metamorphic rocks is accompanied by granitic intrusions along shear zones, which might have triggered rapid exhumation during late-orogenic extension (Tiwari and Biswal 2019b). There is indication from other orogens that emplacement of granitoids into shear zones have facilated the uplift of crustal blocks (Scheuvens and Zulauf 2000; Lexa et al. 2011). The age of these intrusions (860 Ma U/Pb dating on zircon, Singh et al. 2010) give information on timing of this process. Continuing deformation is recorded in several episodes of thrusting at 810 ± 6 Ma (Chatterjee et al. 2017), and 775 ± 26 Ma (Just et al. 2011).

The structural evolution seen here is typical for doubly-vergent orogens (e.g. the Pyrenees, Ford et al. 2016). Considering an eastward directed subduction along the western flank of the orogeny as can be inferred from the architecture of the Sendra–Ajmer arc terrain, the deformation along the eastern flank of the Aravalli-Delhi Mobile Belt can be explained as a retro-wedge setting. Our model includes foreland-directed thrusts and reactivated major pre-existing basement faults, initiated during rifting of the BGC and active during the depositional history.

6 Conclusions

A prograde mono-metamorphic history has been deciphered for the Upper Aravalli Jharol Formation. Along the eastern proximity the deformation occurred under higher metamorphic conditions (500 °C/7 kbar, from petrological study; 500 °C from microstructural analyses) and higher strain has to be inferred from the intense and complex folding pattern.

We relate structural imprint and metamorphism in the Aravalli Supergroup to the ~980 Ma Grenvillian crustal convergence. Compression under changing stress fields imposed during amalgamation of the Northern Indian blocks and the Central Indian Craton blocks might account for the complex superimposed fold pattern in the Aravallis. In contrast to the mono-metamorphic history of the Aravalli Supergroup, the northern poly-metamorphic complexes (Sandmata Complex, Mangalwar Complex, and North Delhi Fold Belt) have been constituted during formation of

Columbia Supercontinent and reworked during the Genvillian event. Care should be given not to relate older metamorphic records, seen in the northern sector of the ADMB (e.g. Sandmata Complex, Mangalwar Complex, and North Delhi Terrane), with age of rifting, deformation and metamorphism in the Aravalli Supergroup. The deformation along the eastern flank of Aravalli-Delhi Mobile Belt can be explained in terms of a retro-wedge setting.

Acknowledgements G. Kodl acknowledges a Promos grant by the University of Erlangen. We thank M. K. Pandit for his company and for discussions in the field. S. Gilbricht assisted during SEM and EPMA work at the Division of Economic Geology and Petrology at TU Freiberg/Saxony. We thankfully appreciate the detailed comments of an anonymous reviewer to improve the manuscript.

References

Absar N, Sreenivas B (2015) Petrology and geochemistry of greywackes of the ~1.6 Ga Middle Aravalli Supergroup, northwest India: evidence for active margin processes. Int Geol Rev 57:134–158

Abu-Hamatteh ZSH (2005) Geochemistry and petrogenesis of mafic magmatic rocks of the Jharol Belt, India: geodynamic implication. J Asian Earth Sci 25:557–581

Abu-Hamatteh ZSH, Raza M, Ahmad T (1994) Geochemistry of early Proterozoic mafic and ultramafic rocks of the Jharol Group, Rajasthan, Northwestern India. J Soc India 44:141–156

Acharyya SK, Roy A (2000) Tectonothermal history of the Central Indian tectonic zone and reactivation of major faults/shear zones. J Geol Soc India 55:239–256

Ahmad T, Dragusanu C, Tanaka T (2008) Provenance of Proterozoic Basal Aravalli mafic volcanic rocks from Rajasthan, Northwestern India: Nd isotopes evidence for enriched mantle reservoirs. Precambr Res 162:150–159

Bakliwal PC, Ramasamy SM (1987) Lineament fabric of Rajasthan and Gujarat, India. Geol Surv India (Jaipur) 113(7):54–64

Bhattacharya A, Mohanty L, Maji A, Sen SK, Raith M (1992) Non-ideal mixing in the phlogopite-annite binary: constraints from experimental data on Fe-Mg partitioning and a reformulation of the garnet-biotite geothermometer. Contrib Miner Petrol 111:87–93

Bhowmik SK, Dasgupta S (2012) Tectonothermal evolution of the Banded Gneissic Complex in Central Rajasthan, NW India: present status and correlation. J Asian Earth Sci 49:339–348

Bhowmik SK, Bernhardt H-J, Dasgupta S (2010) Grenvillian age high-pressure upper amphibolite-granulite metamorphism in the Aravalli-Delhi Mobile Belt, Northwestern India: new evidence from monazite chemical age and its implication. Precambr Res 78:168–184

Bhowmik SK, Wilde SA, Bhandari A, Pal T, Pant NC (2012) Growth of the Greater Indian Landmass and its assembly in Rodinia: geochronological evidence from the Central Indian Tectonic Zone. Gondwana Res 22:54–72

Bhowmik SK, Dasgupta S, Baruah S, Kalita D (2018) Thermal history of a Late Mesoproterozoic paired metamorphic belt (?) during Rodinia assembly: new insight from medium-pressure granulites from the Aravalli-Delhi Mobile Belt, Northwestern India. Geosci Front 9:335–354

Biju-Sekhar S, Yokoyama K, Pandit MK, Okudaira T, Santosh M (2003) Paleoproterozoic magmatism in Delhi Fold belt, NW India and its significance: evidence from EPMA chemical ages of zircons. J Asian Earth Sci 22:189–207

Biswal TK, Gyanib KC, Parthasarathyc R, Pantc DR (1998) Implications of the geochemistry of the Pelitic Granulites of the Delhi Supergroup, Aravalli Mountain Belt, Northwestern India. Precambr Res 87:75–85

Bose S, Seth P, Dasgupta N (2017) Meso-Neoproterozoic mid-crustal metamorphic record from the Ajmer-Shrinagar section, Rajasthan, India and its implication to the assembly of the Greater Indian Landmass during the Grenvillian-age orogenesis. Geol Soc London Spec Publ 457:291–318

Budzyń B, Harlov DE, Williams ML, Jercinovic MJ (2011) Experimental determination of stability relations between monazite, fluorapatite, allanite, and REE-epidote as a function of pressure, temperature, and fluid composition. Am Miner 96:1547–1567

Buick IS, Allen C, Pandit M, Rubatto D, Hermann J (2006) The Proterozoic magmatic and metamorphic history of the banded gneiss complex, Central Rajasthan, India: LA-ICP-MS U-Pb zircon constraints. Precambr Res 151:119–142

Buick IS, Clark C, Rubatto D, Hermann J, Pandit M, Hand M (2010) Constraints on the proterozoic evolution of the Aravalli-Delhi Orogenic belt (NW India) from monazite geochronology and mineral trace element geochemistry. Lithos 120:511–528

Chatterjee SM, Roy Choudhury M, Das S, Roy A (2017) Significance and dynamics of the Neoproterozoic (810 Ma) Phulad Shear Zone, Rajasthan, NW India. Tectonics 36:1432–1454

Chattopadhyay N, Gangopadhyay S (1984) Petrological studies of the ultramafic rocks of Rajasthan. Spec Publ Geol Surv India 12:17–24

Chauhan NK, Sharma BL, Mohemmad SA (1996) Structural Geometry and Strain History of Early Proterozoic Aravalli Rocks of Gorimari, Udaipur District, Rajasthan. J Geol Soc India 47:59–74

D'Souza J, Prabhakar N, Xu Y, Sharma KK, Sheth H (2019) Mesoarchean to Neoproterozoic (3.2–0.8 Ga) crustal growth and reworking in the Aravalli Craton, northwestern India: insights from the Pur-Banera supracrustal belt. Precambr Res 332:105383

de Wall H, Pandit MK, Chauhan NK (2012) Paleosol occurrences along the Archean-Proterozoic contact in the Aravalli craton, NW India. Precambr Res 216:120–131

de Wall H, Pandit MK, Sharma KK, Schöbel S, Just J (2014) Deformation and granite intrusion in the Sirohi area, SW Rajasthan-constraints on Cryogenian to Pan-African crustal dynamics of NW India. Precambr Res 254:1–18

de Wall H, Pandit MK, Donhauser I, Schöbel S, Wang W, Sharma KK (2018) Evolution and tectonic setting of the Malani—Nagarparkar igneous suite: a neoproterozoic silicic-dominated large igneous province in NW India-SE Pakistan. J Asian Earth Sci 160:136–158

Deb M (1999) Metallic mineral deposits of Rajasthan. In: Kataria P (ed) Proceedings of seminar on geology of Rajasthan: status and perspective, Udaipur, India. Geology Department, Mohanlal Sukhadia University, pp 213–237

Deb M, Sarkar SC (1990) Proterozoic tectonic evolution and metallogenesis in the Aravalli-Delhi orogenic complex, Northwestern India. Precambr Res 46:115–137

Deb M, Thorpe RI, Krstic D, Corfu F, Davis DW (2001) Zircon U-Pb and galena Pb isotope evidence for an approximate 1.0 Ga terrane constituting the western margin of the Aravalli-Delhi orogenic belt, northwestern India. Precambr Res 108:195–213

Deb M, Thorpe RI, Krstic D (2002) Hindoli Group of rocks in the eastern fringe of the Aravalli-Delhi orogenic belt-Archean secondary greenstone belt or Proterozoic supracrustals? Gondwana Res 5:879–883

Derez T, Pennock G, Drury M, Sintubin M (2015) Low-temperature intracrystalline deformation microstructures in quartz. J Struct Geol 71:3–23

Dey B, Das K, Dasgupta N, Bose S, Hidaka H, Ghatak H (2019) Zircon U–Pb (SHRIMP) ages of the jahazpur granite and mangalwar gneiss from the Deoli-Jahazpur sector, Rajasthan, NW India: a preliminary reappraisal of stratigraphic correlation and implications to crustal growth. In: Mondal M (ed) Geological evolution of the precambrian Indian shield. Society of Earth Scientists Series. Springer, Cham

Dharma Rao CV, Santosh M, Purohit R, Wang J, Jiang X, Kusky T (2011) LA–ICP–MS U-Pb zircon age constraints on the Paleoproterozoic and Neoarchean history of the Sandmata Complex in Rajasthan within the NW Indian Plate. J Asian Earth Sci 42:286–305

Dharma Rao CV, Santosh M, Kim SW, Li S (2013) Arc magmatism in the Delhi Fold Belt: SHRIMP U-Pb zircon ages of granitoids and implications for neoproterozoic convergent margin tectonics in NW India. J Asian Earth Sci 78:83–99

Din SU, Iqbaluddin (2000) Thrust controlled lineaments of the Aravalli-Delhi Fold Belt, Rajasthan, India mapped from Landsat TM image. J Indian Soc Remote Sens 28:47–58
Fandrich R, Gu Y, Burrows D, Moeller K (2007) Modern SEM-based mineral liberation analysis. Int J Miner Process 84:310–320
Fareeduddin, Kröner A (1998) Single zircon age constraints on the evolution of Rajasthan granulite. In: Paliwal BS (ed) The Indian precambrian. Scientific Publishers, India, Jodhpur, pp 547–566
Ford M, Hemmer L, Vacherat A, Gallagher K, Christophoul F (2016) Retro-wedge foreland basin evolution along the ECORS line, eastern Pyrenees, France. J Geol Soc 173:419–437
Ganguly J, Cheng W, Tirone M (1996) Thermodynamics of aluminosilicate garnet solid solution: new experimental data, an optimized model, and thermometric applications. Contrib Miner Petrol 123:137–151
Ghosh SK, Mandal N, Khan D, Deb SK (1992) Modes of superposed buckling in single layers controlled by initial tightness of early folds. J Struct Geol 14:381–394
Gopalan K, Macdougall JD, Roy AB, Murali AV (1990) Sm–Nd evidence for 3.3 Ga old rocks in Rajasthan, northwestern India. Precamb Res 48:287–297
Goswami-Banerjee G, Robyr M (2015) Pressure and temperature conditions for crystallization of metamorphic allanite and monazite in metapelites: a case study from Miyar Valley (high Himalayan Crystalline of Zanskar, NW India). J Metamorph Geol 33:535–556
Grasemann B, Wiesmayr G, Draganits E, Fusseis F (2004) Classification of refold structures. J Geol 112:119–125
Gregory LC, Meert JG, Pradhan V, Pandit MK, Tamrat E, Malone SJ (2006) Apaleomagnetic and geochronologic study of the Majhgawan Kimberlite, India: implications for the age of the Vindhyan Supergroup. Precambr Res 149:65–75
Groshong RH (1988) Low-temperature deformation mechanisms and their interpretation. Geol Soc Am Bull 100(9):1329–1360
Grujic D (1993) The influence of initial fold geometry on type 1 and type 2 interference patterns: an experimental approach. J Struct Geol 15(3–5):293
Gupta SN, Arora YK, Mathur RK, Iqballuddin BP, Sahai TN, Sharma SB (1980) Lithostratigraphic map of Aravalli region. Geological Survey of India, Hyderabad
Gupta SN, Arora YK, Mathur RK, Iqballuddin PB, Sahai TN, Sharma SB (1997) The Precambrian geology of the Aravalli Region, Southern Rajasthan and Northeastern Gujarat. Memoir Geol Surv India 123:262
Handy MR, Hirth G, Bürgmann R (2007) Continental fault structure and rheology from the frictional-to-viscous transition downward. In: Handy MR, Hirth G, Hovius N (eds) Tectonic faults: agents of change on a dynamic earth. MIT press, Cambridge, Massachusetts, London, pp 139–181
Hazarika P, Upadhyay D, Mishra B (2013) Contrasting geochronological evolution of the Rajpura-Dariba and Rampura-Agucha metamorphosed Zn-Pb deposit, Aravalli-Delhi belt, India. J Asian Earth Sci 73:329–339
Hazarika P, Mishra B, Ozha MK, Pruseth KL (2017) An improved EPMA analytical protocol for U-Th–Pbtotal dating in xenotime: age constraints from polygenetic Mangalwar complex, Northwestern India. Chem Erde Geochem 77:69–79
Heron AM (1953) The geology of central Rajputana. Mem Geol Surv India 79:389
Holdaway MJ (2001) Recalibration of the GASP geobarometer in light of recent garnet and plagioclase activity models and versions of the garnet-biotite geothermometer. Am Miner 86:1117–1129
Holland TJB, Powell R (1998) An internally-consistent thermodynamic dataset for phases of petrological interest. J Metamorph Geol 16:309–344
Holland TJB, Powell R (2011) An improved and extended internally-consistent thermodynamic dataset for phases of petrological interest, involving a new equation of state for solids. J Metamorph Geol 29:333–383

Janots E, Brunet F, Goffé B, Poinssot C, Burchard M, Cemic L (2007) Thermochemistry of monazite-(La) and dis-sakisite-(La): implications for monazite and allanite stability in metapelites. Contrib Miner Petrol 154:1–14
Jessel MW (1987) Grain-boundary migration microtextures in a naturally deformed quartzite. J Struct Geol 9(8):1007–1014
Just J, Schulz B, de Wall H, Jourda F, Pandit MK (2011) Monazite CHIME/EPMA dating of Erinpura granitoid deformation: Implications for Neoproterozoic tectono-thermal evolution of NW India. Gondwana Res 19(2):402–412
Kaur P, Chaudhari N, Raczek I, Kröner A, Hofmann AW (2007) Geochemistry, zircon ages and whole-rock Nd isotopic systematic for Paleoproterozoic A-typegranitoids in the northern part of the Delhi belt, Rajasthan, NW India: implications for late Paleoproterozoic crustal evolution of the Aravalli craton. Geol Mag 144:361–378
Kaur P, Chaudhri N, Raczek I, Kröner A, Hofmann AW (2009) Record of 1.82 Ga Andean-type continental arc magmatism in NE Rajasthan, India: insights from zircon and Sm–Nd ages, combined with Nd–Sr isotope geochemistry. Gondwana Res 16:56–71
Kaur P, Zeh A, Chaudhri N, Gerdes A, Okrusch M (2011) Archaean to Palaeoproterozoic crustal evolution of the Aravalli mountain range, NW India, and its hinterland: the U-Pb and Hf isotope record of detrital zircon. Precambr Res 187:155–164
Kaur P, Zeh A, Chaudhri N, Gerdes A, Okrusch M (2013) Nature of magmatism and sedimentation at a Columbia active margin: insights from combined U-Pb and Lu-Hf isotope data of detrital zircons from NW India. Gondwana Res 23:1040–1052
Kaur P, Zeh A, Chaudhri N, Eliyas N (2017) Two distinct sources of 1.73–1.70 Ga A-type granites from the northern Aravalli orogen, NW India: constraints from in situ zircon U-Pb ages and Lu-Hf isotopes. Gondwana Res 49:164–181
Kruhl JH, Nega M (1996) The fractal shape of sutured quartz grain bound-aries: application as a geothermometer. Geol Rundsch 85:38–43
Lexa O, Schulmann K, Janoušek V, Štípská P, Guy A, Racek M (2011) Heat sources and trigger mechanisms of exhumation of HP granulites in Variscan orogenic root. J Metamorphic Geology 29:79–102
Mamtani MA, Greiling RO, Karanth RV, Merh SS (1999) Orogenic deformation and its relationship to AMS fabric—an example from the southern margin of the Aravalli Mountain Belt, India. In: Radhakrishna T, Piper JD (eds) The Indian subcontinent and Gondwana: a Palaeomagnetic and rock magnetic perspective, vol 44. Geological Society of India Memoir, pp 9–24
Mancktelow NS, Pennacchioni G (2004) The influence of grain boundary fluids on the microstructure of quartz-feldspar mylonites. J Struct Geol 26:47–69
McKenzie NR, Hughes NC, Myrow PM, Sharma M (2011) Correlation of Precambrian-Cambrian sedimentary successions across northern India and the utility of isotopic signatures of Himalayan lithotectonic zones. Earth Planet Sci Lett 312:471–483
McKenzie NR, Hughes NC, Myrow PM, Banerjee DM, Deb M, Planavsky NJ (2013) New age constraints for the Proterozoic Aravalli-Delhi successions of India and their implications. Precambr Res 238:120–128
Meert JG, Pandit MK, Pradhan VR, Banks J, Sirianni R, Stroud M, Newstead B, Gifford J (2010) Precambrian crustal evolution of Peninsular India: 3.0 billion year odyssey. J Asian Earth Sci 39:483–515
Mohanty S (2012) Spatio-temporal evolution of the Satpura Mountain Belt of India: a comparison with the Capricorn Orogen of Western Australia and implication for evolution of the supercontinent Columbia. Geosci Front 3:241–267
Mondal MEA, Goswami JN, Deomurari MP, Sharma KK (2002) Ion microprobe 207Pb/206Pb ages of zircons from the Bundelkhand massif, northern India: implications for crustal evolution of the Bundelkhand-Aravalli protocontinent. Precambr Res 117:85–100
Oriolo S, Wemmer K, Oyhantçabal P, Fossen H, Schulz B, Siegesmund S (2018) Geochronology of shear zones—a review. Earth Sci Rev 185:665–683

Ozha MK, Mishra B, Hazarika P, Jeyagopal AV, Yadav GS (2016) EPMA monazite geochronology of the basement and supracrustal rocks within the Pur-Banera basin, Rajasthan: evidence of Columbia breakup in Northwestern India. J Asian Earth Sci 117:284–303

Ozha MK, Pal DC, Mishra B, Desapati T, Shaji TS (2017) Geochemistry and chemical dating of uraninite in the Samarkiya area, central Rajasthan, northwestern India—implication for geochemical and temporal evolution of uranium mineralization. Ore Geol Rev 88:23–42

Pandit MK, Carter LM, Ashwal LD, Tuckerd RD, Torsvike TH, Jamtveitf B, Bhushang SK (2003) Age, petrogenesis and significance of 1 Ga granitoids and related rocks from the Sendra area, Aravalli Craton, NW India. J Asian Earth Sci 22:363–381

Pandit MK, de Wall H, Chauhan NK (2008) Paleosol at the Archean-Proterozoic contact in NW India revisited: evidence for oxidizing conditions during paleo-weathering? J Earth Syst Sci 117(3):201–209

Perchuk LL, Lavrenteva LV (1983) Experimental investigation of exchange equilibria in the system cordierite-garnet-biotite. In: Saxena SK (ed) Kinetics and equilibrium in mineral reactions. Springer, New York, pp 199–239

Pfiffner OA (2017) Thick-skinned and thin-skinned tectonics: a global perspective. Geosciences 7:71

Powell R, Holland TJB (1993) On the formulation of simple mixing models for complex phases. Am Miner 78:1174–1180

Powell R, Holland TJB (1994) Optimal geothermometry and geobarometry. Am Miner 79:120–133

Powell R, Holland TJB (2008) On thermobarometry. J Metamorph Geol 26:155–179

Purohit R, Bhu H, Sarkar A, Ram R (2015) Evolution of the ultramafic rocks of the Rakhabdev and Jharol Belts in Southeastern Rajasthan, India: new evidences from imagery mapping, petro-minerological and O-H stable isotope studies. J Geol Soc India 85:331–338

Ramsay JG (1967) Folding and fracturing of rocks. McGraw-Hill, New York, p 568

Rasmussen B, Bose PK, Sakar S, Banerjee S, Fletcher IR, McNaughton NJ (2002) 1.6 Ga U-Pb zircon age for the Chorhat Sandstone, lower Vindhyan, India: possible implications for the early evolution of animals. Geology 20:103–106

Rogers JJW, Santosh M (2002) Configuration of Columbia, a Mesoproterozoic supercontinent. Gondwana Res 5:5–22

Roy A, Jakhar SR (2002) Geology of Rajasthan: precambrian to recent: Jodhpur. Scientific Publishers (India), India, p 421

Roy AB, Kröner A (1996) Single zircon evaporation ages constraining the growth of the Archaean Aravalli craton, northwestern Indian shield. Geol Mag 133:333–342

Roy AB, Prasad MH (2003) Tectonothermal events in Central Indian Tectonic Zone (CITZ) and its implications in Rodinian crustal assembly. J Asian Earth Sci 22:115–129

Roy AB, Sharma B, Paliwal B, Chauhan N, Nagori D, Golani P, Bejarniya B, Bhu H, Sabah MA (1993) Lithostratigraphy and tectonic evolution of the Aravalli supergroup: a protogeosynclinal sequence. In: Cassyap SM (ed) Rift basins and aulacogens. Nainital, Gyanodaya Prakashan, pp 73–90

Roy AB, Kröner A, Laul V (2001) Detrital zircons constraining basement age in a late Archaean greenstone belts of southeastern Rajasthan, India. Curr Sci 81:407–410

Roy AB, Kröner A, Bhattacharyya PK, Rathore S (2005) Metamorphic evolution and zircon geochronology of early Proterozoic granulites in the Aravalli Mountains of northwestern India. Geol Mag 142:287–302

Roy AB, Kagami H, Yoshida M, Roy A, Bandyopadhyay BK, Chattopadhyay A, Khan AS, Huin AK, Pal T (2006) Rb–Sr and Sm–Nd dating of different metamorphic events from the Sausar Mobile Belt, central India: implications for proterozoic crustal evolution. J Asian Earth Sci 26:61–76

Roy AB, Kröner A, Rathore S, Laul V, Purohit R (2012) Tectono-metamorphic and geochronologic studies from Sandmata Complex, northwest Indian shield: implications on exhumation of Late-Palaeoproterozoic granulites in an Archaean-early Palaeoproterozoic granite gneiss terrane. J Geol Soc India 79:323–334

Ruj T, Dasgupta NJ (2014) Tectonic imprints within a granite exposednear Srinagar, Rajasthan. J Earth Syst Sci 123:1361–1374

Saha D, Chakraborty S (2003) Deformation pattern in the Kurnool and Nallarnalai Groups in the Northeastern Part (Palnad Area) of the Cuddapah Basin, South India and its Implication on Rodinia/Gondwana Tectonics. Gondwana Res 6:573–583

Saini P, Singh S, Pandit MK (2006) Angular relationship between rocks of the Aravalli and Delhi Supergroup in southeastern Rajasthan—a possible unconformity. Curr Sci 91:432–434

Sarangi S, Gopalan K, Kumar S (2004) Pb–Pb age of earliest megascopic, eukaryotic alga bearing Rhotas formation, Vindhyan Super Group, India: implications for Precambrian atmospheric oxygen evolution. Precambr Res 121:107–121

Sarkar G, Barman T, Corfu F (1989) Timing of continental arc magmatism in northwest India: evidence from U-Pb Zircon geochronology. J Geol 97:607–612

Scheuvens D, Zulauf G (2000) Exhumation, strain localization, and emplacement of granitoids along the western part of the Central Bohemian shear zone (Bohemian Massif). Int J Earth Sci 89:617

Schulz B (2017) Polymetamorphism in garnet micaschists of the Saualpe Eclogite Unit (Eastern Alps, Austria), resolved by automated SEM methods and EMP-Th-U-Pb monazite dating. J Metamorph Geol 35:141–163

Sen S (1981) Proterozoic palaeotectonics in the evolution of crust and location of metalliferous deposits, Rajasthan. Quart J Geol Min Metall Soc India 53:162–185

Sharma RS (1988) Patterns of metamorphism in the Precambrian rocks of the Aravalli mountain belt. In: Roy AB (ed) Precambrian of the Aravalli Mountain Rajasthan, vol 7. Memoir Geological Society India, India, pp 33–76

Sharma KK, Rahman A (2000) The early Archaean-Paleoproterozoic crustal growth of the Bundelkhand craton, northern Indian shield. In: Deb M (ed) Crustal evolution and metallogeny in the Northwestern Indian shield. Narosa Publishing House, New Delhi, pp 51–72

Singh YK, De Waele B, Karmarkar S, Sarkar S, Biswal TK (2010) Tectonic setting of the Balaream-Kui-Surpagla-Kengora granulites of the South Delhi Terrane of the Aravalli Mobile Belt, NW India and its implication on correlation with the East African Orogen in the Gondwana assembly. Precambr Res 183:669–688

Sinha-Roy S, Malhotra G, Mohanty MK (1998) Geology of Rajasthan. Geological Society of India, p 278

Spear FS (1993) Metamorphic phase equilibria and pressure-temperature-time paths. Mineralogical Society of America Monograph Series, Washington

Spear FS (2010) Monazite-allanite phase relations in metapelites. Chem Geol 279:55–62

Spear FS, Pyle JM (2010) Theoretical modeling of monazite growth in a low-Ca metapelite. Chem Geol 273:111–119

Spruzeniece L, Piazolo S, Maynard-Casely HE (2017) Deformation-resembling microstructure created by fluid-mediated dissolution–precipitation reactions. Nat Commun 8:1–9

Stipp M, Stünitz H, Heilbronner R, Schmid SM (2002) The eastern Tonale fault zone: a 'natural laboratory' for crystal plastic deformation of quartz over a temperature range from 250 to 700 °C. J Struct Geol 24:1861–1884

Sugden T, Deb M, Windley B (1990) The tectonic setting of mineralisation in the Proterozoic Aravalli Delhi Orogenic belt, NW India. Dev Precambr Geol 8:367–390

Tiwari SK, Biswal TK (2019a) Dynamics, EPMA Th-U-total Pb monazite geochronology and tectonic implications of deformational fabric in the lower-middle crustal rocks, a case study of Ambaji granulite, NW India. Tectonics 38:2232–2254

Tiwari SK, Biswal TK (2019b) Paleostress and magma pressure measurement of granite veins in the Neoproterozoic Ambaji granulite, South Delhi terrane, Aravalli Delhi mobile belt, NW India: Implications towards the extension-driven exhumation of the moddle-lowe crustal rocks. J Earth Syst Sci 128:150

Venkatachala BS, Sharma M, Shukla M (1996) Age and life of the Vindhyans—facts and conjectures. In: Bhattacharyya A (ed) Recent advances in Vindhyan geology, vol 36. Geological Society India Memoir, pp 137–165
Verma P (1996) Evolution and age of the great boundary fault of Rajasthan 36:197–212
Verma PK, Greiling RO (1995) Tectonic evolution of the Aravalli orogen (NW India): an inverted Proterozoic rift basin? Geol Rundsch 84:683–696
Vernooij MGC, Langenhorst F (2005) Experimental reproduction of tectonic deformation lamellae in quartz and comparison to shock-induced planar deformation features. Meteorit Planet Sci 40:1353–1361
Volpe AM, Macdaugall JD (1990) Geochemistry and isotopic characteristics of mafic (Phulad Ophiolite) and related rocks in the Delhi Supergroup Rajasthan, India: implications for rifting in the Proterozoic. Precambr Res 48:167–191
Wang W, Cawood PA, Zhou MF, Pandit MK, Chen WT (2017) Zircon U-Pb age and Hf isotope evidence for an Eoarchean crustal remnant, and episodic crustal reworking in response to supercontinental cycles in NW India. J Geol Soc 174(4):759–772
Wang W, Cawood PA, Pandit MK, Zhu M-F, Zhou JH (2018) Evolving passive- and active-margin tectonics of the Paleoproterotoic Aravalli Basin. NW India. Geol Soc Am Bull 131(3/4):426–443
Wiedenbeck M, Goswami JN (1994) High-precision $^{207}Pb/^{206}Pb$ zircon geochronology using a small ion microprobe. Geochim Cosmochim Acta 58:2135–2141
Wiedenbeck M, Goswami JN, Roy AB (1996) Stabilisation of the Aravalli craton of the northwestern India at 2.5 Ga.: an ion-microprobe zircon study. Chem Geol 129:325–340
Wu CM (2015) Revised empirical garnet-biotite-muscovite-plagioclase (GBMP) geobarometer in metapelites. J Metamorph Geol 33:167–176
Zhao G, Sun M, Wilde SA, Li S (2003) Assembly, Accretion and Breakup of the Paleo- Mesoproterozoic Columbia Supercontinent: Records in the North China Craton. Gondwana Res 6:417–434

Evolution of the "Central Indian Tectonic Zone": A Critique Based on the Study of the Sausar Belt

Sarada P. Mohanty

Abstract Detailed structural mapping carried out in different parts of the Central Indian Tectonic Zone (CITZ) combined with an analysis of spatio-temporal evolution of different domains have shown evolution in three tectono-thermal events. The felsic gneisses, referred as Tirodi Gneiss, belong to four generations. The oldest gneissic unit, Tirodi Gneiss—I (>2450 Ma), has development of a paleosol horizon at the top, which was involved in anoxic weathering, and forms the basement of the Sausar Group. The Sausar Group was deposited during the period ca. 2400–2250 Ma and contains glaciogenic sediments of Paleoproterozoic "snowball earth" event and ore-grade manganese deposits. The first deformation marked by the development of isoclinal folds with axial planar cleavage and coeval granulite facies metamorphism of the Sausar Group took place at ~2100 Ma, when grey gneisses and migmatites, Tirodi Gneiss—II, were developed. The upright to steeply inclined isoclinal folds in the Sausar Group with regional ~EW axial traces were developed during the second deformation (~1600 Ma), which was also associated with granulite facies metamorphism and development of grey gneisses, Tirodi Gneiss—III, during the second cycle of the Satpura orogeny. The Tirodi Gneiss—IV (~1450 Ma) were developed during the waning phase of the Satpura orogeny. Low intensity deformation in CITZ marked by open folds with ~NS axial traces and thermal event (~950 Ma) were possibly related to the amalgamation of India and Australia to form East Gondwana along ~NS trending Pinjarra Orogen (~1100–900 Ma) on the western margin of Western Australia.

Keywords Satpura orogenic trend · Great oxidation event · Paleoproterozoic glaciation · Anoxic paleosol · Pinjarra orogeny

1 Introduction

The Peninsular India has outcrops of Precambrian rocks in the central part, transected by linear grabens containing Paleozoic-Mesozoic sediments (Gondwana

S. P. Mohanty (✉)
Department of Applied Geology, Indian Institute of Technology (Indian School of Mines), Dhanbad, Dhanbad 826004, India

T. K. Biswal et al. (eds.), *Structural Geometry of Mobile Belts of the Indian Subcontinent*, Society of Earth Scientists Series,
https://doi.org/10.1007/978-3-030-40593-9_3

Supergroup), and a cover of Cretaceous-Paleocene volcanics (Deccan Trap); the Mesozoic-Cenozoic basins occupy areas encircling/cross-cutting the Precambrian cratons (Fig. 1a). The geological diversity of region provides opportunity to examine the geological events throughout the geological time span and reconstruct the evolution of Earth. The oldest Paleoarchean Tonalite-Trondhjemite-Granodiorite (TTG) gneisses are reported from five blocks: the Bastar Craton (3610 ± 336 Ma Rb-Sr whole rock isochron age by Sarkar and Gupta 1990; 3582 ± 4 Ma U-Pb zircon age by Rajesh et al. 2009), the Dharwar Craton (~3580 Ma detrital zircons ages by Nutman et al. 1992 and 3358 ± 66 Ma Rb-Sr isochron age by Beckinsale et al. (1980), the Singhbhum Craton (3775 ± 85 Ma Sm–Nd whole rock age Basu et al. 1981), the Bundelkhand Craton (3503 ± 99 Ma Rb–Sr WR Sarkar et al. 1996) and the Western Indian Craton (3307 ± 65 Ma Sm–Nd whole rock ages by Gopalan et al. 1990). These proto-continental nuclei are surrounded by Paleoproterozoic-Mesoproterozoic metamorphic rocks and sedimentary cover rocks.

In the earliest attempt to understand tectonic evolution of India, Krishnan (1943) identified four Precambrian orogenic belts (Dharwar, Eastern Ghats, Satpura and

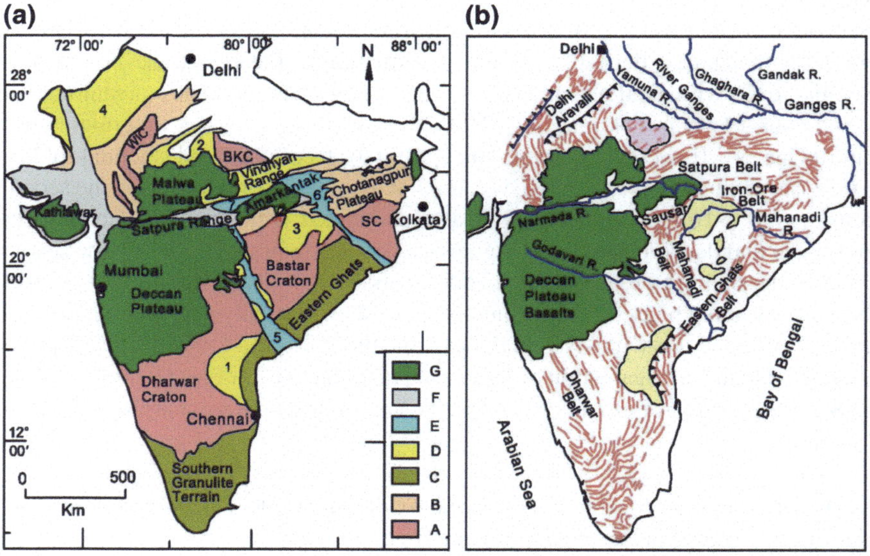

Fig. 1 **a** Generalised geological map of the Peninsular India. A: Archean cratons with Paleoproterozoic metamorphic belts; B: Paleoproterozoic metamorphic belts with Mesoproterozoic overprints; C: Mesoproterozoic metamorphic belts with Neoproterozoic overprints; D: Paleoproterozoic-Neoproterozoic sedimentary basins; E: Paleozoic-Mesozoic sedimentary (Gondwana) basins; F: Mesozoic-Cenozoic sedimentary basins; G: Cretaceous-Paleocene volcanic province (Deccan Traps); SC: Singhbhum Craton; BKC: Bundelkhand Craton; WIC: Western Indian Craton; 1: Cuddapah Basin; 2: Vindhyan Basin; 3: Chhattisgarh Basin; 4: Rajasthan Basin; 5: Godavari Graben; 6: Mahanadi Graben; and 7: Satpura Basin. **b** Generalised tectonic map of the Precambrian orogenic belts of Peninsular India, showing selected Precambrian sedimentary basins and Deccan Trap (after Holmes 1955, 1965)

Delhi-Aravalli) in the Peninsular India (Fig. 1b; Holmes 1955, 1965). On the basis of radiometric dating of rocks intruding into these belts, Holmes et al. (1950) proposed different orogenic cycles to be used for stratigraphic classification of different Precambrian rocks of India. These cycles which separated the Older Gneissic Complex (including Bundelkhand Gneiss) and the younger Precambrian (Purana) sediments of Cuddapah and Vindhyan, include (from older to younger): Dharwar (Aravalli) cycle, Eastern Ghats cycle, Satpura cycle, and Delhi cycle (Holmes et al. 1950). Radhakrishna and Naqvi (1986) identified intersecting "orogenic trends" in the southern part of the Peninsular India. The older NW-SE to N-S trends in the "Early Proterozoic" rocks (referred as Dharwar trend) is truncated by the "Middle Proterozoic" mobile belts: NE-SW trending Eastern Ghats and ENE-WSW trending Satpura Belt (Fig. 2a). In contrast with the ideas of Holmes et al. (1950) and Holmes (1955), Radhakrishna and Naqvi (1986) proposed that the Eastern Ghats trend joins with the Satpura trend, and also joins with the Aravalli-Delhi mobile belt. On the basis of these observations, the proto-continental nuclei of India were divided into two blocks (Fig. 2b): the North Indian Block (NIB) and the South Indian block (SIB) (Naqvi and Rogers 1987). The former comprises of the Archean nuclei of the Dharwar, Bastar and Singhbhum cratons (Dharwar-Singhbhum proto-continent of Radhakrishna and Naqvi 1986; the Southern Peninsular Block of Radhakrishna

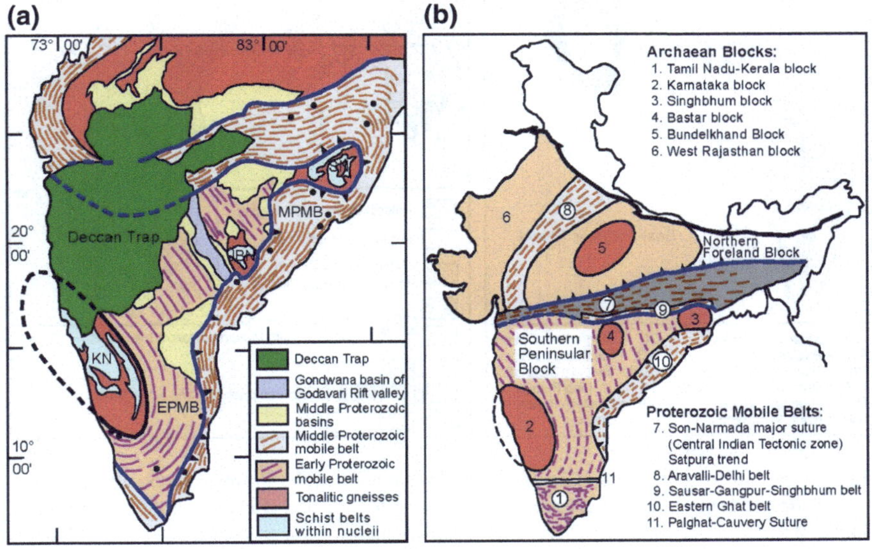

Fig. 2 **a** Generalised geological map of India showing different proto-continental nuclei and tectonic trends in different Proterozoic 'mobile belts' (after Radhakrishna and Naqvi 1986). **b** Sketch map showing diagrammatically the disposition of Archean blocks and the Proterozoic mobile belts of India (after Radhakrishna and Ramakrishnan 1988). The oval character of the Archean provinces and a progressively linear disposition during the Proterozoic was assumed. Dip directions of regional thrusts at the boundaries of different blocks are shown by black triangles

and Ramakrishnan 1988), and the latter includes the Bundelkhand and Aravalli cratons (Bundelkhand proto-continent of Radhakrishna and Naqvi 1986; the Northern Foreland Block of Radhakrishna and Ramakrishnan 1988). The boundary of the two blocks is marked by a zone bound by two prominent lineaments (Son-Narmada in the north and Tapti in south) trending ~ENE-WSW in the central part of India, presently marked by high topography and forming the Satpura Mountains. The Precambrian tectonic fabric of this mountain belt continues eastward through the Chhotanagpur plateau into the Shillong plateau (Figs. 1a, 2b, 3a) and is referred as the "Satpura trend" (Krishnan 1943; Holmes 1955, 1965; Verma 1991). This zone is also referred as the "Central Indian Tectonic Zone (CITZ)" (cf. Radhakrishna and Ramakrishnan 1988; Radhakrishna 1989). The tectonic evolution of this region is a topic of interest for reconstruction of continental assembly of India as well as to understand the relationship between the proto-continental nuclei of India and other blocks of the supercontinent Gondwana.

In the reconstruction model of the East Gondwana, the orogenic trend in the Precambrian rocks of the Satpura Mountain matches with the Albany-Fraser belt of the

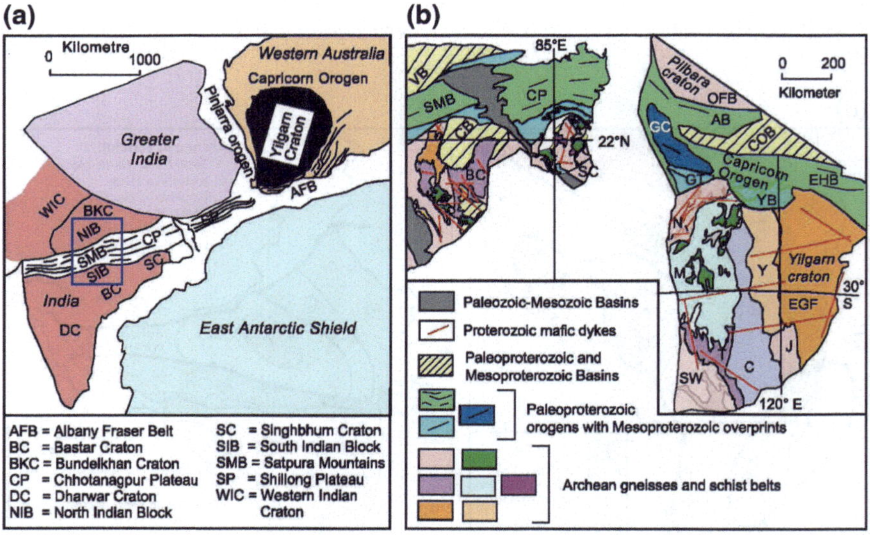

Fig. 3 **a** Reconstructed positions of India and Western Australia in the East Gondwana, showing matched tectonic trends of the Central Indian Tectonic Zone (CITZ) and the Albany-Fraser Mobile Belt (after Harris and Beeson 1993). **b** Schematic reconstructed positions of India and Western Australia after breaking and separation of "SIWA", showing the match of the Satpura orogen and the Capricorn orogen (after Mohanty 2012). AB = Ashburton basin; C = Southern Cross; CB = Chhatisgarh Basin; CGC = Chhotanagpur Gneissic Complex; COB = Collier Basin; EGF = Eastern Goldfields; EHB = Earaheedy Basin; GC = Gascoyne Complex; GT = Glenburgh Terrane; J = Lake Johnson; M = Murchison; N = Narryer Gneiss; OFB = Opthalmian Fold Belt; SW = South West; T = Toodyay Lake Grace; VB = Vindhyan Basin; Y = Yeelirrie; and YB = Yerrida Basin. The blue rectangle in (**a**) is referred to be CITZ, and CP and SP are mentioned to be different tectonic domains (Bhowmik et al. 2011)

Western Australia (Fig. 3a; Harris 1993; Harris and Beeson 1993). The "Satpura orogenic trend" was believed to be developed at ~1000 Ma (Holmes et al. 1950; Aswathanarayana 1956; Mahadevan and Aswathanarayana 1962; Sarkar et al. 1964; Roy and Prasad 2003; Bhowmik et al. 2005; Roy et al. 2006). However, Mohanty (2003a) suggested a polyphase evolution of the Satpura orogen with ages ranging from Paleoproterozoic to Neoproterozoic; the Neoproterozoic low intensity deformation during the last (waning) phase evolution of the mountain belt was overprinted on the earlier phases of pervasive deformation, which defines the "Satpura trend". Through detailed comparison of the spatio-temporal data, Mohanty (2010, 2012) demonstrated continuity of the Satpura orogenic trend into the Capricorn orogen of Western Australia at ~2000 Ma (Fig. 3b). Detailed analysis of the geological history of the Central Indian Shield shows similarity with the evolution of the Western Australia and South African cratons, and raises the possibility for the assembly with "Vaalbara". The purpose of this paper is to provide a critical appraisal of the tectonic evolution of the CITZ, which can be useful in understanding tectonic evolution of this region and used for supercontinent reconstructions.

2 Review of Ideas Regarding Evolution of the "CITZ" and Significance of the "Satpura Orogenic Trend"

2.1 Spatio-Temporal Relations

The Satpura Range extends in WSW-ENE direction from Rajpipla in the west (~73° E) to Daltonganj (at the margin of Amarkantak plateau) in the east (~84° E) for a distance of 1200 km and is located to the south of the Vindhyan Range (Fig. 1a). The Satpura Range has outcrops of Precambrian rocks, Gondwana sediments, Deccan Traps and Cenozoic sediments. The regional WSW-ENE extension of this mountain rage is believed to reflect the trends of the rocks in the basement to the younger sediments and Deccan Traps, and is referred as the "Satpura trend" (Krishnan 1943; Holmes et al. 1950; Holmes 1955; Verma 1991). Similar trends are found in the Chhotanagpur plateau to the east of the Satpura range and in the Shillong plateau of Meghalaya (Figs. 1, 2b, 3a). It may be noted that the significance of the "orogenic trends" in India was critically analysed by Naha (1964), who identified that the "tectonic trends" reflect either the physiographic trends or the generalized strikes of foliation planes, not trends of fold axes. Naha (1964) appreciated the use of "tectonic trends" for correlation of different provinces, but used caution for use of this parameter only after detailed studies of superposed folds in Precambrian terrains and radiometric dating of events related to different tectonic deformations. This was found to be correct for the "Satpura orogenic trend" (Das 1966). Thus, contrary to the idea of one orogenic cycle for the evolution of the Precambrian rocks of the Satpura Range, studies in the eastern part of the Satpura region identified different tectonic evolutions of the northern and southern parts, and lead to the suggestion that the northern region constitutes an younger orogenic belt (Rajgir orogeny) of pre-Vindhyan age (Das 1966).

Radhakrishna and Naqvi (1986) suggested that the collision between the Dharwar-Singhbhum protocontinent in the south and the Bundelkhand protocontinent in the north took place along the central part of India during "Mid-Proterozoic" (Fig. 2a). This zone of collision, named as the "Central Indian Tectonic Zone" (Radhakrishna and Ramakrishnan 1988), was shown to be a suture zone with both southern and northern boundaries dipping northward and the northern block interpreted to be the foreland belt (Fig. 2a, b).

On the basis of detailed structural studies in the Sausar belt (southern part of the CITZ) and compilation of data related to Proterozoic granites of the region, Mohanty (2000) identified five clusters of ages for granites associated with the tectonothermal events in the region (1850 ± 50 Ma, 1600 ± 50 Ma, 1300 ± 50 Ma, 1100 ± 50 Ma and 900 ± 100 Ma) and correlated the ~1850 Ma event to be associated with the "Satpura Orogeny". Yoshida et al. (2001) identified two major tectonothermal events in the CITZ, Paleoproterozoic (~2200–2100 Ma) and Mesoproterozoic (~1700–1500 Ma), overprinted by a minor event at ~1100–600 Ma. However, regional mapping and metamorphic studies Bhowmik and his co-workers (Bhowmik et al. 1999, 2000; Bhowmik and Roy 2003; Bhowmik and Spiering 2004) and Roy and his coworkers (Acharyya and Roy 2000; Roy and Prasad 2003) proposed a different timeframe for the collisional process and suggested the amalgamation of the North Indian Block (NIB) and the South Indian Block (SIB) at ~1000 Ma (a Grenvillian event). The lacunae in the time frame proposed in the above model was discussed by Mohanty (2006a, 2010, 2012), whose model of poly-phase evolution of the Satpura Mountain Belt (in at least three orogenic episodes) indicated amalgamation of the NIB and SIB during the second deformation in the Sausar Group at ~1700–1800 Ma and ~1500–1600 Ma (cf. Mohanty 2003a, 2006a, 2010, 2012, 2015) following the earliest deformation in the Sausar Group at ~2100 Ma. An updated correlation between the deformation and thermal events associated with igneous intrusions, migmatisation and metamorphic growth history in CITZ will be discussed later in Sects. 4 and 5.1.

2.2 Location and Geometry of the Subduction Zone

Based on the discordance of structural trends of the Bundelkhand protocontinent (having an ENE-WSW trend) and the Deccan protocontinent (with NS trend), the southern margin of the Bundelkhand protocontinent (Figs. 1b, 2a, b) was suggested to be the site of collision between the NIB and SIB, and was named as the CITZ, with two positions defining the suture (Radhakrishna and Naqvi 1986; Radhakrishna and Ramakrishnan 1988). The southern suture marking the boundary between the "Early Proterozoic Mobile Belt (EPMB)" and the "Middle Proterozoic Mobile Belt (MPMB)" was shown to have the dip of subduction zone towards MPMB (Fig. 2a), and the northern suture defined at the contact between the Bundelkhand block and the northern limit of the "Satpura Trend" or CITZ had dip towards the Bundelkhand block as a foreland block (Fig. 2b). A ~500 km shear zone coinciding with the position of the southern suture was termed as the "Central Indian Suture (CIS)", with the dip of subduction towards south (Yedekar et al. 1990). The occurrence of a high grade granulite terrain to the north of CIS and a low grade island-arc tholeiite suite (Sakoli and Nandgaon volcanic) in the Deccan protocontinent to the south were cited as evidence of southward subduction of the Bundelkhand protocontinent below the Deccan protocontinent (Yedekar et al. 1990). The tectonic model for CITZ proposed by Yedekar et al. (1990) envisaged initiation of the subduction process at ~2300 Ma leading to intrusion of Malanjkhand and Dongargarh plutons and calc-alkaline magmatism, later development of island-arc system (Sakoli and Nandgaon volcanic) during the culminating stage of the collision at ~2100 Ma and the terminal suturing at ~1600 Ma. In this model the development of the Sausar fold belt and thrusting of lower crustal granulites north of CIS were suggested to be at 1700–1500 Ma. The back-arc extensional basin development (Khairagarh Group) was thought to be a later event associated with the cessation of collision process between 1000 and 700 Ma.

On the basis of the results from deep seismic surveys indicating offset of the Moho and intersecting lower crustal reflectors, Divakara Rao et al. (1998) also interpreted the CIS to be a suture between the Satpura mobile belt and Kotri-Dongargarh mobile belt. Mishra et al. (2000) computed crustal models from gravity-magnetic studies, and interpreted the CIS to be a suture zone between the Bundelkhand craton and the Bhandara craton, wherein the Bundelkhand craton was considered to be the subducting block containing the granulitic rocks of the region, the Sausar Group and the Tirodi Gneiss, and northward thrusting of different zones. The model proposed by Mishra et al. (2000) was contested by Acharyya et al. (2001) on the basis of the time frame for different tectonic events in the region and the nature of the CIS as a zone of ductile shear, not a suture. Acharyya et al. (2001) argued for formation of granulites and suturing to be pre-Sausar event, which was in contrast with the observations that the high-grade (up to granulite facies) metamorphism and migmatisation of the Sausar Group to form gneisses similar to the Tirodi Gneiss took place during the collisional event (Brown and Phadke 1983).

The model proposed by Acharyya (2003) envisaged subduction of the northern crustal province below the southern crustal province during 1700–1800 Ma, leading to the suturing of both the provinces during 1500–1700 Ma and development of northward directed thrust sequences. This model also proposed coeval deposition of the sediments in a foreland basin forming the Vindhyan Supergroup on the northern province and the Sausar Group in the southern sides of the suture zone in a back-arc basin set-up during 1200–1400 Ma. The collision process in the CITZ was thought to be completed at 1200–1000 Ma with development of southward directed thrusts in the southern province (Acharyya 2003).

A completely different tectonic model proposed by Roy and Prasad (2003) envisaged subduction of the Bastar craton (southern province) below the Bundelkhand craton (northern province) during 2100–1800 Ma, with the development of the Mahakoshal belt and the Betul belt in back-arc tectonic set-up. The model identified the granulites of the Bastar craton to be older than the subduction process. The collision process during 2100–1800 Ma led to the development of northward directed thrusts in the northern block and granulites (of the Makrohar region) occurring south of the Mahakoshal belt. The collision was thought to be completed at 1500 Ma with the suture zone marked at the margin of the Bastar craton, development of the Ramakona granulite belt as an accretionary terrane by southward thrusting, and deposition of the Vindhyan Supergroup. The Sausar Group was thought to be deposited during 1100–1500 Ma in a collisional sag basin. The closing of the collision process was identified to be at 900–1100 Ma with deformation of the Sausar Group. Interestingly, the suture between the Bastar craton and the Bundelkhand craton was marked at the southern boundary of the Ramakona granulite belt, north of the Sausar belt.

A modified model of Acharyya (2003) was proposed by Sarbadhikari and Bhowmik (2008), in which the southward subduction of the North Indian Block below the South Indian Block was thought to be at 2100–2670 Ma and the granulites of the South Indian Block were reported to be older than the subduction process. A series of orogenic uplift and collapse were identified in the overriding plate during 1400–1590 Ma with the development of a series of thrust directed towards south, antithetic (backward) with respect to the subduction margin, and deposition of the sediments in the Sausar Group. The period between 1000 and 1400 Ma was identified to be the period of extension with emplacement of mafic plutons and continued deposition of the Sausar Group. The collision was identified to be complete at ~1000 Ma. Interestingly, the suture marking the boundary between the two crustal blocks was marked at a place far north of the Ramakona block (at the margin of the Vindhyan Basin). A drastically different position of the suture (at southern boundary of the Sausar Group with the granulites and gneisses of the southern domain) has been proposed by Bhowmik et al. (2011).

Deep seismic data across CIS shows different reflectivity characters in the northwest (southerly dip of reflection planes) and southeast (northerly dip) and indicate presence of a zone of suture (Mall et al. 2008). The subduction model proposed by Mall et al. (2008) shows subduction of the Bastar craton below the Bundelkhand craton, in which the latter includes the rocks of the Sausar Group, Tirodi Gneiss and associated granulites. A model similar to this has also been proposed by Naganjaneyulu and Santosh (2010). On the basis of aeromagnetic data, Rajaram and Anand (2003) marked the margin of the Bastar craton far north of the Central Indian Shear, matching with the Tan shear.

2.3 Summary

A review of the ideas regarding the position of the suture zone, the dip of the subducting plate and the time of subduction indicates conflicting models. In all the models proposed before 2010 identified CIS as the suture zone, although this zone is a ductile shear zone separating felsic gneisses on the north from those in the south. The CIS has poor signature in gravity anomaly and aeromagnetic maps. Moreover, all the models (except those of Divakara Rao et al. 1998 and Mishra et al. 2000) show subduction of the Bastar craton below the Bundelkhand craton along CIS. Such models cannot explain the occurrences of granulite facies rocks in the overriding plate unless such high-grade terranes are exotic blocks of the subducting plate accreted to the overriding plate; in such a situation, the suture is likely to be present to the north of the high-grade terranes (cf. Rajaram and Anand 2003). The time of collision between the NIB and SIB was thought to be 'middle' Proterozoic (~1600 Ma; Radhakrishna and Ramakrishnan 1988; Radhakrishna 1989; Yedekar et al. 1990), but the models of Bhowmik et al. (1999, 2000), Acharyya and Roy (2000), Bhowmik and Roy (2003), Roy and Prasad (2003), Bhowmik and Spiering (2004) and Roy et al. (2006) demonstrated a continuous orogenic event in the region leading to the amalgamation of blocks during 'Grenville orogeny' at ~1000 Ma.

A multi phase evolutionary model for the CITZ has been proposed in which the earliest orogeny is identified to be at ~2200–2100 and the subsequent orogenies at ~1800–1700 Ma and ~1600–1500 Ma resulted in the amalgamations of the NIB and SIB along the Satpura Orogen; the ~1000 Ma event resulted in low intensity deformation in CITZ giving rise to broad folding with ~N-S axial planes overprinted on the dominant ~E-W tectonic fabric of CITZ (Mohanty 2000, 2003a, 2006a, 2010, 2012, 2015).

3 Deformation Pattern of Different Domains of CITZ and Its Tectonic Implication

The CITZ has Precambrian rocks exposed in two domains: the western domain (covering the Betul belt and the Sausar belt in the southern flank) and the eastern domain (the Mahakoshal belt in the northern flank and the Chhotanagpur Gneissic Complex in the south). Eastward continuation of the CITZ is found in the Shillong plateau. Structural studies carried out in the Sausar belt (Fig. 4) and the Chhotanagpur Gneissic Complex combined with a compilation of data related to the Proterozoic granites associated with different deformation and metamorphic events provide clues to understand tectonic evolution of the region.

3.1 Deformation in the Sausar Belt in the Western Domain of the CITZ

The western domain covers the Satpura region of the CITZ and has been the focus for interpreting the "Satpura orogenic trend" (discussed in the Sect. 1). Oldham (1856), Blanford (1872) and Bose (1888) carried out the first systematic mapping in the Central India and reported the occurrences of manganese ore horizons in the region (Fermor 1906, 1909, 1911). On the basis of regional mapping, the name "Sausar Series" (henceforth referred as Sausar Group) and the stratigraphic divisions for

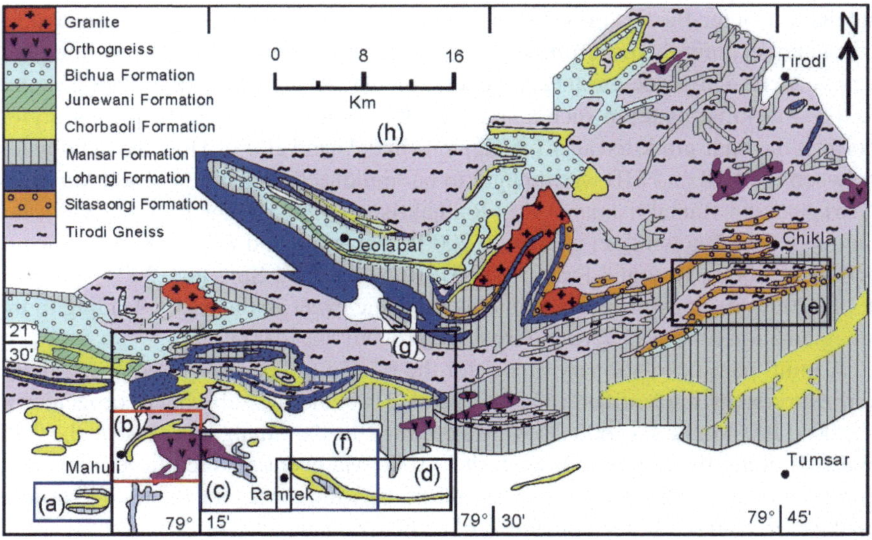

Fig. 4 Geological map of the central part of the Sausar belt, central India (after Narayanaswami et al. 1963). The formations (Sitasaongi-Bichua) constitute the Sausar Group. Outcrops of rocks of the Sausar Group in this map area of relatively low relief show curved/closed geometries, indicating presence of regional folds of two generations. A broad swing in regional strike from NW-SE to SW-NE near Deolapar defines a regional fold of last generation having ~NS axial trace. Framed rectangles: (a) Fig. 5a, (b) Fig. 6a, (c) Fig. 7a, (d) Fig. 9a, (e) Fig. 10a, (f) Fig. 11 and (g) Fig. 12. The position marked as (h) is the location for Fig. 13. Position of Fig. 4 falls on the margin of SMB in the blue rectangle in Fig. 3a. The dimensions of the framed rectangles are approximate

the succession were proposed by Fermor (in Pascoe 1926). West (1936) reported nappe structures from the northern part of the Sausar belt. The first generation mapping identified complex deformation history of the region and associated high-grade metamorphism producing granulites from the sediments of the Sausar Group, and migmatites-gneisses (referred as "streaky gneisses" by West 1933). The maps prepared for the region recorded presence of thin bands of metasediments of the Sausar Group as relicts (resistites) within the gneisses. West (1933) analysed different fabric elements within the "streaky gneiss" in conjunction with the map pattern of the region and concluded that the granulite facies metamorphism of the region took place during deformation and metamorphism of the sediments of the Sausar Group. The second generation mapping in the Sausar belt (Straczek et al. 1956; Narayanaswami et al. 1963) redesignated the "streaky gneiss" as "Tirodi Gneiss" (Fig. 4), being a part of the Sausar Group, and preferred the status of the unit to be the basement of the Sausar Group. The regional map prepared for the Sausar belt indicated presence of isoclinal folds with ~EW regional axial traces, showing variations from NW-SE to SW-NE due to late phase deformation (Fig. 4).

Presence of superposed folds of three generations have been recorded from the Sausar belt by Basu and Sarkar (1966), Agrawal (1975), Sarkar et al. (1977), Mohanty (1988, 1993, 2002), Mohanty and Mohanty (1996), and Mohanty et al. (2000). The early folds (F_1) are isoclinal with an axial planar cleavage (S_1) and show variable orientation of axial surface and folds axes because of overprinting and reorientation by later folds. The folds of second generation (F_2) are tight to isoclinal, with axial planar crenulation cleavage (S_2) and congruous pucker axis lineations; these folds have ~EW striking axial planes. The late folds (F_3) are open upright with ~NS striking axial planes. The folds of the first two generations define the regional E-W alignment of the rock units of the Sausar Group and its basement (referred together as the Sausar belt). The isoclinal folds of first and second generation in the Sausar Group were reported to be coaxial, giving rise to hook-shaped map patterns at number places (cf. Basu and Sarkar 1966; Agrawal 1975; Sarkar et al. 1977). On the other hand, detailed structural mapping carried out in the Sausar belt by Mohanty and his coworkers (Mohanty 1988, 1993, 2002; Mohanty and Mohanty 1996; Mohanty et al. 2000) have established non-coaxial geometry of the early folds, which have given rise to Type 2 ("mirror-image") interference patterns (discussed later in this section).

Contrary to the structural patterns proposed by all the previous workers, the third generation mapping in the Sausar belt (Bandyopadhyay et al. 1995; Chattopadhyay et al. 2003a, b, 2015; Khan et al. 2003) suggested that the earliest deformation in the Sausar Group gave rise to the development of regional thrusts and associated small scale folds with axial planar schistosity; progressive rotation of these structures during a single continuous deformation related to the "Sausar Orogeny" developed the folds of second generation with ~EW axial traces. The presence of several thrusts (some with cm scale spacing) to explain the complex map pattern of the Deolapar area by Chattopadhyay et al. (2003a) has been debated by Mohanty (2003b) for lack of evidence of thrusting, interpretation of the map pattern and folded geometry of the thrust planes (further discussed in Sect. 3.1.6).

3.1.1 Detailed Structural Mapping in the Parseoni Area

The metasediments of the Sausar Group, comprising of quartzite, mica schist with manganese ore and marble, in the Parseoni area have an antiformal fold closure with plunge to the east (Fig. 5a; Mohanty 2002). The core of the fold does not have any surface outcrops; a quartzite, with micaceous character at many places, defines a ridge around the core. The northern limb of this unit has pebble horizons with pebbles of grey and white quartzite as well as amphibolites. The girdle distribution of the poles of S_1 schistosity indicates the regional fold of the area to be of second generation with a plunge of 13°/N089° (Fig. 5b). The second generation linear structures are concentrated around this β-axis. The upright attitude of the regional F_2 fold is suggested by the point maxima pattern of S_2 surfaces.

3.1.2 Detailed Structural Mapping in the Mahuli Area

The area around Mahuli (Fig. 6a) has continuous outcrops of a granite gneiss east of Ghuksi and north of Potgowari, but outcrops of the Sausar Group are partial and discontinuous. The granite gneiss is the westward continuation of the granite

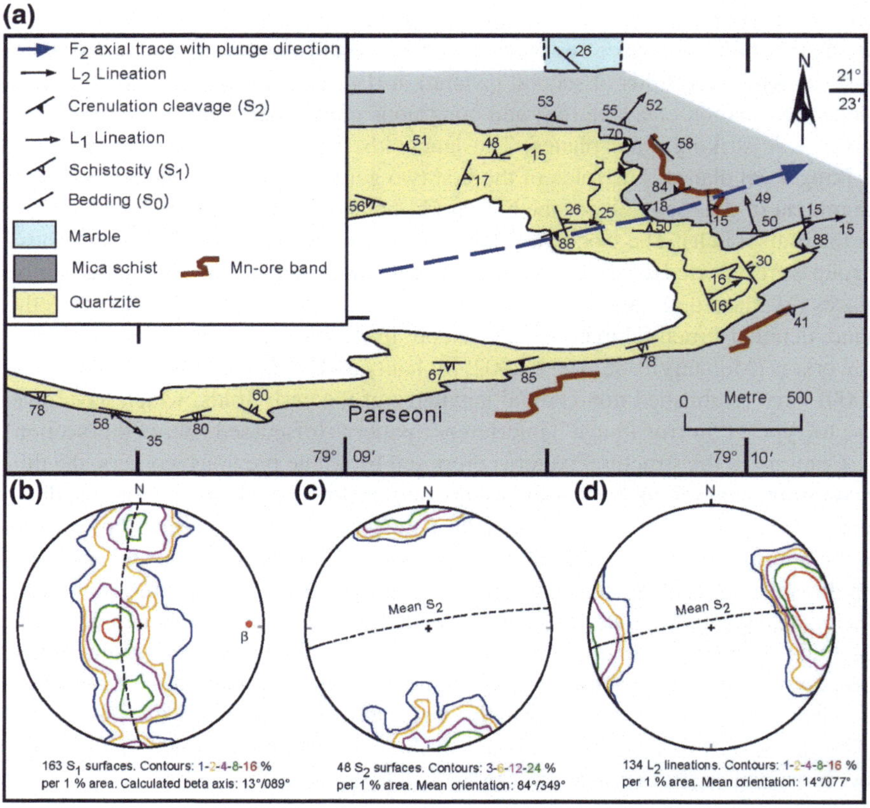

Fig. 5 **a** Geological map of the Parseoni area, showing orientations of planar and linear structures (after Mohanty 2002). **b** Lower hemisphere equal area projection of the attitudes of planar and linear structures of the Parseoni area (after Mohanty 2002)

near Mansar (Tirodi Gneiss—I) and is considered to be the basement for the Sausar Group. A conglomerate horizon of the Sausar Group is exposed at the contact of the granite west of Mahuli and SW of Potgowari, corroborating the status of basement for the granite gneiss. Outcrops of dolomitic marble are found in the areas between Ghuksi and Potgowari Dolomite Mines Office, Mahuli and Gundri, Chichbhuwan and Gundri, and Kalapatha and Palora (Fig. 6a). Manganiferous rocks are found in the central part of the area, north of Gundri. However, scattered outcrops and smaller sizes of outcrops do not permit to prepare a detailed lithological map of the Sausar Group. On the basis of the geometry of the contact between the granite gneiss and the Sausar Group and the structural data, regional folds of two generations can be identified from the area. The folds with ~EW straight axial traces are found to deform the axial traces showing NE-SW to NW-SE. Therefore, the folds with ~EW axial traces are considered to be later generation (F_2) and the other set as the early

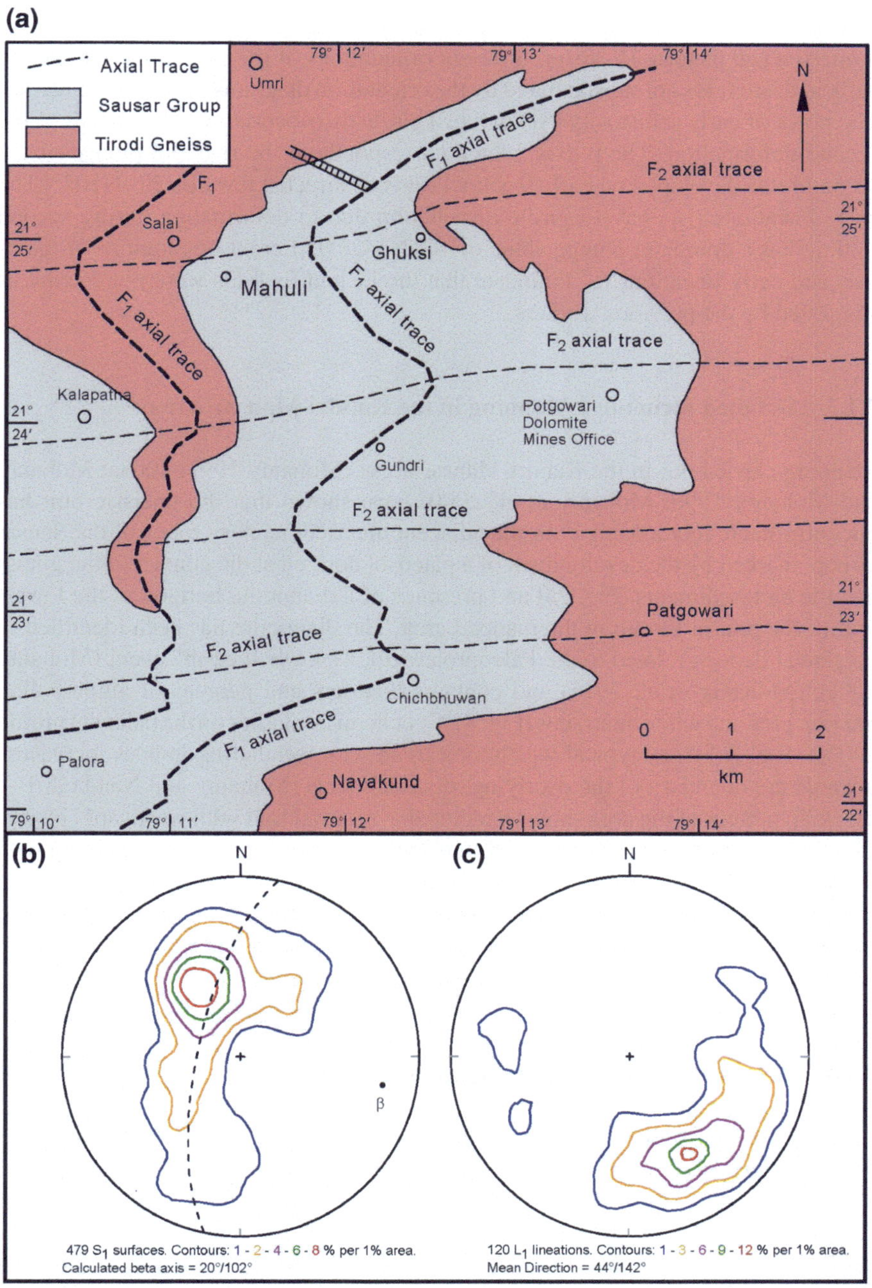

Fig. 6 **a** Simplified geological map of the Mahuli area, showing large-scale folding of the axial traces of early folds around later folds with ~E-W axial trace, defining a "zig-zag" map pattern. **b** Lower hemisphere equal-area projection of attitudes of different planar and linear structures of the Mahuli area. The map and projection diagrams are compiled from unpublished work of the author

folds (F_1). The overall map pattern is showing "zig-zag" geometry, apparently Type-3 interference pattern. However, different orientations of the axial directions of the early and late folds are substantiated by the orientation diagrams of the area (Fig. 6b). The poles of early schistosity (S_1) define a girdle distribution, which is result of the second deformation. The β-axis, which corresponds to the regional plunge of the second folds developed on early isoclinal folds, is directed towards 20°/N102°. The early lineations (L_1) have a girdle distribution due to deformation during second folding has a dominant plunge value of 44°/N142°. Different orientations of the β-axis and early lineations (L_1) indicate that the F_1 and F_2 folds were not coaxial as suggested by the previous workers.

3.1.3 Detailed Structural Mapping in the Kandri-Mansar Area

Mapping carried out in the Kandri-Mansar areas (Mohanty 1993, 2003a; Mohanty and Mohanty 1996; Mohanty et al. 2000) have shown that the gneissic unit has unconformable relationship with the adjacent metasedimentary rocks of the Sausar Group, marked by the development of a paleosol horizon at the contact of the gneiss and the metasediments (Fig. 7a) and presence of a diamictite horizon as the lowest unit of the Sausar Group in the mapped area. The diamictite has been identified to be glacial deposit related to the Paleoproterozoic "snowball earth" event (Mohanty 2006b; Mohanty et al. 2015) and contains boulders and pebbles of amphibolite, granite, grey and white quartzites (Fig. 8). Geochemical studies on the paleosol profile (WP in Fig. 7a) show typical weathering trend with weathering indices increasing towards the boundary of the overlying metasediments (Mohanty and Nanda 2016). The map pattern of the area shows a regional synformal fold with westward closure. The girdle distribution of early foliation planes (S_1), with the β-axis coinciding with the second generation lineations (L_2), indicate this fold to be of second generation (Fig. 7b). The correlation of discontinuous Mn-ore occurrences at Mansar Mines and Kandri Mines can be made on the basis of the structural geometry of a dolomitic marble horizon in the core of the regional fold.

3.1.4 Detailed Structural Mapping in the Ramtek Area

The map patterns showing "mirror-image" pattern of fold interference, which develop when the late (second) axial surfaces and fold axes are at high angle to the axial surfaces and fold axes of the early (first) folds, are common in the Sausar belt. Such an interference patterns are mapped in the Ramtek area (Fig. 9a; Mohanty 1988). In the Ramtek area different horizons of quartzite of the Sausar Group show a typical "boomerang-shaped" map pattern with the axial surface of second generation folds (F_2 with average strike of N123° and dip of 82° towards NNE) as the symmetry plane. The projection diagrams of linear structures of the Ramtek area (Fig. 9b–e) show a girdle distribution pattern for L_1 lineations but a point maximum pattern for the L_2 lineations, with different positions of maxima (L_1 at 21°/N128° and L_2 at 11°/N120°),

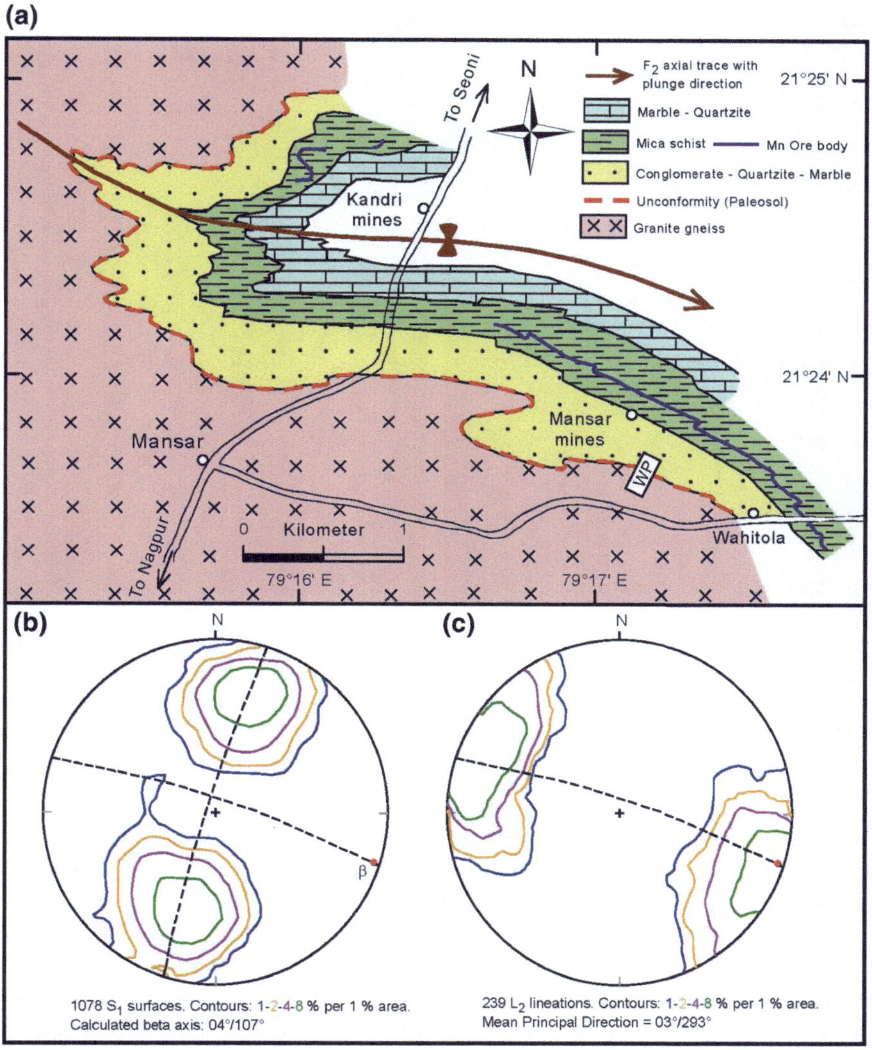

Fig. 7 **a** Geological map of the Kandri-Mansar area (modified after Mohanty and Mohanty 1996; Mohanty et al. 2000) showing the positions of a paleosol horizon and a conglomerate horizon between the Tirodi Gneiss and the Sausar Group, indicating the gneissic unit to be the basement for the Sausar Group. **b** Lower hemisphere equal-area projection of attitudes of different planar and linear structures of the Kandri-Mansar area (after Mohanty and Mohanty 1996; Mohanty et al. 2000)

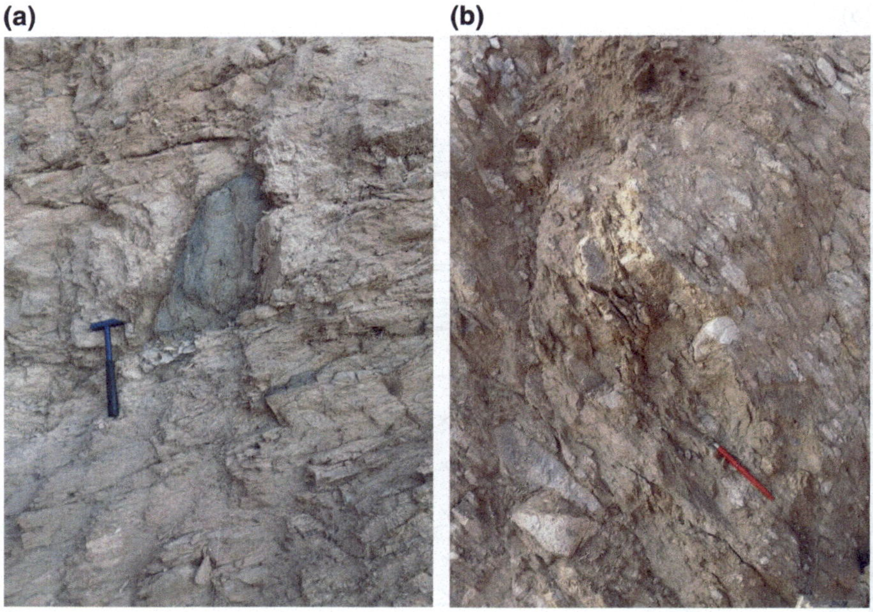

Fig. 8 Diamictite unit of the Mansar Mines area (21° 23′ 51″ N: 79° 17′ 02″ E): **a** A boulder of amphibolite in the matrix of arkosic composition (E-W vertical section, view towards north), and **b** clasts of different shape, size, angularity and composition (mostly granite and grey and white quartzites) in a matrix of arkosic composition (N-S vertical section, view towards west). The outcrop location is close to the marked site WP in Fig. 7

and corroborate non-coaxial geometry of the early folds in the Sausar belt. Extreme flattening associated with the folds of second generation rotated the early lineations to make low angle with the second fold axes (Mohanty 1988). The stratigraphic younging directions in the limb areas across the early (F_1) axial trace indicate the mica schist to be the younger stratigraphic unit than the different lithological variants of quartzite, which has raised questions regarding the correlations of quartzite with the Chorbaoli Formation and mica schist with the Mansar Formation (Mohanty 1988).

3.1.5 Detailed Structural Mapping in the Chikla-Dongribuzurg Area

In the Chikla-Dongribuzurg area, the outcrops of the basement gneiss (Tirodi Gneiss—I) surrounded by the Sausar metasediments show a typical "mirror-image" pattern with symmetry around the F_2 axial surface (Fig. 10a), having the strike of N094° and dip of 72° towards S). Non-coaxial geometry of F_1 and F_2 folds is corroborated by different positions of maxima for the linear structures related to these folds (Fig. 10b). In the Ramtek area as well as the Chikla-Dongribuzurg area, the effects of late folds (F_3) with nearly NNE-SSW axial traces have caused changes in the regional strikes and plunge variations of F_2 folds.

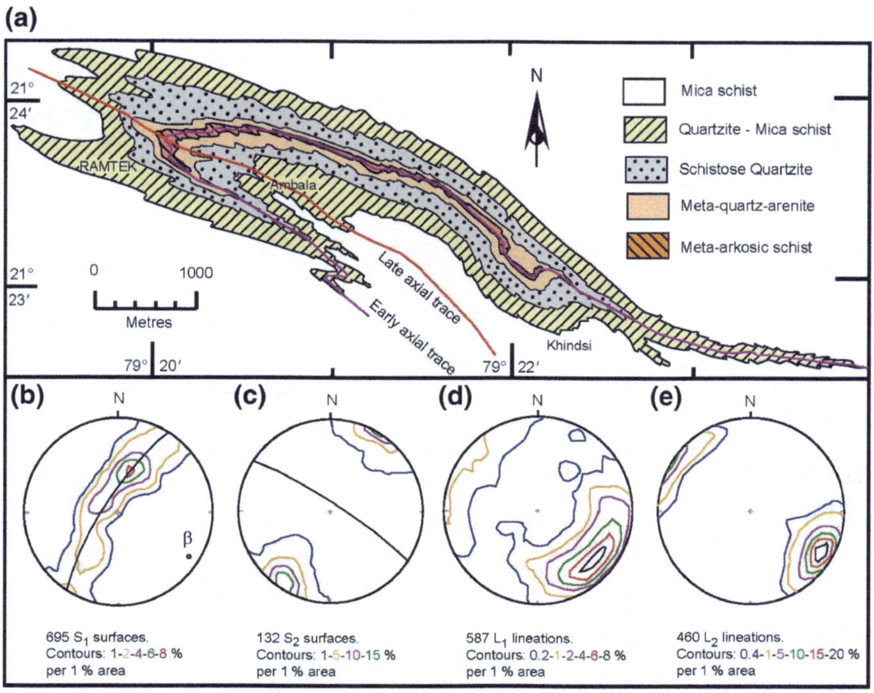

Fig. 9 **a** Simplified geological map of the Ramtek area, showing large-scale axial folding of an early fold around a later isoclinal fold with WNW-ESE axial trace and southeastward plunge, defining a “boomerang-shaped” map pattern. **b** Lower hemisphere equal-area projection of attitudes of different planar and linear structures of the Ramtek area (after Mohanty 1988, 2010). Non-coaxial geometry of the F_1 and F_2 folds are substantiated by the Type 2 interference pattern and diverse orientations of deformed L_1 lineations compared to the point maxima pattern of L_2 lineations

3.1.6 Interpretations of Map Patterns of the Third-Generation Mapping in the Sausar Belt

The third-generation mapping of the Sausar belt (Bandyopadhyay et al. 1995; Chattopadhyay et al. 2003a, b, 2015; Khan et al. 2003) ruled out presence of any mappable fold related to the first deformation in the Sausar Group; the earliest deformation structures (F_1) in the Sausar Group were reported to be thrust planes and associated minor folds and schistosity surfaces developed during the collision of the NIB and SIB, with continuous monocyclic evolution of the Sausar belt giving rise to regional (mappable) isoclinal folds with ~EW axial surfaces during the second deformation (F_2). It may be noted that crescent-shaped map patterns for the Kandri Mines and the Ramtek areas (Fig. 11) were shown by Chattopadhyay et al. (2003b), but these were interpreted to be results of a single deformation. The geometrical interpretation of such a map pattern by coaxial-refolding was questioned by Mohanty (2003c), but the authors avoided the problem by stating that the area is having soil covers (though the lithological units were shown in the map and had points where three different

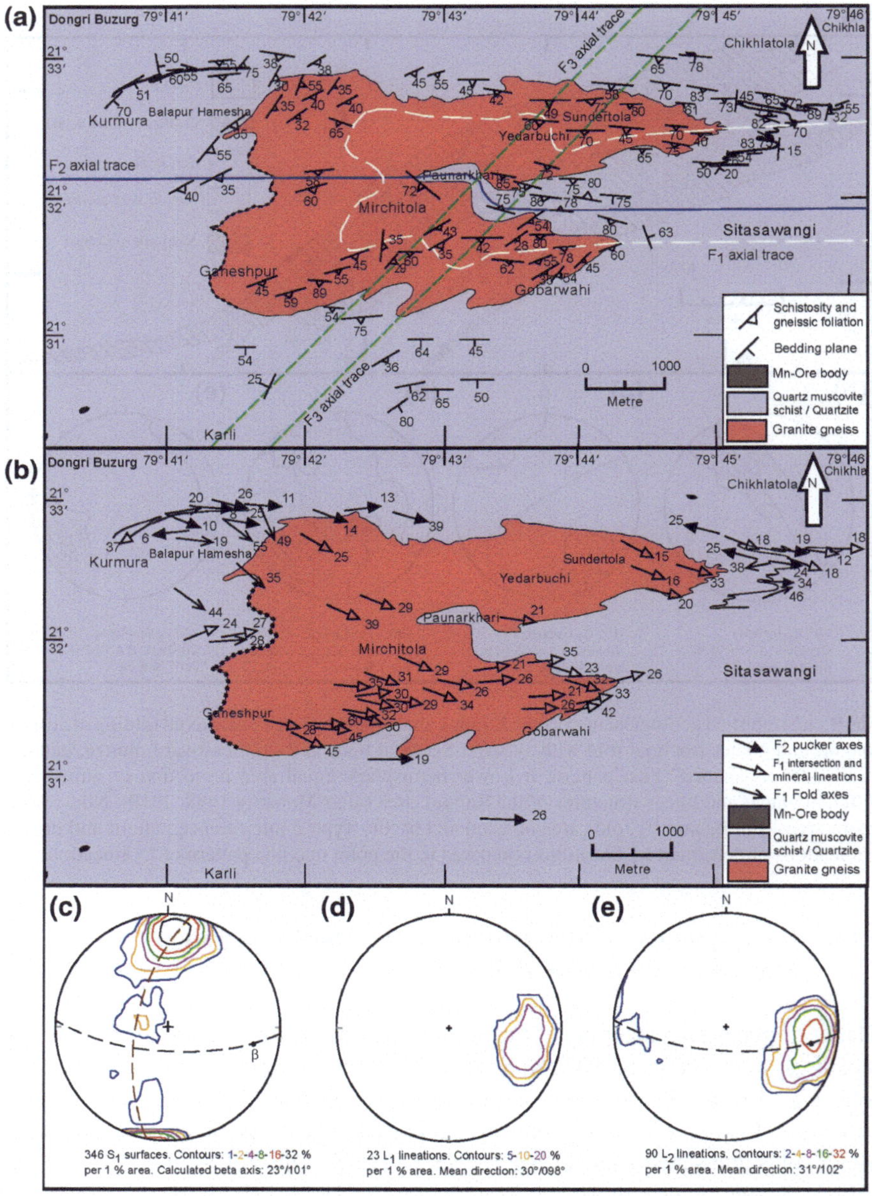

Fig. 10 **a** Simplified geological map of the Chikla-Dongribuzurg area, showing large-scale axial folding of an early fold around a later isoclinal fold with ~E-W axial trace and eastward plunge, defining a "boomerang-shaped" map pattern. **b** Lower hemisphere equal-area projection of attitudes of different planar and linear structures of the Chikla-Dongribuzurg area. Non-coaxial geometry of the early folds is substantiated by the map pattern. The map and projection diagrams are compiled from unpublished work of the author

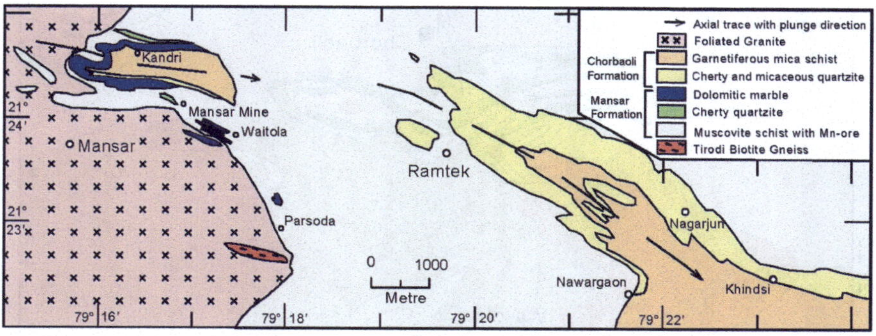

Fig. 11 Geological map of the Kandri-Mansar-Ramtek area (from Chattopadhyay et al. 2003b), showing a southeasterly plunging synformal fold on the Sausar Group. Crescent-shaped map pattern of the Chorbaoli Formation (showing symmetry around the WNW-ESE axial trace) near Kandri and down plunge repetition of the Mansar Formation can be noted (see discussion by Mohanty 2003c). The regional axial traces can be traced in the foliated granite, which is considered to be syn-tectonic intrusion with respect to the second deformation in the Sausar Group (Chattopadhyay et al. 2015). These features indicate the deposition of the rocks of the Sausar Group and granite intrusion (?) must be pre-F_2 age (~1600 Ma)

lithological units meet). The status of the gneissic unit mapped near Mansar was assigned to be an intrusion syn-tectonic with respect to the second deformation in the Sausar Group (Chattopadhyay et al. 2003b, 2015); the curved boundary between the gneissic unit and the Sausar Group was having similar geometry of the folds in the Sausar Group; the synformal axial trace between Kandri-Ramtek, and antiformal axial trace between Parsoda-Nawargaon pass through the foliated granite (Fig. 11). The significance of the fold geometry vis a vis the status of the gneissic unit near Mansar will be analysed in a later part of this section.

A close scrutiny of the map (Fig. 11) prepared by Chattopadhyay and his coworkers (Chattopadhyay et al. 2015 and references therein) shows that the map for the central part of the Sausar fold belt has been updated by removing the Chorbaoli Formation from the core of the synformal fold near Kandri possibly to avoid the questions raised by Mohanty (2003c), but a similar "boomerang-shaped" pattern, with down-plunge repetition of the Mansar Formation, still remains in the area near Ramtek (please compare Figs. 11 and 12). The overall geometry (i.e. westward closure and symmetry of the fold) in Fig. 12 (Chattopadhyay et al. 2015) matches with that in Fig. 9a (Mohanty 1988). As discussed in Sect. 3.1.4, such a pattern supports the observation of Mohanty (1988) that mappable F_1 are preserved in the Sausar belt.

Furthermore, along strike terminations of different formations of the Sausar Group inside the "Biotite Gneiss/Migmatite (TBG)" can be observed in the map of the south-central part of the Sausar belt (Fig. 12). Although no explanations are available for such patterns, these can happen under any one of the two conditions: (a) the migmatisation leading to the formation of the so called basement (TBG) later than the deposition and deformation of the Sausar Group, leading to termination of different lithology in a migmatite as suggested by West (1933), or (b) the terminations are fold

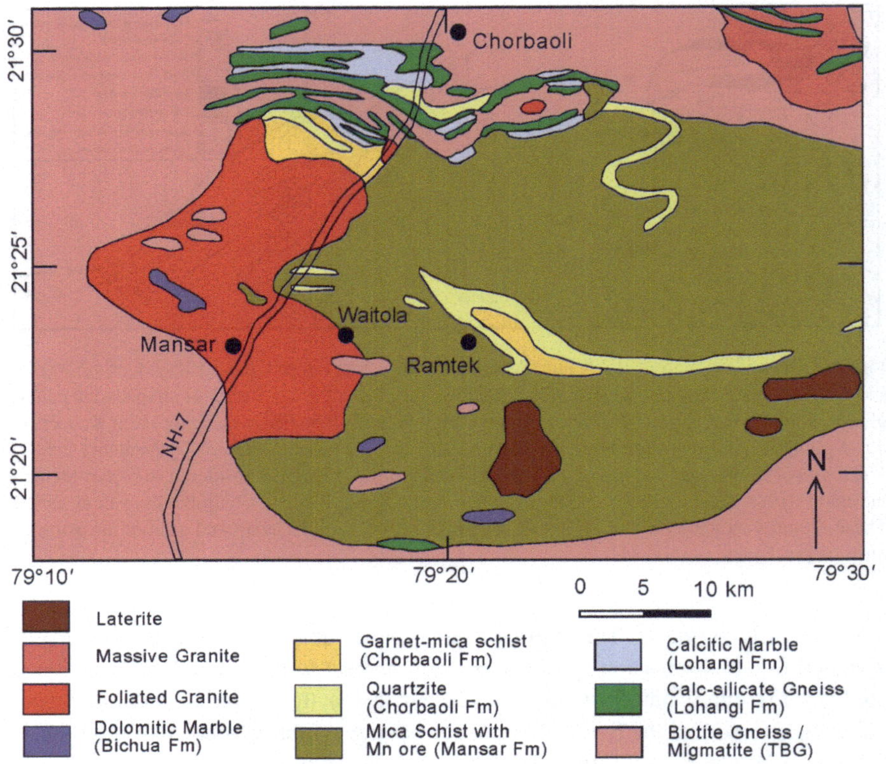

Fig. 12 Geological map of the Kandri-Mansar-Ramtek area prepared during the third generation mapping in the Sausar belt (from Chattopadhyay et al. 2015; please refer to the original source therein). It may be noted that the crescent-shaped pattern near Kandri (Fig. 11) has been removed to avoid the questions raised by Mohanty 2003c, but another crescent-shaped map pattern of the Chorbaoli Formation (showing symmetry around the WNW-ESE axial trace) and down plunge repetition of the Mansar Formation can be noted near Ramtek

closures defining "mirror-image type" interference patterns (west of the area near Chorbaoli; Fig. 12). The third option regarding presence of thrust sheets proposed for the area north of Deolapar by Chattopadhyay et al. (2003a) is not considered for reasons discussed in Sect. 3.1. Furthermore, the thrusts, in general, place older rocks of the hanging wall block over younger rocks of the footwall block; the reverse situation of younger rocks overlying older rocks can be explained by an unconformable relationship. In the map prepared for the area north of Deolapar, Chattopadhyay et al. (2003a) have shown the younger rocks of the Sausar Group in the core of a synform near Manegaon having the basement TBG below the Sausar Group and several closed outcrops of Sausar Group surrounded by the TBG; the map pattern was explained by invoking thrusts at the contact of TBG and the Sausar Group. This interpretation has been questioned for lack of evidence of thrusting (such as presence of cataclastic rocks with shear sense indicators) and the stratigraphic relationship which can be explained by unconformable relationship (Mohanty 2003b).

Interestingly, the outcrop map of the foliated granite near Mansar shows a "boomerang-shaped" pattern with symmetry around the EW axial trace (Fig. 12). The status of this granite as the basement of the Sausar Group was suggested by Mohanty (1993) on the basis of the presence of a paleosol horizon and a conglomerate at the boundary of the gneiss and the Sausar Group (discussed earlier), but Chattopadhyay et al. (2015) have mapped this as an intrusion in the Sausar Group, syn-tectonic with the second deformation. Monazite chemical dating of this granite exposed near Waitola (Fig. 12) provided an average age of 944 ± 5 Ma, but grains with core and rim structures yield 1609 ± 32 Ma; monazite grains from the Tirodi Gneiss (basement?) from the Chorbaoli area (Fig. 12) define age of 948 ± 4 Ma; and a massive granite emplaced post-tectonically into the Sausar Group provide an age of 928 ± 4 Ma (Chattopadhyay et al. 2015). All these age data can be reconciled only when we consider that the ages provide record of post-tectonic thermal equilibrium at ~950 Ma and the second deformation in the Sausar Group, which imparted foliation in the granite of the Mansar-Waitola area was associated with the thermal event at 1609 ± 32 Ma (core age of monazite at Waitola).

3.1.7 Summary of Structural-Tectonic Studies in the Sausar Belt

Analysis of the structural studies in the Sausar Group indicates that the regional E-W alignment of the lithological units in the Sausar belt resulted during the second deformation (F_2) in the region and was associated with the second episode of the "Satpura Orogeny" (~1600 Ma, Mohanty 2015). These observations go against the model proposed by the previous workers (Bhowmik et al. 1999, 2000; Acharyya and Roy 2000; Bhowmik and Roy 2003; Roy and Prasad 2003; Bhowmik and Spiering 2004; Roy et al. 2006; Chattopadhyay et al. 2015) regarding the tectonic evolution of the CITZ vis-à-vis the deposition of the sediments of the Sausar Group in late Mesoproterozoic (~1600–1000 Ma) and its deformation and metamorphism at ~1000 Ma.

3.2 Structural Evolution of the Chhotanagpur Complex in the Eastern Domain of CITZ

The Chhotanagpur Gneissic Complex (CGC) is a collage of high-grade (amphibolite-granulite facies) gneisses, migmatites and metsedimentary-metavolcanic rocks with ~EW trend covering ~500 km length and ~200 km width. Three sedimentary basins in the southern part (Purulia-Ranchi, Parasnath and Kodarma belts), aligned in ENE-WSW direction parallel to the "Satpura orogenic trend" show medium to high grade metamorphism, and the Munger-Rajgir-Gaya basin on the north shows a relatively lower grade of metamorphism. Three distinct deformational episodes have been identified from this terrain (Bhattacharyya 1975; Roy 1977; Sarkar and Jha 1985; Mazumdar 1988; Sarangi and Mohanty 1998). The first generation folds (F_1) are isoclinal with diverse orientation of axial planes; the second generation structures (F_2) are represented by isoclinal folds with EW-striking upright axial planes and the third generation folds (F_3) are open to tight with nearly NS-striking axial planes (Roy 1977; Sarangi and Mohanty 1998). The dominant ~EW trending fabric in CGC was developed during the second deformation and matches with the "Satpura Orogenic trend" (Sarangi and Mohanty 1998). The CGC is intruded by mafic plutons, anorthosite plutons, hypersthene granites (charnockites) and granites. The mafic and anorthosite plutons were emplaced after the first deformation and were syn- to post-tectonic with respect to the second deformation (Roy 1977; Sarangi and Mohanty 1998). Rb-Sr whole rock ages from the CGC indicate 1599 $\pm$ 33 Ma (hypersthene granites; Mallik et al. 1991), and 1457 $\pm$ 63 Ma (Dumka Syenite and granulite metamorphism; Ray Barman et al. 1993, 1994) for these intrusions. Thus, in spite of limitations of the Rb-Sr methods, the high-grade granulite event related to the second deformation is constrained to be ~1600 Ma (Ray Barman et al. 1994). Based on U-Pb zircon age of 1550 $\pm$ 12 Ma for the Bengal anorthosite, Chatterjee et al. (2008) interpreted the intrusion of anorthosite to be 50–150 Ma after the metamorphism related to the collision of the north Indian and south Indian blocks.

4 Spatio-Temporal Correlation of Different Domains of the CITZ

Based on the available geological and geochronological data from different domains of the CITZ (the western domain covering the Betul and Sausar belts, the central domain of the Mahakoshal belt, and the eastern domain of the Chhotanagpur region) and adjacent cratonic blocks, and eastward continuation of the "Satpura orogenic trend" in the Shillong plateau and adjacent regions, an event stratigraphy was established (cf. Mohanty 2010, 2012). The event stratigraphy has been updated with new age data from all the domains (Table 1). The blocks to the south of the CITZ in the western and eastern domains and the Shillong plateau have >2400 Ma felsic gneisses with record of granulite facies metamorphism at ~2600 Ma; the Bundelkhand Complex to the north of CITZ also contains granites and gneisses of >2400 Ma but does not have any record of granulite facies metamorphism. All the domains of the CITZ have a volcano-sedimentary succession overlying the Archean basement (felsic gneiss of >2400 Ma), which was involved in deformation and metamorphism at ~2200 Ma. These data provide the stratigraphic age of 2400–2200 Ma for the rocks of the Betul belt, Sausar Group, Lower Mahakoshal Group, Lower Singhbhum Group and Lower Bomdila Group. The metamorphism in the western domain was of ultra-high temperature (UHT) type leading to longer cooling and monazite growth history up to 2100 Ma. A decompressional event following this first orogenic episode in the CITZ was associated with intrusion of mafic dykes and development of basins in which

Table 1 Events stratigraphy of the Central Indian Tectonic Zone (after Mohanty 2010, 2012)

Time (Ma)	Western domain (Sausar)	Central domain (Mahakoshal)	Eastern domain (Chhotanagpur)	Shillong Plateau
<900	Betul (Navegaon) Granite (850 ± 15 Ma), monazite growth rims in Sausar granite (785 ± 15 Ma)	Upper Vindhyan sedimentation (Bhander Formation): Limestone (866 ± 90 Ma, 908 ± 72 Ma)	Raigarh (Kunkuri) Granite (803 ± 49 Ma)	Sindhuli Granite (881 ± 39 Ma)
900–1000	Low intensity deformation (SD_3) Monazite age of post-Sausar granite (928 ± 4 Ma, 944 ± 5 Ma) Monazite age from Tirodi gneiss (948 ± 4 Ma)	Upper Vindhyan sedimentation (Rewa Formation) Thermal imprints on Majhgawan kimberlite (974 ± 30 Ma)	Deformation of Munger Group Monazite growth age from CGC (957 ± 17 Ma) Zircon growth age from CGC (1009 ± 24 Ma)	Low intensity deformation of Shillong Group and Dirang Formation
1000–1300	Decompressional event; Dolerite dyke in Sausar Group (1112 ± 77 Ma); Mandla granite intrusion (1147 ± 16 Ma) Deposition of Chhatisgarh Supergroup (upper part)	Kimberlite intrusion (1073 ± 14 Ma; 1170 ± 45 Ma) Upper Vindhyan sedimentation (Upper Kaimur Formation): Bijaigarh Shale (1210 ± 52 Ma)	Jajawal Granite intrusion (1100 ± 20 Ma) Deposition of Munger Group	Mikir Hill granite (1110 ± 15 Ma) Deposition of Dirang Formation Deposition of Shillong Group
1300–1450	Decompressional event; intrusion of Dolerite dykes Thermal resetting ages of mafic granulites of Bhandara craton (1407 ± 11 Ma); metamorphic imprints in Sausar Group (1415 ± 23 Ma) and Tirodi Gneiss (1454 ± 5 Ma)	Bundelkhand mafic dykes (1407 ± 11 Ma)	Post-CD_2 intrusion: Raigarh (Nagam) granite (1331 ± 42 Ma), Dumka syenite (1457 ± 63 Ma) Post-collisional ferroan granite batholiths of CGC (1446 ± 6 Ma)	Longavalli granite (1430 ± 10 Ma)
1450–1500	Deposition of Chhatisgarh Supergroup (lower part)	Upper Vindhyan sedimentation (Lower Kaimur Formation)	Deposition of Munger Group and Chandil Formation (1487 ± 34 Ma)	Deposition of Shillong Group
1500–1550	SD_{2b} deformation and UHT Metamorphic overprint of Sausar Group (1525 ± 70 Ma), Sausar granulite (1553 ± 19 Ma), metamorphic overprint on Tirodi Gneiss (1534 ± 15 Ma), Re-metamorphism of BBG belt (1525 ± 13 Ma)	Syn-MD_3 Dudhi granite emplacement (1576 ± 76 Ma)	Chhotanagpur/Simultala Orogeny CD_2 folding, metamorphism, and granite emplacement (1590 ± 30 Ma) Mor valley granulites (1515 ± 5 Ma) and granites Monazite growth age (1547 ± 18 Ma)	BD_{2b} deformation, Ziro high-Ca granite (1536 ± 60 Ma) Thermal imprint on mafic rocks of Garo Hills (1596 ± 15 Ma)
1550–1600	Decompression: Betul (Morkha) granite (1550 ± 50 Ma)	Deposition of Upper Semri Group (1599 ± 48 Ma, 1602 ± 10 Ma)	Intrusion of Bengal anorthosite (1550 ± 12 Ma) Deposition of Upper Kodarma Group	Deposition of Shillong Group
1600–1780	Decompression: Mafic dyke of Bastar (1776 ± 13 Ma) Thermal imprints on Tirodi Gneiss (1618 ± 8 Ma) and Dongargarh granite (1640 ± 60 Ma)	Bundelkhand mafic dykes (1670 ± 11 Ma) Lower Vindhyan (Lower Semri Group) sedimentation (1631 ± 5 Ma, 1721 ± 90 Ma) Rihand granite intrusion (1731 ± 36 Ma) Jhirgadandi post-orogenic granite (1753 ± 9 Ma)	Isobaric cooling (IBC) of hypersthene gneiss of Mor Valley (1624 ± 5 Ma) Daltonganj granite (1741 ± 65 Ma) Newer Dolerite Dyke IV (1765 ± 1 Ma) Ultramafic and mafic intrusions in Dalma Group (1619 ± 38 Ma) Deposition of Middle Kodarma Group	Rongjeng granite of Garo Hills (1778 ± 37 Ma) Masang Kang mafic igneous suite of Bhutan (1742 ± 39 Ma) Tonalite of Dinajpur (1722 ± 6 Ma) Shillong Group: Conglomerate, shale and carbonaceous shale
1780–1850	Satpura Orogeny (SD_{2a} deformation) Spot chemical ages of monazite cores in BBG (1794 ± 44 Ma)	Satpura Orogeny (MD_2)	Chhotanagpur/Simultala Orogeny CD_{2a} deformation, granites, granulite event, and migmatisation Dalma Group (volcanics)	BD_{2a} deformation, porphyritic and biotite granite
1850–1950	Decompressional event: Post-SD_1 (Unfoliated) Harda granite intruding the Betul Belt (1856 ± 68 Ma)	Bundelkhand mafic dykes (1863 ± 9 Ma), Granite intrusions post-tectonic with respect to MD_1 episode: Sidhi granite (1850 ± 40 Ma) Deposition of Gwalior Group	Deposition of Lower Kodarma Group	Bomdila augen gneiss (1914 ± 25 Ma)
1950–2050	Sausar Orogeny Ib (SD_{1b}) and UHT metamorphism (amalgamation of NIB and SIB) Monazite rim in felsic granulite (2040 ± 17 Ma)	Mahakoshal Orogeny Ib (amalgamation of NIB and SIB)	Singhbhum Orogeny Ib Deformation, granites, migmatisation, and granulite event in CGC (CD_1)	BD_{1b} deformation
2050–2100	Decompressional event: High-Mg mafic dyke of Bastar (2100 ± 11 Ma) Deposition of Chilpi Group Khairagarh Group	Bundelkhand mafic dykes (2063 ± 45 Ma) Deposition of Upper Mahakoshal Group	Newer Dolerite III (2105 ± 38Ma) Deposition of protoliths of pelitic schists and gneisses of CGC Deposition of Upper Singhbhum Group (Kolhan Formation)	Deposition of Upper Bomdila Group (Tenga Formation)
2100–2250	Sausar Orogeny Ia (SD_{1a} deformation) and UHT metamorphism Thermal imprint on Bijli rhyolite (2180 ± 25 Ma) Monazite growth in BBG (2089 ± 14 Ma)	Mahakoshal Orogeny Ia (MD_{1a} deformation) Thermal imprint on Mahakoshal volcanics (2110 ± 20 Ma, 2150 ± 90 Ma)	Singhbhum Orogeny Ia Soda Granite (2200 ± 54 Ma))	Deformation BD_{1a} and metamorphism of felsic gneisses of Meghalaya (2230 ± 13 Ma)
2250–2400	Deposition of volcano-sedimentary rocks, diamictites and Mn-bearing siliciclastics in the Sausar Group/Betul belt	Deposition of BIF-bearing volcano-sedimentary units of Lower Mahakoshal Group (2362 ± 90 Ma)	Deposition of Lower Singhbhum Group (Chaibasa Formation), Jagannathpur volcanics (2250 ± 81 Ma) and BIF and Mn-bearing Koira Group	Deposition of the volcano-sedimentary units of Lower Bomdila Group (Khetabari Formation)
>2400	Amgaon gneiss (2432 ± 5 Ma) Malanjkhand granite (2477 ± 10 Ma) Dongargarh granite (2465 ± 22 Ma) Granulite grade metamorphism and migmatisation in the basement (2672 ± 54 Ma)	Mafic dykes (2420 ± 77 Ma) Bundelkhand granite (2492 ± 10 Ma)	Newer Dolerite II dykes Granulite grade metamorphism and migmatisation in CGC metapelites (2569 ± 18 Ma) Newer Dolerite I dykes (2752 ± 1 Ma, 2762 ± 2 Ma, 2800 ± 1 Ma) Singhbhum Granite	Granulite metamorphism Older Metamorphic Group (2637 ± 55 Ma)

sedimentation continued till 2050 Ma; the final closing of this orogenic cycle took place at ~1950–2050 Ma, giving rise to the first deformation in the Chhotanagpur region. The closing age this cycle is constrained from post-orogenic granite intrusions with ages of 1900–1850 Ma; the deposition of the Lower Kodarma Group took place during this period. The period 1850–1780 is marked by the second orogenic movement in the CITZ, marked by a major tectono-thermal event in the central domain, followed by the intrusions of post-orogenic granites and deposition of the Lower Vindhyan Supergroup. Depositions of the Middle Kodarma Group in the eastern domain and the Shillong Group took place during the decompression (orogenic collapse) between 1780 and 1550 Ma. The period 1550–1500 Ma was marked by the closing of the second deformation in the CITZ, associated with ultra-high temperature (UHT) metamorphism. Post-orogenic period (1500–1000 Ma) is marked by the deposition of sedimentary successions of the Vindhyan Supergroup, Chhatishgarh Supergroup, Munger Group and Shillong Group, and intrusions of post-collisional (undeformed) granite plutons in all the domains. The crustal stability in the CITZ during 1500–1000 Ma was punctuated by a low-intensity deformation resulting in open folds with ~NS axial traces. Granite intrusions (900–800 Ma) with no deformational fabric present in all the domains represent magmatic activities during a period in which the CITZ had formed a stable cratonic domain.

5 Discussion

5.1 Tectonic Evolution of CITZ

The models proposed for the tectonic evolution of the CITZ by the previous workers (Bhowmik et al. 1999, 2000; Acharyya and Roy 2000; Bhowmik and Roy 2003; Roy and Prasad 2003; Bhowmik and Spiering 2004; Roy et al. 2006; Chattopadhyay et al. 2015) had key elements related to the granulite facies metamorphism in the CITZ and age of the Sausar Group. The main arguments include (i) the granulite facies metamorphism was earlier than the deposition of the Sausar Group, (ii) all the grey gneisses (the Tirodi Gneiss) are older than the Sausar Group, and (iii) the Sausar Group was deposited, deformed and metamorphosed during "the Grenville Orogeny". Based on the detailed structural studies in the Sausar belt and synthesis of tectono-thermal data from the CITZ, these opinions were questioned (Mohanty 2003a, 2006a, 2010, 2012). It has been demonstrated that the gneissic rocks of the Sausar belt (the Tirodi Gneiss) belong to four generations; while some of the gneisses (Tirodi Gneiss—I) form basement for the Sausar Group, some of the gneisses (Tirodi Gneiss—II–IV) are later than the Sausar Group (Mohanty 2012, 2015). These divisions are corroborated by the observations of West (1933), who had noted intersecting gneissic bands with three different orientations on a single outcrop of "streaky gneiss" (Fig. 13), and interpreted that "these streaky gneisses bear a discordant relation towards the rest of the Sausar series" (p. 351 in West

Fig. 13 Three generations of gneissic bands intersecting each other recorded from the "streaky gneiss" in the Teliya River, between Manegaon and Khawasa (from West 1933). Approximate location is marked as (h) in Fig. 4, in which the rock is redesignated as Tirodi Gneiss by Narayanaswami et al. (1963)

1933) and "*Thus both the field evidence and the microscopic evidence are strongly in favour of the streaky gneisses which occur in these two localities being composite rocks, formed by the intimate penetration of the granulite by the aplite*" (p. 352 in West 1933). The Neoproterozoic age for the Sausar Group proposed in the previous tectonic models (Bhowmik et al. 1999, 2000; Acharyya and Roy 2000; Bhowmik and Roy 2003; Roy and Prasad 2003; Bhowmik and Spiering 2004; Roy et al. 2006) failed to note that the cryptomelane age of 953 ± 10 Ma was interpreted to be the alteration related to post-amphibolite facies metamorphism (Lippolt and Hautmann 1995) and the analysed sample (L3 in Lippolt and Hautmann 1995) had a peak at ~1700 Ma. Brown and Phadke (1983) had also shown that the rocks of the Sausar Group in the Ramakona area were deformed and metamorphosed to granulite facies and produced gneisses which were classified as "Tirodi Biotite Gneiss", and Phadke (1990) recommended for the change of the status of the "Tirodi Biotite Gneiss" as an unit younger than the Sausar Group. Furthermore, Rb-Sr whole rock isochron age of 1525 ± 70 Ma for the Tirodi Gneiss was also interpreted to be age of regional metamorphism of the Sausar Group and the Tirodi Gneiss (Sarkar et al. 1986).

The spatio-temporal evolution of the CITZ was synthesized to understand the overall tectono-thermal evolution of the region. The Proterozoic age data from the region (Fig. 14) show three peaks at ~2100, 1650 and 950 Ma, in order of decreasing relative probability. These three peaks correspond to the three deformation episodes

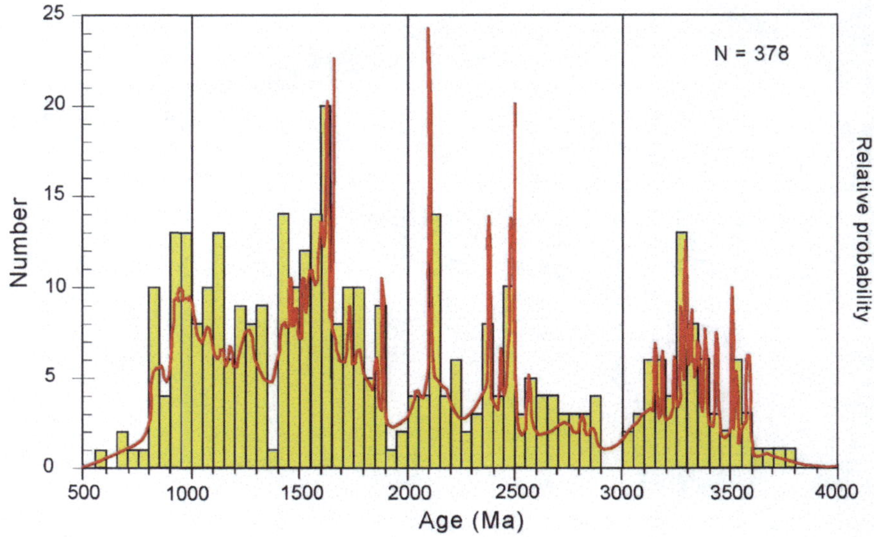

Fig. 14 Distribution of Precambrian age data from the Satpura Mountains and adjoining cratons. Gaussian probability plot of ages are shown by red colour (after Mohanty 2012)

recorded in the rocks of the Sausar Group as well as the Chhotanagpur Gneissic Complex. The first deformation was associated with UHT metamorphism in CITZ having pressure of ~9 kbar and temperature ~900 °C (Bhowmik et al. 2005; Bhowmik 2006). The second deformation, giving rise to the regional ~EW alignment of CITZ, was also associated with UHT metamorphism at P ~6–8 kbar and T ~650–850 °C in the Sausar belt and the Chhotanagpur Gneissic Complex. Both these deformation and metamorphic events resulted in partial melting of the rocks of the basement and the sedimentary cover rocks in all the domains of CITZ. The resultant gray gneisses, commonly referred as the Tirodi Gneiss in the western domain, are considered to be the basement for the sediments of the Sausar Group (Bhowmik et al. 1999, 2000; Acharyya and Roy 2000; Bhowmik and Roy 2003; Roy and Prasad 2003; Bhowmik and Spiering 2004; Roy et al. 2006). However, close scrutiny of deformation fabrics within these gray gneisses show truncations of earlier fabric by the later ones (Fig. 13), suggesting development of the grey gneisses at least in three different episodes of migmatisation; some of which are younger than the Sausar Group (discussed in Sect. 3.1.6). The Tirodi Gneiss developed during the regional deformation defining the "Satpura orogenic trend" have SHRIMP U-Pb zircon age of 1618 ± 8 Ma (Bhowmik et al. 2011). Ages close to this were also obtained for the grey coloured Tirodi Gneiss (1534 ± 13 Ma) and for the granulites (1569 ± 15 Ma) of the Ramtek area, which were interpreted to be age of peak metamorphism and generation of the Tirodi Gneiss (Ahmad et al. 2009). Interestingly, U-Pb zircon ages of the Tirodi Gneiss indicate protolith age of 2503–2118 Ma, with different components of 2056 ± 7 Ma and 1506 ± 11 Ma (Mishra et al. 2010); these ages correspond to the depositional age of the rocks of the Sausar Group, first and second deformations in the Sausar Group (Mohanty 2015). Considering all these tectono-thermal events, the Tirodi Gneiss of the CITZ has been proposed to be distinguished as four variants (Mohanty 2015); the oldest forming the basement of the Sausar Group is Tirodi Gneiss—I (>2450 Ma), the grey gneisses and granites developed during the first deformation form Tirodi Gneiss—II (~2100 Ma), the migmatised gneisses and granites associated with the second deformation of the Sausar Group constitute Tirodi Gneiss—III (~1600 Ma), and the post-granulite facies granites are Tirodi Gneiss—IV (~1450 Ma). The youngest tectonic event in the CITZ resulted in the development of open folds with ~NS axial planes and associated granite emplacements and thermal effects at ~950 Ma.

In the reconstruction model of East Gondwana during Proterozoic, the tectono-thermal evolution of the CITZ can be correlated with the Capricorn Orogen (Fig. 3), both showing multiple events of orogenic movements and separation during the period 2100–1500 Ma and matching tectonic fabric in the reconstructed positions (Mohanty 2010, 2012). It is to be noted that the Pinjarra orogen and its southward counterpart in the Prydz Bay region of Antarctica mark the collisional boundary between the Australo-Antarctic and Indo-Antarctic domains at ~550–500 Ma, but presence of Neoproterozoic gneissic terranes (1100–900 Ma) in the southern segment of the Pinjarra Orogen truncating the CITZ and the Albany-Fraser-Wilkes orogen suggest Neoproterozoic collision along the Pinjarra orogen (Fitzsimons 2003; Markwitz et al. 2017). Reworking of the Capricorn Orogen at ~1030–950 Ma is also interpreted to be the effect of Grenvillian orogeny (Sheppard et al. 2007) Therefore, the development of open folds with ~NS axial traces at ~950 Ma and associated granite emplacements in the CITZ are possibly related to the amalgamation of India-Australia-Antarctica along the Pinjarra Orogen. This tectonic model rules out the amalgamation of the NIB and SIB along CITZ during the Grenvillian Orogeny at ~1000 Ma suggested for the evolution of CITZ by the previous workers (Bhowmik et al. 1999, 2000; Acharyya and Roy 2000; Bhowmik and Roy 2003; Roy and Prasad 2003; Bhowmik and Spiering 2004; Roy et al. 2006; Chattopadhyay et al. 2015).

5.2 Dip of Subduction Boundary and Position of the Suture

Review of the position of the suture between NIB and SIB (discussed earlier in Sect. 2.2) indicates that most of the models have shown the suture to be coincident with the CIS in spite of the debate that the CIS represents a zone of ductile deformation separating felsic gneisses of identical characters (Acharyya et al. 2001). The dip of the subduction boundary has also controversy regarding northward or southward models. The position of suture along CIS and northward subduction of SIB below the NIB cannot explain the granulite facies metamorphism recorded in the Ramakona block as the overriding plate will not undergo the required pressure temperature condition. Therefore, the position of the suture between the NIB and SIB can be explained by two possible models: (a) the suture located to the north of the Sausar belt having northward dip of the southern province or (b) the suture located along CIS having southward dip of the subduction.

6 Conclusions

First generation mapping in the southern flank of the Satpura Mountain belt (referred as the "Central Indian Tectonic Zone (CITZ)" had identified complex tectono-thermal evolutional of the region. Although the gneissic rocks of the area, mapped as "streaky gneisses", were shown to have fabric development in three phases, and were considered to be younger than the Sausar Group and post-dating the granulite facies metamorphism of the region (West 1933), the third generation mapping in the region (Bhowmik et al. 1999, 2000; Acharyya and Roy 2000; Bhowmik and Roy 2003; Roy and Prasad 2003; Bhowmik and Spiering 2004; Roy et al. 2006) overlooked these interpretations and assigned the status of all the grey gneisses (renamed as Tirodi Gneiss; Straczek et al. 1956; Narayanaswami et al. 1963) to be the basement of the Sausar Group. Views similar to that of West (1933) regarding the status of granulite facies metamorphism of the Sausar Group and migmatisation forming "Tirodi Biotite Gneiss" were documented by Brown and Phadke (1983) and Phadke (1990). Based on these controversial status and local mapping of granulite facies rocks with improper correlation with the spatio-temporal evolution of the region led the third generation mapping of the Sausar belt to propose a single continuous subduction model ending up during "Grenville orogeny" at ~1000 Ma for the amalgamation of NIB and SIB and deformation and metamorphism of the Sausar Group (Bhowmik et al. 1999, 2000; Acharyya and Roy 2000; Bhowmik and Roy 2003; Roy and Prasad 2003; Bhowmik and Spiering 2004; Roy et al. 2006; Chattopadhyay et al. 2015). This model also ignored the earlier interpretation of second generation mapping data, which indicated the age of regional metamorphism of the Sausar Group and the Tirodi Gneiss to be 1525 ± 70 Ma. The misinterpreted evolutions outlined for the CITZ have proposed to place the Sausar Group stratigraphically above the Lower Vindhyan Supergroup (Bhowmik et al. 1999, 2000; Acharyya and Roy 2000; Bhowmik and Roy 2003; Roy and Prasad 2003; Bhowmik and Spiering 2004; Roy et al. 2006; Chattopadhyay et al. 2015).

Detailed structural mapping carried out in different parts of the CITZ combined with an analysis of spatio-temporal evolution of different domains have corroborated the ideas developed during the first generation mapping that the earliest deformation in the Sausar Group preceded the granulite facies metamorphism in the CITZ, and the idea of development of grey gneisses (Tirodi Gneiss) of four generations. The oldest gneissic unit of the region was of Archean age, Tirodi Gneiss—I (>2450 Ma), and formed the basement of the Sausar Group. This oldest gneiss shows development of a paleosol horizon formed by anoxic weathering at Archean-Paleoproterozoic transition period. The metasedimentary rocks forming the Sausar Group were deposited during 2400–2250 Ma. The glaciogenic sediments reported from the Sausar Group are stratigraphic markers of Paleoproterozoic "snowball earth" event and the ore-grade manganese deposits present discontinuously over a strike length of ~200 km are related to the Paleoproterozoic "giant manganese fields" formed during the Great Oxidation Event (GOE). The first deformation and granulite facies metamorphism of the Sausar Group at ~2100 Ma formed grey gneisses, Tirodi Gneiss—II, younger than the Sausar Group. The second deformation of the Sausar Group giving rise to the development of isoclinal folds with regional ~EW axial trace, regional granulite facies metamorphism and development of grey gneisses, Tirodi Gneiss—III (~1600 Ma), was associated with the Satpura orogeny ("Satpura orogenic trend"). The post-orogenic gneisses, Tirodi Gneiss—IV (~1450 Ma), were developed during the waning phase of the Satpura orogeny. Low intensity deformation in CITZ developing open folds with ~NS axial traces and thermal event (~950 Ma) were possibly related to the Mesoproterozoic amalgamation of India and Australia to form the East Gondwana during the formation of ~NS trending Pinjarra Orogen (~1100–900 Ma) on the western margin of Western Australia.

References

Acharyya SK (2003) The nature of Mesoproterozoic Central Indian Tectonic Zone with exhumed and reworked older granulites. Gondwana Res 6:197–214

Acharyya SK, Roy A (2000) Tectonothermal history of the Central Indian Tectonic Zone and reactivation of major fault/shear zones. J Geol Soc India 55:239–256

Acharyya SK, Bandyopadhyay BK, Roy A (2001) Comment on 'Two cases of continental collisions and related tectonics during the Proterozoic period in India—insight from gravity modeling constrained by seismic and magnetotelluric studies' by Mishra DC, Singh B, Tiwari VM, Gupta SB, Rao MBSV (2000) Precambrian Res 99:149–169. Precambrian Res 108:335–338

Agrawal VN (1975) Fold interference patterns in the Sausar Group, northern Nagpur district, Maharashtra, India. J Geol Soc India 16:176–187

Ahmad T, Kaulina TV, Wanjari N, Mishra MK, Nitkina EA (2009) U-Pb zircon chronology and Sm–Nd isotopic characteristics of the Amgaon and Tirodi Gneissic Complex, Central Indian Shield: constraints on Precambrian crustal evolution. In: Singh VK, Chandra R (eds) Precambrian continental growth and tectonism. Excel India Publishers, New Delhi, pp 137–138

Aswathanarayana U (1956) Absolute ages of the Archaean orogenic cycles of India. Am J Sci 254:19–31. https://doi.org/10.2475/ajs.254.1.19

Bandyopadhyay BK, Roy A, Huin AK (1995) Structure and tectonics of a part of the Central Indian shield. In: Sinha-Roy S, Gupta KR (eds) Continental crust of northwestern and central India, vol 31. Memoirs of the Geological Society of India, pp 433–467

Basu NK, Sarkar SN (1966) Stratigraphy and structure of the Sausar Series in the Mahuli-Ramtek-Junawani area, Nagpur district, Maharashtra. Q J Geol Min Metall Soc India 38:77–105

Basu AR, Ray SL, Saha AK, Sarkar SN (1981) Eastern Indian 3800 million years old crust and early mantle differentiation. Science 212:1502–1505

Beckinsale RD, Drury SA, Holt RW (1980) 3360 m.y. old gneisses from the South Indian Craton. Nature 283:469–470

Bhattacharyya BP (1975) Structural evolution in the central part of Santhal Parganas district Bihar. Bull Geol Min Metall Soc India 48:41–47

Bhowmik SK (2006) Ultra high temperature-metamorphism and its significance in the Central Indian Tectonic Zone. Lithos 92:484–505

Bhowmik SK, Roy A (2003) Garnetiferous metabasites from the Sausar Mobile Belt: petrology, P-T path and implications for the tectonothermal evolution of the Central Indian Tectonic Zone. J Petrol 44:387–420

Bhowmik SK, Spiering B (2004) Constraining the prograde and retrograde P-T paths of granulites using decomposition of initially zones garnets: an example from the Central Indian Tectonic Zone. Contrib Miner Petrol 147:581–603

Bhowmik SK, Pal T, Roy A, Pant NC (1999) Evidence of pre-Grenvillian high-pressure granulite metamorphism from the northern margin of the Sausar mobile belt in central India. J Geol Soc India 53:385–399

Bhowmik SK, Pal T, Pant NC, Shome S (2000) Implication of Ramakona cordierite gneiss in the crustal evolution of Sausar Mobile Belt in central India. Geol Surv India Spec Publ 57:131–150

Bhowmik SK, Sarbadhikari AB, Spiering B, Raith M (2005) Mesoproterozoic reworking of Palaeoproterozoic ultrahigh-temperature granulites in the Central Indian Tectonic Zone and its implications. J Petrol 46:1085–1119

Bhowmik SK, Wilde SA, Bhandari A (2011) Zircon U-Pb/Lu–Hf and monazite chemical dating of the Tirodi biotite gneiss: implication for latest Palaeoproterozoic to Early Mesoproterozoic orogenesis in the Central Indian Tectonic Zone. Geol J 46:574–596

Blanford WT (1872) Description of the geology of Nagpur, and its neighbourhood. Mem Geol Surv India 9:295–330

Bose PN (1888) Notes on the igneous rocks of the districts of Raipur and Balaghat, Central Provinces. Rec Geol Surv India 21(2):56–61

Brown M, Phadke AV (1983) High temperature retrograde reactions in pelitic gneiss from the Precambrian Sausar metasediments of the Ramakona area, Chhindwara district, Madhya Pradesh (India): definition of the exhumation P-T path and the tectonic implications. In: Phadke AV, Phansalkar VG (eds) Professor K.V. Kelkar Memorial Volume. Indian Society of Earth Scientists, Pune, pp 61–96

Chatterjee N, Crowley JL, Ghose NC (2008) Geochronology of the 1.55 Ga Bengal anorthosite and Grenvillian metamorphism in the Chotanagpur gneissic complex, eastern India. Precambr Res 161:303–316

Chattopadhyay A, Huin AK, Khan AS (2003a) Structural framework of Deolapar Area, Central India and its implications for Proterozoic nappe tectonics. Gondwana Res 6:107–117

Chattopadhyay A, Khan AS, Huin AK, Bandyopadhyay BK (2003b) Reinterpretation of stratigraphy and structure of Sausar Group in Ramtek-Mansar-Kandri Area, Maharashtra, central India. J Geol Soc India 61:75–89

Chattopadhyay A, Das K, Hayasaka Y, Sarkar A (2015) Syn- and post-tectonic granite plutonism in the Sausar Fold Belt, central India: age constraints and tectonic implications. J Asian Earth Sci 107:110–121

Das B (1966) A new orogeny in the Pre-Cambrian tract of the northern Satpura–Rajgir orogeny. Proc Indian Natl Sci Acad 32-A:202–209

Divakara Rao V, Gupta HK, Gupta SB, Mall DM, Mishra DC, Murthy PRK, Narayana BL, Reddy PR, Tewari HC (1998) A geotransect in the central Indian shield, across the Narmada-Son lineament and the central Indian suture. Int Geol Rev 40:1021–1037

Fermor LL (1906) Notes on the petrology and manganese-ore deposits of the Sausar Tahsil, Chhindwara district, Central Provinces. Rec Geol Surv India 33:159–220

Fermor LL (1909) The manganese-ore deposits of India. Mem Geol Surv India 37:1–610

Fermor LL (1911) On the age and continuation in depth of the manganese-ores of the Nagpur-Balaghat area, Central Provinces. Rec Geol Surv India 41(1):1–11

Fitzsimons ICW (2003) Proterozoic basement provinces of southern and southwestern Australia, and their correlation with Antarctica. In: Yoshida M, Windley BE, Dasgupta S (eds) Proterozoic East Gondwana: supercontinent assembly and breakup. Geol Soc Lond Spec Publ 206:93–130

Gopalan K, Macdougall JD, Roy AB, Murali AV (1990) Sm-Nd evidence for 3.3 Ga old rocks in Rajasthan, northwestern India. Precambr Res 48:287–297

Harris LB (1993) Correlations between the Central Indian Tectonic zone and the Albany Mobile Belt of Western Australia: evidence for a continuous Proterozoic orogenic belt. In: Findlay RH, Unrug R, Banks MR, Veevers JJ (eds) Gondwana 8: assembly, evolution and dispersal. A. A. Balkema, Rotterdam, pp 165–180
Harris LB, Beeson J (1993) Gondwanaland significance of Lower Palaeozoic deformation in central India and SW Western Australia. J Geol Soc Lond 150:811–814
Holmes A (1955) Dating the Precambrian of Peninsular India and Ceylon. Proc Geol Assoc Can 7:81–106
Holmes A (1965) Principles of physical geology. Thomas Nelson and Sons Ltd., London, 1288 p
Holmes A, Leland WT, Nier AO (1950) Age of uraninite from a pegmatite near Singar, Gaya district, India. Am Miner 35:19–28
Khan AS, Mukherjee MK, Sarkar RL (2003) Reinterpretation of geology of Ramakona-Sausar-Mogra area of Sausar fold belt in Madhyay Pradesh. In: Roy A, Mohabey DM (eds) Advances in Precambrian of Central India. Gondwana Geol Mag Spec Publ 7:119–128
Krishnan MS (1943) Geology of India and Burma. Higginbothams, Madras, 518 p
Lippolt HJ, Hautmann S (1995) $^{40}Ar/^{39}Ar$ ages of Precambrian manganese ore minerals from Sweden, India and Morocco. Min Depos 30:246–256
Mahadevan C, Aswathanarayana U (1962) Age levels of Precambrian orogenic cycles of India. Bull Volcanol 24:233–248. https://doi.org/10.1007/BF02599350
Mall DM, Reddy PR, Mooney WD (2008) Collision tectonics of the Central Indian Suture zone as inferred from a deep seismic sounding study. Tectonophysics 460:116–123
Mallik AK, Gupta SN, Ray Barman T (1991) Dating of early Precambrian granite greenstone complex of the Eastern Indian Precambrian shield with special reference to the Chotanagpur Granite Gneiss Complex. Rec Geol Surv India 124(2):20–21
Markwitz V, Kirkland CL, Evans NJ (2017) Early Cambrian metamorphic zircon in the northern Pinjarra Orogen: implications for the structure of the West Australian Craton margin. Lithosphere 9:3–13. https://doi.org/10.1130/L569.1
Mazumdar SK (1988) Crustal evolution of the Chhotanagpur gneissic complex and the mica belt of Bihar. In: Mukhopadhyay D (ed) Precambrian of the Eastern Indian shield. Geol Soc India Mem 8:49–83
Mishra DC, Singh B, Tiwari VM, Gupta SB, Rao MBSV (2000) Two cases of continental collisions and related tectonics during the Proterozoic period in India—insight from gravity modeling constrained by seismic and magnetotelluric studies. Precambr Res 99:149–169
Mishra MK, Ahmad T, Wanjari N, Kaulina TV, Nitkina EA (2010) Geochemical and isotopic characteristics of the Amgaon Gneissic Complex and Tirodi Gneissic Complex of Central Indian Shield: Constraints on Precambrian crustal evolution. In: Proceedings of 7th annual meeting of Asia Oceania Geosciences Society (AOGS), 5–9 July 2010, Hyderabad, SE12-A012
Mohanty S (1988) Structural evolution of the southern part of the Sausar belt near Ramtek, Nagpur district, Maharashtra. Indian J Geol 60:200–210
Mohanty S (1993) Stratigraphic position of the Tirodi gneiss in the Precambrian terrane of central India: evidence from the Mansar area, Nagpur district, Maharashtra. J Geol Soc India 42:55–60
Mohanty S (2000) Proterozoic granites of Madhya Pradesh and Bihar: a compilation. In: Fist Meeting of IGCP—426 "Proterozoic granites and crustal evolution", Jaipur, pp 226–264
Mohanty S (2002) Structural evolution of Sausar Group around Parseoni, Nagpur district, Maharashtra: its implication for stratigraphy. J Geol Soc India 60:309–315
Mohanty S (2003a) Structural-stratigraphic relations in Precambrian rocks of Sausar belt, central India. In: Roy A, Mohabey DM (eds) Advances in Precambrian of Central India. Gondwana Geol Mag Spec Publ 7:109–117
Mohanty S (2003b) Structural framework of Deolapar area, Central India and its implication for Proterozoic nappe tectonics: comment. Gondwana Res 6(4):934–935. https://doi.org/10.1016/S1342-937X(05)71039-X

Mohanty S (2003c) Reinterpretation of stratigraphy and structure of Sausar Group in Ramtek-Mansar-Kandri area, Maharashtra, Central India, by Chattopadhyay A, Khan AS, Huin AK, Bandyopadhyay BK (2003) J Geol Soc India 61:75–89. Comments: J Geol Soc India 61(6):743–745
Mohanty S (2006a) Structural evolution of the Sausar belt, central India: Implications for Tectonic evolution of the Central Indian Tectonic Zone. Indian J Geol 78:135–142
Mohanty S (2006b) Evidence of volcanism and glaciation from the Sausar Group, Central India. J Geol Soc India 68:764–768
Mohanty S (2010) Tectonic evolution of the Satpura Mountain Belt: a critical evolution and implication on supercontinent assembly. J Asian Earth Sci 39:516–526
Mohanty S (2012) Spatio-temporal evolution of the Satpura Mountain Belt of India: a comparison with the Capricorn Orogen of Western Australia and implication for evolution of the supercontinent Columbia. Geosci Front 3:241–267
Mohanty SP (2015) Palaeoproterozoic supracrustals of the Bastar Craton: Dongargarh Supergroup and Sausar Group. In: Mazumder R, Eriksson PG (eds) Precambrian Basins of India: stratigraphic and tectonic context. Geological Society, London. Memoirs 43:151–164. https://doi.org/10.1144/m43.11
Mohanty AK, Mohanty S (1996) Structural patterns in the Sausar Group around Mansar, Nagpur district, Maharashtra. J Geol Soc India 48:559–565
Mohanty SP, Nanda S (2016) Geochemistry of a paleosol horizon at the base of the Sausar Group, Central India: implications on atmospheric conditions at the Archean-Paleoproterozoic boundary. Geosci Front 7:759–773. https://doi.org/10.1016/j.gsf.2015.10.002
Mohanty S, Mohanty AK, Nayak PK (2000) Structure of manganese-bearing metasedimentary rocks of the Sausar Group around Kandri, Nagpur district, Maharashtra. J Geol Soc India 56:79–87
Mohanty SP, Barik A, Sarangi S, Sarkar A (2015) Carbon and oxygen isotope systematics of a Paleoproterozoic cap-carbonate sequence from the Sausar Group, Central India. Palaeogeogr Palaeoclimatol Palaeoecol 417:195–209. https://doi.org/10.1016/j.palaeo.2014.10.036
Naganjaneyulu K, Santosh M (2010) The Central India Tectonic Zone: a geophysical perspective on continental amalgamation along a Mesoproterozoic suture. Gondwana Res 18:547–564
Naha K (1964) A critique of "orogenic trends" in Archaean correlation in India. Tectonophysics 1(5):431–438
Naqvi SM, Rogers JJW (1987) Precambrian Geology of India. Oxford University Press, Oxford, 223 p
Narayanaswami S, Chakravarty SC, Vemban NA, Shukla KD, Subramanyam MR, Venkatesh V, Rao GV, Anandalwar MA, Nagarajaiah RA (1963) The geology and manganese-ore deposits of the manganese belt in Madhya Pradesh and adjoining parts of Maharashtra. Part I—general introduction. Bull Geol Surv India A-22(I):1–69
Nutman AP, Chadwick B, Ramakrishnan M, Viswanatha MN (1992) SHRIMP U-Pb ages of detrital zircons from Sargur supracrustal rocks in Western Karnataka, Southern India. J Geol Soc India 39:367–374
Oldham T (1856) Remarks on the classification of rocks of central India resulting from the investigation of the Geological Survey. J Asiat Soc Bengal 25:224–256
Pascoe EH (1926) General report of the Geological Survey of India for the year 1925. Rec Geol Surv India 59:1–114
Phadke AV (1990) Genesis of the granitic rocks and the status of the "Tirodi Biotite Gneiss" in relation to the metamorphites of the Sausar Group and the regional tectonic setting. Geol Surv India Spec Publ 28:287–302
Radhakrishna BP (1989) Suspect tectono-stratigraphic terrane elements in the Indian subcontinent. J Geol Soc India 34:1–24
Radhakrishna BP, Naqvi SM (1986) Precambrian continental crust of India and its evolution. J Geol 94:145–166
Radhakrishna BP, Ramakrishnan M (1988) Archaean-Proterozoic boundary in India. J Geol Soc India 32:263–278

Rajaram M, Anand SP (2003) Central Indian tectonics revisited using aeromagnetic data. Earth Planets Space 55:e1–e4
Rajesh HM, Mukhopadhyay J, Beukes NJ, Gutzmer J, Belyanin GA, Armstrong RA (2009) Evidence for an early Archaean granite from Bastar craton, India. J Geol Soc London 166:193–196
Ray Barman T, Bishui PK, Mukhopadhyay K, Ray JN (1993) Geochronology of high-grade rocks of Jamua-Dumka sector of Chotanagpur Gneissic Complex, Bihar. Rec Geol Surv India 126(2):25–26
Ray Barman T, Bishui PK, Mukhopadhyay K, Ray JN (1994) Rb-Sr geochronology of the high-grade rocks from Purulia, West Bengal and Jamua-Dumka sector, Bihar. Indian Miner 48:45–60
Roy AK (1977) Structural and metamorphic evolution of the Bengal Anorthosite and associated rocks. J Geol Soc India 18:203–223
Roy A, Prasad MH (2003) Tectonothermal events in Central Indian Tectonic Zone (CITZ) and its implications in Rodinian crustal assembly. J Asian Earth Sci 22:115–129
Roy A, Kagami H, Yoshida M, Roy A, Bandyopadhyay BK, Chattopadhyay A, Khan AK, Pal T (2006) Rb–Sr and Sm–Nd dating of different metamorphic events from the Sausar Mobile Belt, central India: implication for Proterozoic crustal evolution. J Asian Earth Sci 26:61–76
Sarangi S, Mohanty S (1998) Structural studies in the Chhotanagpur Gneissic Complex near Gomoh, Dhanbad district, Bihar. Indian J Geol 70:73–80
Sarbadhikari AB, Bhowmik SK (2008) Constraining the metamorphic evolution of a cryptic hot Mesoproterozoic orogen in the Central Indian Tectonic Zone, using P-T pseudosection modeling of mafic intrusions and host reworked granulites. Precambr Res 162:128–149
Sarkar A, Paul DK, Potts PJ (1996) Geochronology and geochemistry of Mid-Archaean trondhjemitic gneisses from the Bundelkhand craton, central India. Rec Res Geol Geophys Precambrians 16:76–92
Sarkar AN, Jha BN (1985) Structure, metamorphism and granitic evolution of the Chhotanagpur granite gneiss complex around Ranchi. Rec Geol Surv India 113:1–12
Sarkar G, Gupta SN (1990) Dating of early Precambrian granite complex of Bastar district, Madhya Pradesh. Rec Geol Surv India 123(2):31–33
Sarkar SN, Polkanov AA, Gerling EK, Chukrov FV (1964) Geochronology of the pre-Cambrians of Peninsular India: a synopsis. Sci Cult 30:527–537
Sarkar SN, Gautam KVVS, Roy S (1977) Structural analysis of a part of the Sausar Group rocks in Chikla-Sitekesa area, Bhandara district, Maharashtra. J Geol Soc India 18:627–643
Sarkar SN, Trivedi JR, Gopalan K (1986) Rb-Sr whole rock and mineral isochron ages of the Tirodi gneiss, Sausar Group, Bhandara district, Maharashtra. J Geol Soc India 27:30–37
Sheppard S, Rasmussen B, Muhling JR, Farrell TR, Fletcher IR (2007) Grenvillian-aged orogenesis in the Palaeoproterozoic Gascoyne Complex, Western Australia: 1030–950 Ma reworking of the Proterozoic Capricorn Orogen. J Metamorph Geol 25:477–494
Straczek JA, Subramanyam MR, Narayanaswami S, Shukla KD, Vemban NA, Chakravarty SC, Venkatesh V (1956) Manganese ore deposits of Madhya Pradesh, India. In: Proceedings of 20th international geological congress, Mexico, Symposium Sobre Yacimientos de Manganeso, Tomo IV, pp 63–96
Verma RK (1991) Geodynamics of the Indian Peninsula and the Indian Plate Margin. Oxford and IBH Publishing Company, New Delhi, 357 p
West WD (1933) The origin of the streaky gneisses of the Nagpur district. Rec Geol Surv India 67:344–356
West WD (1936) Nappe structure in the Archaean rocks of the Nagpur district. Trans Natl Inst Sci India 1(6):93–102
Yedekar DB, Jain SC, Nair KKK, Dutta KK (1990) The central Indian collision suture. In: Dutta KK, Choudhary SV, Rajurkar ST, Deshmukh SS (eds) Precambrian of Central India. Geol Surv India Spec Publ 28:1–43
Yoshida M, Divi RS, Santosh M (2001) Precambrian Central India and its role in the Gondwanaland-Rodinia Context. Gondwana Res 4:208–211

Tectonic Development of the Bengal Basin in Relation to Fold-Thrust Belt to the East and to the North

Md. Sakawat Hossain, Md. Sharif Hossain Khan, Rashed Abdullah and Khalil R. Chowdhury

Abstract Tectonic development of the Bengal Basin is related to the complex interplay among the Himalaya orogen to the north, the Indo-Burma orogen to the east, and Stable Indian Craton to the west. To the north, convergence tectonic loading presumed to transfer from the deformation front, along the Oldham and Dauki faults to the south and shaped not only the Shillong Plateau and Assam Basin, but also the northern part of the Bengal Basin. To the east, oblique subduction related transpressional tectonics splits along different morphotectonic units separated by large-scale transpressive dextral strike-slip faults and produces fold thrust belt, thrust front, and emerging fold belt, whose western limit is marked by the deformation front to the east. The Stable Shelf part of the basin to the west is characterized by the presence of passive margin rift faults with numerous graben and half-graben structures, which show sign of tectonic reactivation in response to the ongoing N-S collision and E-W subduction of the Indian Plate.

Keywords Bengal Basin · Deformation front · Fold-thrust belt · Locked thrust · Active faults

1 Introduction

Fold-thrust belts and their associated deformation fronts are directly related to the geodynamic evolution and tectonic development of the peripheral foreland basins (Mitra and Fisher 1992; DeCelles 2012; Wang et al. 2014). On the outer margins of the most collisional orogenic fold-thrust belts, generally a wide deformation zone of thick prism of sedimentary rock (i.e. accretionary wedge) is developed (Cawood et al. 2009). The sediments were mainly deposited at the stable margin of the craton, and thicken rapidly away from the craton margin towards the core of the orogenic belt (Roeder et al. 1978; Price 1981; Yassaghi 1998; Sihua et al. 2009). These sedimentary rocks are commonly compressed into long parallel fold belt, which are generally cut by parallel thrust-shear zones (Rodgers 1991). The fold-thrust belts vergence is

Md. Sakawat Hossain (✉) · Md. Sharif Hossain Khan · R. Abdullah · K. R. Chowdhury
Department of Geological Sciences, Jahangirnagar University, Dhaka 1342, Bangladesh
e-mail: sakawat@juniv.edu

T. K. Biswal et al. (eds.), *Structural Geometry of Mobile Belts of the Indian Subcontinent*, Society of Earth Scientists Series,
https://doi.org/10.1007/978-3-030-40593-9_4

dominantly directed to the stable craton. Towards the basin, the frontal margin of the fold-thrust belts, in general, indicates by surficial projection of the subsurface décollement level or basal detachment where slip diminishes to zero. The surficial projection of the subsurface décollement, where slip gradually diminishes to zero is termed as deformation front.

The Bengal Basin lies at the edge of Indian Craton in the northeastern part of the Indian Plate (Fig. 1) and is one of the largest peripheral collisional foreland basins in South Asia (DeCelles 2012; DeCelles et al. 2014). The geodynamic development of this basin is related to the collision of the Indian Plate with the Eurasian and the Burmese Plate, building the Himalaya and the Indo-Burma Orogen in the north and east, respectively (Hossain et al. 2014; Vernant et al. 2014; Wang et al. 2014). The basin is largely overlain by the Ganges-Brahmaputra-Meghna (GBM) delta and accommodated the huge thickness of fluvio-deltaic to shallow marine sedimentary successions. Thickness of the sediments within the basin gradually increases to the east and southeast with a maximum thickness reported is ~18–20 km (Hiller and Elahi 1984; Uddin and Lundberg 2004; Singh et al. 2016). These thick sedimentary successions are being accreted as accretionary prism to form the Indo-Burman orogen as a result of eastward subduction of the Indian Plate beneath the Burmese Plate. The Chittagong-Tripura Fold Belt (CTFB) formed from this accreted sediments in the eastern part of the basin, which is the western prolongation of the Indo-Burman accretionary prism.

To the north, part of the Bengal Basin extends up to the Main Frontal Thrust (MFT), which separates the Sub-Himalayan Zone from the Indo-Gangetic Alluvial Plain. The MFT defines the southernmost margin of Himalayan orogen across an entire ~2500 km-long arc, generates and maintains the active, present Himalayan deformation front (Srivastava et al. 2018). This Himalayan deformation front represented by MFT is thought to be the near-surface expression of the main basal décollement called the Main Himalayan Thrust (MHT) (Bilham and England 2001). Therefore, the MFT-MHT deformation boundary is the present-day collisional interface between the Indian and Eurasian plates, which causes strain accumulation along this boundary (Bilham and England 2001; Srivastava et al. 2018). However, the geological gap between eastern Himalayan deformation front and northern edge of the Bengal Basin are occupied by the Assam Basin and Shillong Plateau. Most of the collisional force of the eastern Himalayan frontal belt is accommodated by not only the uplift and convergence along the MFT, but also taken by two other major thrusts namely the Dauki and Oldham faults to the south and the north of the Shillong Plateau, respectively. Rapid development and westward propagation of the CTFB since 2 Ma (Maurin and Rangin 2009) as well as simultaneous uplift of the Shillong Plateau result an intricate relationship between subsidence and concurrent folding in this part of the basin. On the other hand, the western and southwest parts of the basin are characterized by an easterly inclined cratonic passive margin of the Indian Plate (Kumar et al. 2017; Hossain et al. 2019). The basement rock beneath this part of the basin is characterized by tensional tectonics, and tectonically less suffered during orogenic development of the Himalayas. However, the geodynamic significance and

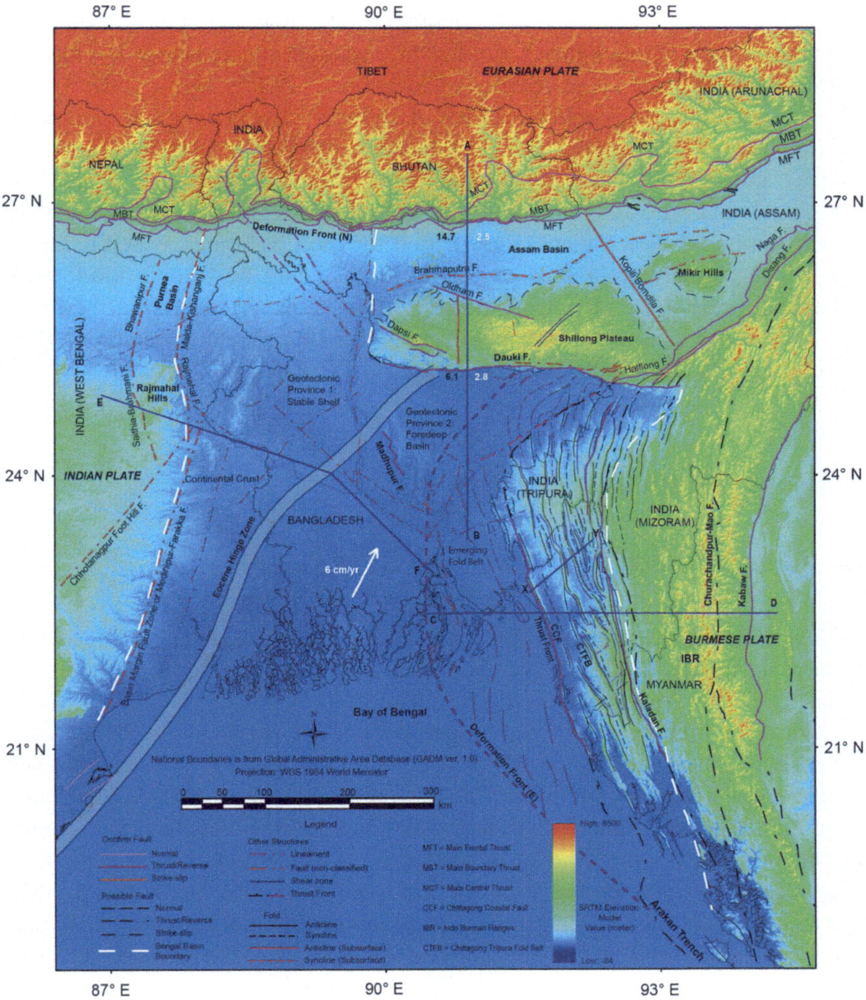

Fig. 1 Conceptual geodynamic map of the Bengal Basin and its surroundings, at scale 1:1,250,000, with focus on the main tectonic features (modified after Mohanty et al. 2014; Wang et al. 2014; Steckler et al. 2016; Betka et al. 2018; Rahman et al. 2019; Hossain et al. 2020, in review). This map relies on the terrestrial map constructed from the Shuttle Radar Topography Mission images (SRTM version 3.0 with 1° × 1° tiles at 1 arc second, i.e., 30 m resolution). Administrative areas, i.e., country boundaries used in the map are taken from Global Administrative Area database (GADM, version 3.6). White arrow indicates the direction of the Indian Plate movement (after Akhter et al. 2018). Dip-slip and strike-slip rate (mm/year) of the MFT (Bhutan Himalaya) and the Dauki Fault are given in black and white digit, respectively (after Lindsey et al. 2018)

complex interplay among the Himalayan orogen to the north, the Indo-Burman orogen to the east and Stable Indian Craton to the west in the geodynamic development of the Bengal Basin are yet to be clearly understood.

The objective of this research is to use of regional-scale gravity data, and extensively compile recent publications and available geological information in order to postulate a comprehensive tectonic model for the basin development. In addition, geological field observation and published 2D seismic sections have been used for the preparation of a geological cross-section for the CTFB area which is largely unknown but can carry a significant value in understanding the tectonic development of the Bengal Basin development. Therefore, this research focuses on the preparation of a conceptual tectonic model of the Bengal Basin and to comprehend the relationship of the fold-thrust belt to the east and to the north.

2 Tectonic Settings

2.1 Geological Settings

The Bengal Basin is a foreland basin consisting of Early Cretaceous–Holocene sedimentary succession (Curray 1991; Curray and Munasinghe 1991). Geographically, the basin lies approximately between 20° 34′–26° 40′ N and 87° 00′–92° 45′ E with its major portion being constituted by Bangladesh and partly by the Indian states of West Bengal, Assam, Tripura, and Mizoram. It is surrounded by four major geotectonic units, which are the Indian Craton to the west and the Shillong Plateau to the north, the Indo-Burman Ranges to the east and the Bay of Bengal to the south. In terms of plate tectonics, the basin itself situated at the northeastern edge of the Indian Plate with close proximity to the Eurasian Plate at the north and Burmese Plate at the east (Fig. 1). Recent geodetic measurement suggests that the Indian plate is moving about 6 cm/year towards northeast direction, collided with the Eurasian Plate with a rate of 45 mm/year, and subducted beneath the Burmese Plate with a rate of ~46 mm/year (Akhter et al. 2018). This convergent collision is taken by the active development of the fold-thrust belt to the east and to the north. Hence, the tectonic evolution of the Bengal Basin is directly related to the development of the Himalayan orogen in the north and the Indo-Burman orogen to the east due to the north and north-eastern collision of the Indian Plate with the Eurasian Plate and the Burmese Plate, respectively (Steckler et al. 2008, 2016; Wang et al. 2014).

2.2 Tectonic Evolution

The tectonic evolution of the basin has involved three major geodynamic episodes. In the first episode, the basin initiated as an intra-cratonic rift basin within Gondwana landmass lies slightly approximately close to the triple junction between Antarctica, Australia, and India (Aitchison et al. 2019). This rift basin received the continental Gondwana sediments till Mid Mesozoic. Widespread volcanism occurred as continental flood basalts (namely the Rajmahal Trap) during the transition from first episode to second episode (Mid-Cretaceous), which covered the Gondwana sediments.

The second geodynamic episode of basin development began in the Late Mesozoic (Late-Cretaceous) with drifting and further break-up of Gondwana. In this episode, the Indian Plate is subjected to drifting, further breakup and isolation, and no significant changes in the basin tectonic framework have taken place. The basin rests on the north-eastern edge of the Indian Plate lithosphere and is underlain by the thick, buoyant continental lithosphere in the west and north and cold-dense oceanic lithosphere in the east and south-east (Bender 1983; Curray 1991; Reimann 1993; Uddin and Lundberg 2004; Singh et al. 2016). The west- and northwestern parts of the basin consists of an easterly dipping shelf of the stable Pre-Cambrian Platform of the Indian Plate. This part of the basin started to develop at the time of rifting of the Gondwana in the Early Mesozoic when graben and half-graben structures were formed, followed by the drifting in the Jurassic-Cretaceous. The later event results the Rajmahal lava flows and is formed the Rajmahal Trap. During the Late Cretaceous to Eocene, carbonate sediments deposited at the outer edge of the shelf (Eocene Hinge Zone; Fig. 1), whereas arenaceous sediments were deposited at the inner edge (Reimann 1993). The basement rock beneath this part of the basin is characterized by tensional tectonics, but is structurally less suffered during this episode. The second geodynamic episode ended at the Late Paleocene.

In the third and final episode of basin development, the Indian Plate is subjected to continent-continent collision along its northern edge during the Eocene-Oligocene transition and ocean-continent subduction along eastern edge during the Early Miocene. Due to continued oblique subduction of the Indian Plate beneath the Burmese Plate, the Bengal Basin evolved into a remnant ocean basin at the beginning of the Miocene (Ingersoll et al. 1995). During this episode, the basin received and accommodated thick pile of sediments mostly from the north and partially from the east (Zhang et al. 2019). As a result of continuous subduction of the oceanic part of the Indian Plate to the east, this thick pile of sediments of the Bengal Basin has been deformed into accretionary wedge since the Mid-Late Pliocene and given rise to fold-thrust belt and large-scale transpressive fault to the east (Alam et al. 2003; Steckler et al. 2008; Maurin and Rangin 2009).

Until the Late Oligocene, the western, central and northern parts of the basin are believed to have undergone similar tectonic and sedimentary evolution (Ingersoll et al. 1995; Alam et al. 2003). From the Late Oligocene onward, central part of the basin has undergone its own tectonic evolution. In the northern part, a major change in sedimentation pattern has most likely occurred in the Mid Pliocene by the major thrust related uplift of the Shillong Plateau along the north dipping Dauki Fault in the south and south dipping Oldham Fault in the north (Bilham and England 2001; Biswas and Grasemann 2005; Yin et al. 2010; Najman et al. 2016). To accommodate the east-west and north-south compression related to the Indian Plate movement during the latest stage of third geodynamic episode, the central part of the basin evolved into several sub-basins (Hossain et al. 2019 and reference there in).

3 Methodology

The regional gravity data from global-1 min grid (Sandwell et al. 2014) are used as an important tool to understand the buried tectonic features. This regional-scale dataset is then superimposed with the crustal thickness dataset (Mitra et al. 2018) to divide the Bengal Basin into two geotectonic provinces as marked in the Fig. 2. Topographic profiles and geological cross-sections across the Chittagong-Tripura Fold Belt (CTFB) were constructed based on detailed field investigation, and published 2D seismic sections (Sikder and Alam 2003; Steckler et al. 2008; Maurin and Rangin 2009; Najman et al. 2012). The profile is created from the Google Earth Pro and then exported and scaled using GRAPHER 347 software. Individual faults identified at the surface are tied with subsurface seismic sections, and then combined with geological map to produce the cross-sections along the line XY (Fig. 3; line location is given in Fig. 1). The conceptual geodynamic model given in this study is mainly based on geological filed investigation, free-air gravity dataset, geological cross-sections, and extensive review of the published literatures.

4 Result and Discussion

4.1 Result

The combined map (Fig. 2) of regional gravity and crustal thickness (Mitra et al. 2018) datasets distinctly divide the Bengal Basin into two geotectonic provinces. We interpreted these two large-scale buried tectonic features as the Geotectonic Province 1 and Geotectonic Province 2. The transition of these two geotectonic provinces is visible in the Fig. 2, which is characterized by almost ENE-WSW gravity anomaly as well as marked variation in the crustal thickness. This transitional structure is known as the Eocene Hinge Zone. In addition, free-air gravity value also confirmed the locations and extent of the major faults in the study area. Some of these major faults are Basin Margin Fault to the west, Dauki Fault to the north, Chittagong Coastal Fault and Kaladan Fault to the east (Fig. 2).

Geological field observations suggest that most of the folds in the CTFB area are intrinsically related to longitudinal thrust faults. Along the line XY (Fig. 3), evidences of longitudinal thrust fault (Fig. 4) are observed in the both flank of the anticlines. The topographic data along the line XY and surface geological observation complement each other. Furthermore, these surface datasets well tied with the subsurface dataset (published 2D seismic sections). These three datasets (surface geology, topography

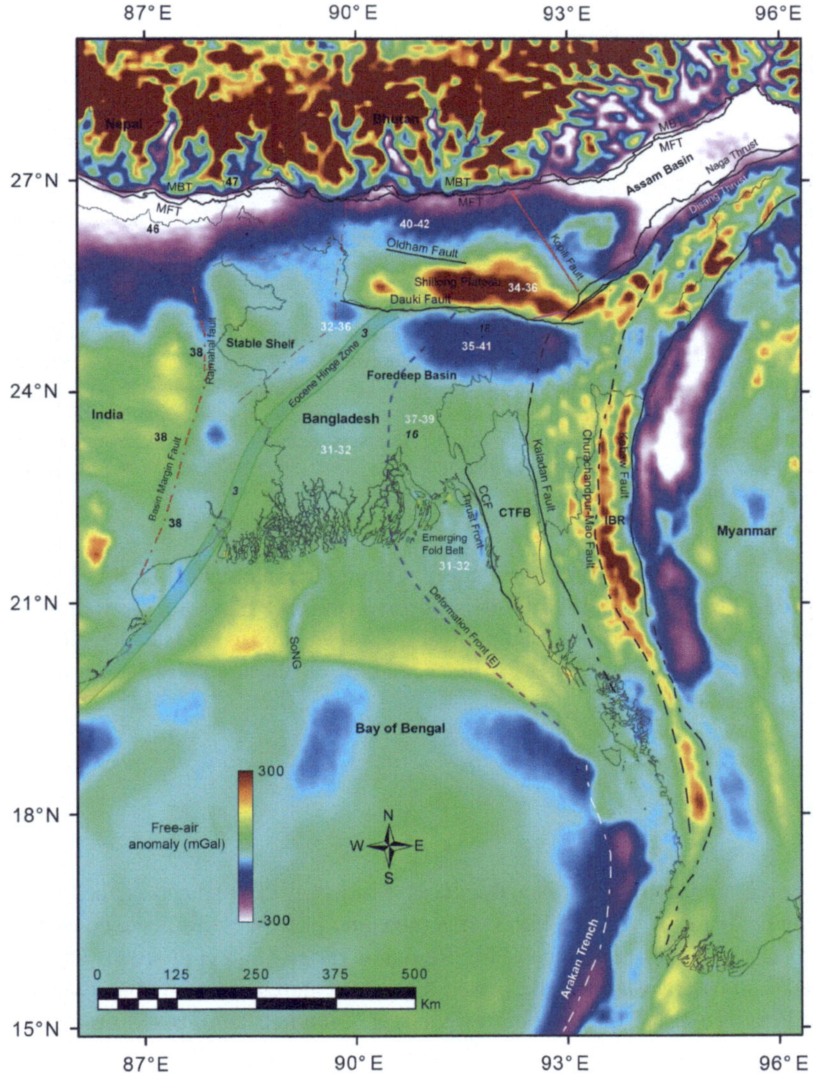

Fig. 2 Free-air gravity map of the Bengal Basin and its surroundings (data taken from Sandwell et al. 2014) with interpreted major tectonic features related to the basin development. The depth of the Moho are shown in bold digit (individual values shown in black colour are taken from Singh et al. 2016 and range values shown in white colour are taken from Mitra et al. 2018). Sedimentary thickness is shown in black italics font (after Singh et al. 2016)

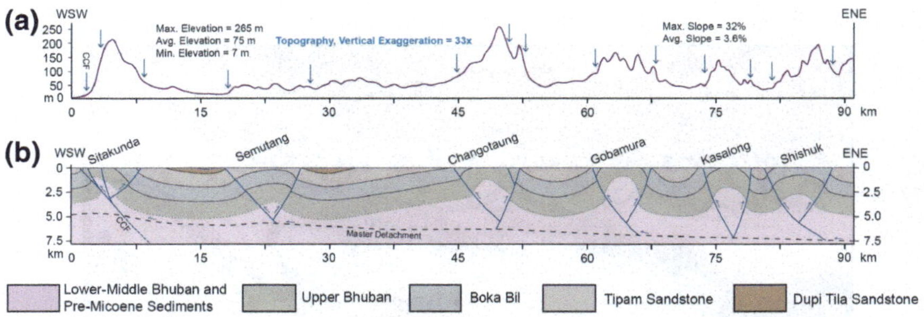

Fig. 3 Topographic profiles and geological cross-sections across the Chittagong-Tripura Fold Belt (CTFB). **a** The topographic profile across the CTFB drawn along the line XY (shown in Fig. 1). The arrows on the topographic profile are used locate the fault positions, which are clearly reflected by the abrupt change in elevation on both side of the structures, except the Semutang Structure. **b** Geological cross-section drawn along the line XY. Datasets are based on several geological field works, published 2D seismic sections (Sikder and Alam 2003; Steckler et al. 2008; Maurin and Rangin 2009; Najman et al. 2012)

and seismic sections) suggest the presence of six anticlines along the line XY (Fig. 3). At least one longitudinal thrust fault is observed in each flank of the individual structure. All west verging and east dipping thrusts are related to a master detachment, which gradually rises up to the west. East flank of all these anticlines are mainly affected by back thrust.

4.2 Discussion

Based on free-air gravity and crustal thickness (Mitra et al. 2018) datasets, the Bengal Basin can be distinctly divided into two geotectonic provinces. To make consistent with the earlier studies (Bakhtine 1966; Reimann 1993; Alam et al. 2003; Hossain et al. 2019), these two large-scale buried tectonic features are named as the Geotectonic Province 1 and Geotectonic Province 2. On other hand, based on *P* wave receiver function analysis of the broadband seismographs data, Mitra et al. (2018) determined the crustal thickness in and around the Bengal Basin, which also shows a distinctive pattern. The minimum crustal thickness observed in the eastern and southeastern parts of the basin (the Foredeep Basin) with transitional crustal types, whereas maximum thickness is observed in the northwestern part (the Stable Shelf) with continental crust type (Fig. 2). Depth of the Moho and thickness of the sediments in the Foredeep Basin suggest that the crustal thickness is approximately half of the Stable Shelf part. To the east, the initial development of the CTFB in the eastern part of the Province 2 is mostly considered to be started from the Pliocene as accretionary prism due to the continuous oblique subduction of the Indian Plate beneath the Burmese Plate (Maurin and Rangin 2009; Najman et al. 2012). As the collision continues, the eastern part of the Province 2 evolved as fold belt and form

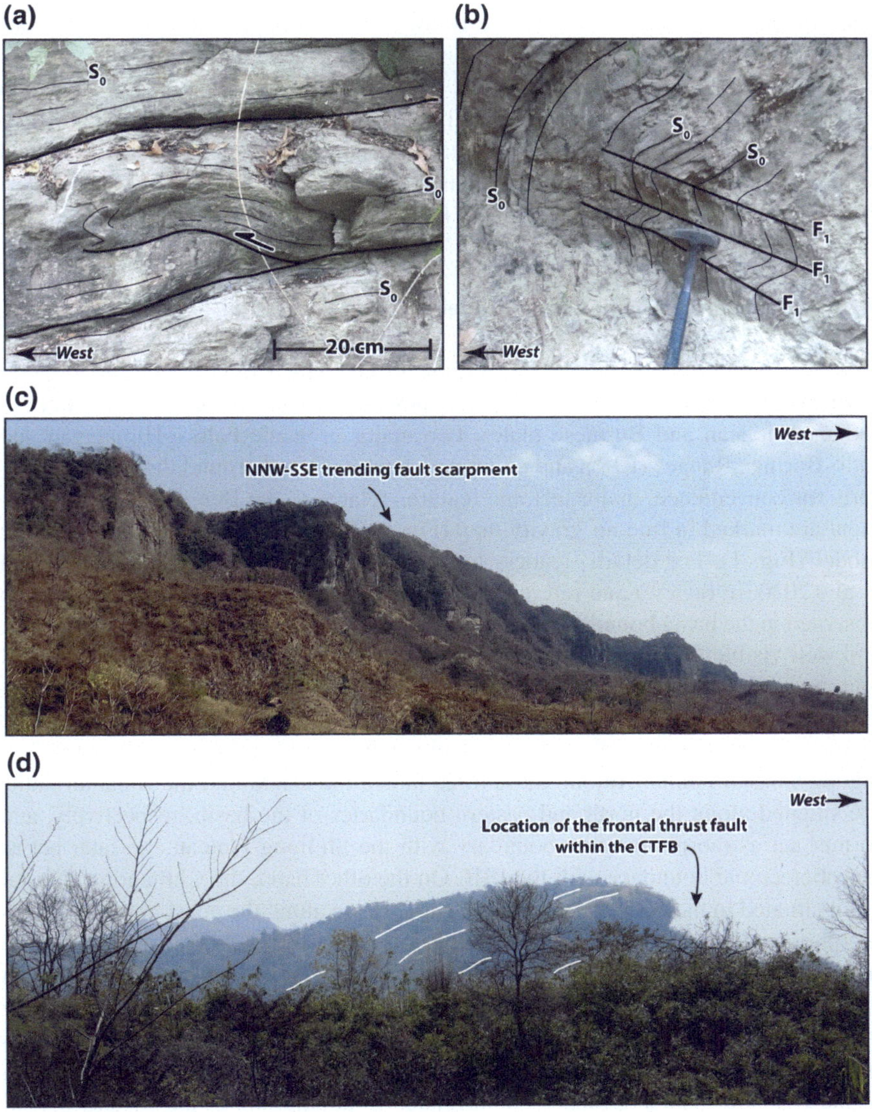

Fig. 4 Field observations from the Sitakund Anticline in the western part of the Chittagong-Tripura Fold Belt. **a** Small scale folding due to westward propagation of bedding (S_0) parallel reverse fault (22° 38′ 51″ N, 91° 39′ 31.7″ E). **b** Small scale recumbent chevron folds (F1) in the western flank (Labanakhya Charra Section) (22° 38′ 44″ N, 91° 38′ 58″ E). **c**, **d** NNW-SSE trending escarpment, possibly developed due to the westward propagation of frontal thrust fault in the Chittagong-Tripura Fold Belt (c: 22° 39′ 08″ N, 91° 39′ 48″ E; d: 22° 41′ 46″ N, 91° 39′ 13″ E)

folded flank of the basin. Gradual westward overprinting of thin-skinned tectonics of the CTFB by thick-skinned tectonics of the IBR is due to rapid propagation of the accretionary wedge westward since last 2 Ma (Maurin and Rangin 2009; Najman et al. 2016). Most of these folds within the CTFB can be interpreted as fault propagating folds that are linked to a major detachment at depth (Fig. 3) is directly related to the subduction of the Indian Plate beneath the Burmese Plate. The westernmost extension of this subsurface detachment is known as deformation front.

4.2.1 Conceptual Geodynamic Model

To comprehend the geodynamic evolution of the Bengal Basin, it is necessary to have insights on the present day spatial positions of the three tectonic plates—Indian, Eurasian and Burmese plates, two major orogenic belts—Himalayan and Indo-Burman Ranges (IBR), and major active faults in and around the basin. Therefore, for convenience, major tectonic features related to the Bengal Basin development are marked in free-air gravity map (Fig. 2) and in the conceptual geodynamic model (Fig. 1). For details, readers are requested to have a look to the Hossain et al. (2020, in review) and reference there in. Although spatially few changes are observed in the basin boundary fault positions in different publications, more precise and well visible fault lines are marked around the Bengal Basin in this study based on topography and gravity datasets (Figs. 1 and 2).

Concerning the tectonic evolution, the Bengal Basin is affected mainly by three major transpressive dextral strike-slip faults, which are Dauki, Kaladan, and Chittagong Coastal Faults. Among these three faults, the Dauki and the Kaladan faults are situated along the north and eastern boundaries of the basin, respectively, and former act as morphotectonic boundary with the Shillong Plateau and later act as morphotectonic boundary with the IBR. On the other hand, the Chittagong Coastal Fault situated in the eastern part of the Bengal Basin along the coast line, also act as morphotectonic boundary that separate the outer Indo-Burmese Wedge (i.e., CTFB) from the Foredeep Basin (Maurin and Rangin 2009; Hossain et al. 2014; Hossain et al. 2019). In addition, another important geodynamic feature that actively shaping the present-day tectonic development of the Bengal Basin is the two deformation fronts (Fig. 1). The deformation front to the east represents the outermost margin of the Indo-Burma accretionary prism related to defamation in the eastern part of the basin. The deformation front to the north is represented by the MFT to the far northwest of the basin, which acts as a morphotectonic boundary that separates the basin from the east Himalayan frontal belt. A brief description of these faults and deformation fronts, and their significance as morphotectonic boundary for the basin development is discussed here after.

The deformation of the Himalayan frontal belt is limited to the south by the MFT. The part of the east Himalayan front to the far northwest of the Bengal Basin is marked by the MFT. Most of the collisional force between India and Eurasia are accommodated by the uplift and convergence along the MFT (Wesnousky et al. 1999). Although an apparent faster convergence rate is observed at the central Himalayan frontal belt (Nepal Himalaya) (Wesnousky et al. 1999; Berthet et al. 2014; Bilham et al. 2017), actually a faster convergence rate observed in the east Himalayas (Bhutan Himalaya), where convergence is divided between the MFT and the Dauki thrust (Lindsey et al. 2018). The distributed slip rate across the MFT in the Bhutan Himalaya is 11.5 mm/year (Panda et al. 2018). In this part of the Himalaya, the slip-deficit is accumulated on a wider fault interface than that of the Nepal Himalaya, highlighting higher seismic risk in the Bhutan Himalaya and its adjacent regions (Li et al. 2018). According to Berthet et al. (2014), the MFT is associated with active low-angle thrusting and associated uplift. South of the MFT (i.e. northwestern most part of the Bengal Basin) represents a zone of active deformation between the Sub-Himalaya and the Indo-Gangetic Alluvial Plain as indicated by the active thrusting, uplift, surface rupture and occurrence of earthquakes (Thakur 2004). In this part of the basin, deformation related to MFT (i.e., deformation front to the north) is slowly migrated to the south (Hossain et al. 2020, in review).

To the south of the MFT, the Bengal Basin is separated from the Shillong Plateau and the Assam Basin by the Dauki Fault (Fig. 5a). The Dauki Fault, which is a continental margin paleo-rift fault that later acts possibly as a lateral dextral ramp combining thrusting and dextral strike-slip faulting. According to Ferguson et al. (2012), it is a regional deeply-rooted north-dipping blind thrust fault with strikes E-W, which can be divided into three segments: western segment within continental India; central segment along the former passive margin; and eastern segment overridden by the west-verging Indo-Burman accretionary prism. In the western segment, the fault penetrates the Indian continent and branched into the WNW-trending Dapsi Fault, with dip in NNE. The uplift of the Shillong Plateau was triggered by north-dipping Dauki Fault and subsequent back thrusting on the south-dipping Oldham Fault, in response to flexural loading of the Eastern Himalaya (Mitra et al. 2018). This interpretation also suggested by the other studies (Bilham and England 2001; Biswas and Grasemann 2005; Biswas et al. 2007; Banerjee et al. 2008; Hossain et al. 2016; Najman et al. 2016). Transition of this part of the basin from passive margin to flexural basin occurred during 3.5 to ~2 Ma, which is related to the flexural loading by the Shillong Plateau (Najman et al. 2016). The distributed slip rate across the Dauki Fault is 8–9 mm/year (Panda et al. 2018). According to Lindsey et al. (2018), dip-slip and dextral strike-slip rate of the Dauki Fault are 6.1 and 2.8 mm/year, respectively. Hence, this northern Foredeep part of the Bengal Basin is subjected to bidirectional compression from north and south-east, and significant dextral strike-slip motion between the Shillong Plateau and the Sylhet Trough. The clastic sediments in this part of the basin are dragged and compressed, actively producing folds with variable intensity and geometric alignment. In addition, downdip locking extent of the MFT and Dauki Fault and high convergence rate suggest that the seismic hazard throughout this area remains high (Lindsey et al. 2018).

To the east of the Bengal Basin, three regional scale active faults constantly accommodate the stress resulting from the Indo-Burma subduction, and therefore are responsible for kinematic evolution, décollement geometry, and geologic deformation, specifically in the eastern half of the Bengal Basin. From east to west, these faults are the Kabaw, Churachandpur-Mao Fault (CMF), Kaladan and Chittagong

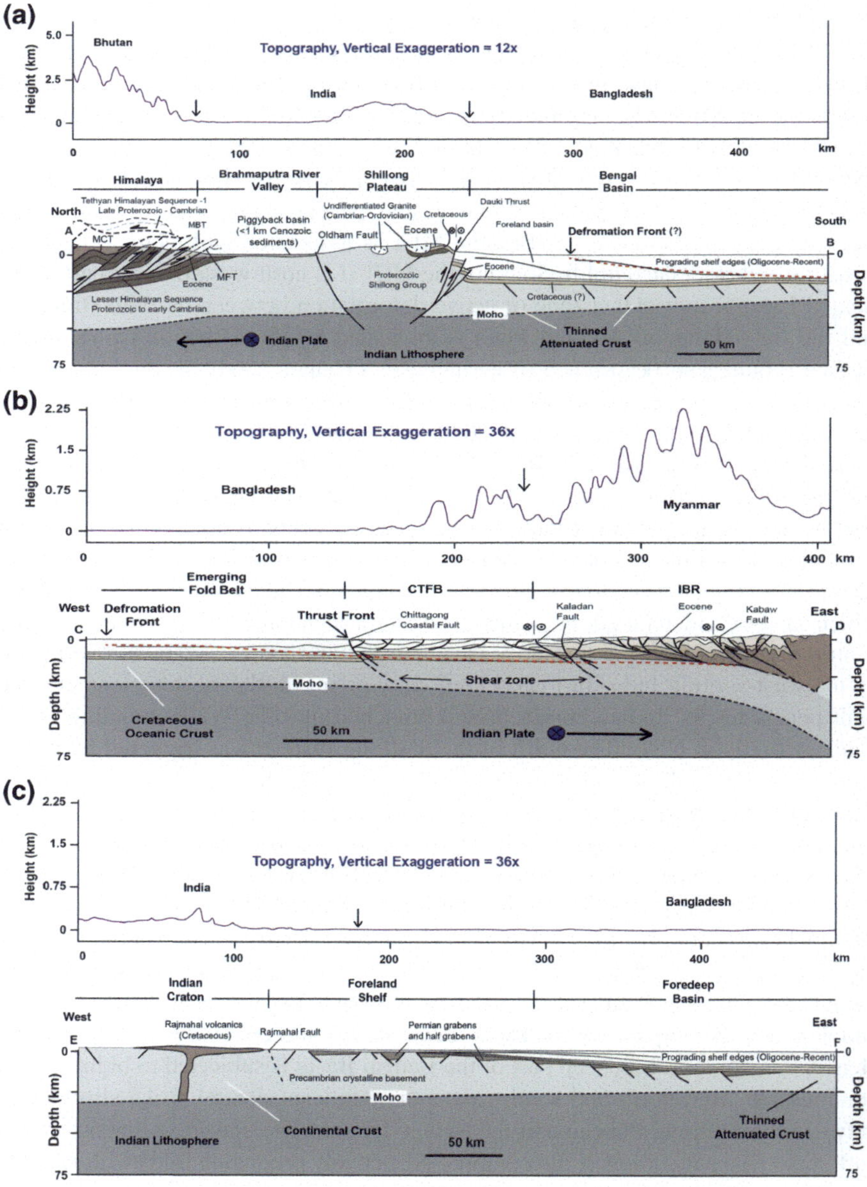

Fig. 5 Topographic profiles and geological cross-sections across the Bengal Basin and its surroundings. **a** Cross-section drawn along the line AB. The arrows on the topographic profile AB indicate the border line of Bhutan, India, and Bangladesh. **b** Cross-section drawn along the line CD. The arrow on the topographic profile CD indicates the border line of Bangladesh and Myanmar. **c** Cross-section drawn along the line EF. The arrow on the topographic profile EF indicates the border line of India and Bangladesh. Locations of the cross-section lines are shown in Fig. 1. The topographic profiles were created from the Google Earth Pro along the above mentioned lines

Coastal Fault (CCF) all of which are running approximately N-S (Fig. 5b). As the CMF and Kabaw Faults are far away from the Bengal Basin, therefore these two faults will not be discussed in this study. According to Wang et al. (2014) and Steckler et al. (2016), a shallow dipping megathrust structurally connects the CCF, Kaladan faults. To the east of the Kaladan Fault, the rate of tectonic convergence is ~13 mm/year; in between the Kaladan Fault and CCF, the rate ~5 mm/year; and between CCF and deformation front, the rate is ~2 mm/year (CDMP II, 2013). Maurin and Rangin (2009) suggest that these rates of convergence result ~200 km advancement of the outer part of the accretionary prism (represented by the CTFB) over the last 2 Ma.

The eastern boundary of the basin with the Indo-Burman Ranges is marked by the ~450 km long and north-south running Kaladan Fault (Fig. 2). The fault mostly follows the course of the Kaladan River and is named after that river (Nandy 2001; Maurin and Rangin 2009; Barman et al. 2016). According to Maurin and Rangin (2009), along 20° 30′ N, the dip of the fault plane is 70° E with an 8°–10° S pitch, and shows a clear right-lateral offset. The orientation of the maximum compression in this area is ~N80° E–N60° E. Seismically, the fault is segmented, and is sparsely active. Active motion is partitioned within a very short distances along this fault, as the right-lateral slip dominates its northern portion, whereas in the middle portion due E-W or WSW-ENE thrusting focal mechanisms are observed. Southward continuation of the Kaladan Fault is most likely represented by the Ramree Thrust (Maurin and Rangin 2009). Among these three portions of the fault, the north and the central portions have significant influences in the tectonic evolution of the eastern part of the Bengal Basin. Conspicuous morphological step between inner wedge to the east and outer wedge to the west has been observed with average elevation of 1300 m, and 250 m, respectively (Fig. 5b). The transpressive active motion along this fault is partitioned on very short distances between N10° W dextral strike-slip, and west verging thrusting and décollement. These west verging thrusts and décollement are responsible in producing fault propagated and detachment en echelon folds in the eastern part of the Bengal Basin (Fig. 3). Orientation of these folds is in accordance with the orientation of maximum compressive stress across the IBR and the CTFB (Angelier and Baruah 2009; Khan et al. 2017, 2018). Due to variable dip and strike-slip convergence rates along the Kaladan and the CCF, the CTFB subjected to shearing. With increasing distance from the Kaladan Fault, the fold amplitude and thrust intensity decrease westward. The western edge of this fold thrust belt (i.e. the CTFB) is marked by the CCF.

Within the Bengal Basin, two major faults actively shaping the geodynamic evolution of some of the tectonic elements and also causes the active tectonics. These are the CCF and the Madhupur Fault. The geological field reconnaissance (Abdullah et al. 2015; Sikder and Alam 2003), multichannel seismic survey (Maurin and Rangin 2009), geomorphic signatures on 30 m resolution SRTM image (Hossain et al. 2019), and free-air gravity data suggest the presence of a ~300 km long major fault along eastern coast of the northern Bay of Bengal. This fault, which stretches between ~19° 40′ N and 23° 30′ N is known as the CCF. As the CCF actively controlled the geomorphic development of the coast, the eastern coast of Bangladesh can be treated as a fault controlled structure (Fig. 2). According to Maurin and Rangin

(2009), the CCF is a crustal transpressive dextral strike-slip fault with an appreciable thrust component, results from thin-skinned and thick-skinned tectonics. In its northern termination, the fault shows splay nature near the Tichna anticline in Tripura, just south of the Surma basin. This fault is probably younger than the Kaladan Fault as the accretionary wedge is progressing westward during the last 2 Ma, and not yet affected by sufficient erosion. Maurin and Rangin (2009) suggest that both the CCF and the Kaladan Faults can be interpreted as deep-seated basement reverse fault (Figs. 3 and 5b) with dextral strike-slip component, but the CCF is seismically less active than the Kaladan Fault. The Foredeep part of the Bengal Basin is divided into two parts along the CCF. The CCF, which marked this transitional boundary, is also known as the 'Thrust Front' (Betka et al. 2018).

To the west of the CCF or 'Thrust Front', recent geological and geophysical studies suggest progressive westward growth of low intensity subsurface folds, towards the deformation front to the east (Figs. 1 and 5b). This young and active blind fold belt exists in between CCF and deformation front to the east, stretches from offshore to the south and onshore to the north (Hossain et al. 2020, in review). Folds within this 'Emerging Fold Belt' are characterized by long-wavelength décollement folds, locally west verging, and subparallel to that of the CTFB. As in the CTFB, their amplitude also seems to decrease westward gradually die out near deformation front to the east (Maurin and Rangin 2009; Hossain et al. 2018, 2019). The east-west compressive tectonic deformation in the Foredeep Basin is mainly accommodated by this 'Emerging Fold Belt', and related faults.

Another major active fault plays significant role in the tectonic framework of the approximately central part of the Bengal Basin. This fault is clearly visible in the 30 m resolution SRTM images (Fig. 1), which appears as a series of en echelon faults with approximately NNW-SSE trend. The fault is assumed to be resulted either from torsion of the region or from the effect of shear along an assumed buried fault or possibly a combination of both (Morgan and McIntire 1959). This fault is known as the Madhupur Fault, which is east- dipping and is considered as an important structure for possible seismic hazard in the central part of the Bengal Basin (Hossain et al. 2019). According to Pickering et al. (2014), the cyclic avulsion history of the Brahmaputra River with a periodicity of about 1800 years is most likely related to slip along the Madhupur Fault.

Beside the above mentioned major active basin-margin faults and deformation fronts, some other faults are present along the western edge of the Bengal Basin (Fig. 5c). In this side, the basin is separated from the Purnea Basin by the approximately N-S running Malda-Kishanganj Fault (north), from the Rajmahal Hill by the Rajmahal Fault (centre), and from the Chhotanagpur Foot Hill by the Basin Margin Fault Zone or Medinipur-Farakka Fault (south) (Hossain et al. 2019). These faults are associated with the rifting of the Indian Plate from the Gondwana. The western part of the basin including the Eocene Hinge Zone is characterized by alternating gravity highs and lows indicate the presence of numerous graben and half-graben structures in most of the Stable Shelf (Khan and Rahman 1992; Khan and Chouhan 1996). These older rift related basement normal faults are currently subjected to positive inversion due to east west compression. Earthquakes with strike-slip source mechanism in this part of the Bengal Basin have been observed, showing the NE-SW nodal plane paralleling the Eocene Hinge Zone. These most likely demarcate the passive margin rift faults, which are tectonically reactivated due to the ongoing N-S compression and E-W subduction (Mitra et al. 2018).

4.2.2 Locked-Thrust/Seismic Gap

The faster convergence rate and presence of seismogenic gap along the East Himalayan deformation belt (Bhutan Himalaya) can produce earthquakes with great magnitudes analogous to those of oceanic subduction zones (Berthet et al. 2014; Li et al. 2018; Lindsey et al. 2018). On the other hand, the deformation to the east along which the Indian Plate subducting beneath the Burmese Plate at the rate of 13–17 mm/year appeared to be loading the locked shallow megathrust that connects CCF, Kaladan and Kabaw Faults (Figs. 1 and 5b). As in the case of other subduction zones, the accumulated strain from this locked thrust will likely be released in future large earthquakes (Steckler et al. 2016, 2018). These locked shallow-mega thrusts to the north and to the east will significantly control the seismic activity within the Bengal Basin and its northern as well as eastern surroundings in the future.

5 Conclusion

The Cenozoic tectonic development of the Bengal Basin is mainly related to the collision of the Indian Plate with the Eurasian Plate to the north and the Burmese Plate to the east. Three major transpressive basin boundary faults: MFT, Dauki and Kaladan Faults, two deformation fronts, thrust front (CCF) and other major active faults play important role in the tectonic development of the basin, specifically the Foredeep Basin part (Province 2). In the Stable Shelf part of the basin, older passive margin rift faults are currently tectonically reactivated and subjected to positive inversion due to east-west compression. All these faults and deformation fronts are directly related to the fold-thrust belt and emerging fold belt within the basin, and actively shaping the present-day geodynamic development of the Bengal Basin.

References

Abdullah R, Yeasmin R, Ameen SMM, Khanam F, Bari Z (2015) 2D Structural modelling and hydrocarbon potentiality of the Sitakund structure, Chittagong Tripura Fold Belt (CTFB), Bengal Basin, Bangladesh. J Geol Soc India 85(6):697–705

Aitchison JC, Ao A, Bhowmik S, Clarke GL, Ireland TR, Kachovich S, Lokho K, Stojanovic D, Roeder T, Truscott N, Zhen Y, Zhou R (2019) Tectonic evolution of the western margin of the Burma microplate based on new fossil and radiometric age constraints. Tectonics 38(5):1718–1741

Akhter SH, Bhuiyan AH, Steckler MS (2018) Geodynamic modeling and hydrocarbon potentiality in Bengal Basin, Bangladesh. AGU Fall Meeting Suppl. Abstract T33-320227

Alam M, Alam MM, Curray JR, Chowdhury MLR, Gani MR (2003) An overview of the sedimentary geology of the Bengal Basin in relation to the regional tectonic framework and basin-fill history. Sed Geol 155:179–208

Angelier J, Baruah S (2009) Seismotectonics in Northeast India: a stress analysis of focal mechanism solutions of earthquakes and its kinematic implications. Geophys J Int 178:303–326

Bakhtine MI (1966) Major tectonic features of Pakistan: Part II. The Eastern Province. Sci Ind 4:89–100

Banerjee P, Burgmann R, Nagarajan B, Apel E (2008) Intraplate deformation of the Indian subcontinent. Geophys Res Lett 35:L18301

Barman P, Jade S, Shrungeshwara TS, Kumar A, Bhattacharyya S, Ray JD, Jagannathan S, Jamir WM (2016) Crustal deformation rates in Assam Valley, Shillong Plateau, Eastern Himalaya, and Indo-Burmese region from 11 years (2002–2013) of GPS measurements. Int J Earth Sci 106:2025–2203

Bender F (1983) Geology of Burma. Borntraeger, Berlin, 260 p

Berthet T, Ritz J-F, Ferry M, Pelgay P, Cattin R, Drukpa D, Braucher R, Hetényi G (2014) Active tectonics of the eastern Himalaya: new constraints from the first tectonic geomorphology study in southern Bhutan. Geology 42:427–430

Betka PM, Seeber L, Thomson SN, Steckler MS, Sincavage R, Zoramthara C (2018) Slip-partitioning above a shallow, weak décollement beneath the Indo-Burman accretionary prism. Earth Planet Sci Lett 503:17–28

Bilham R, England P (2001) Plateau "pop-up" in the great 1897 Assam earth-quake. Nature 410:806–809

Bilham R, Mencin D, Bendick R, Bürgmann R (2017) Implications for elastic energy storage in the Himalaya from the Gorkha 2015 earthquake and other incomplete ruptures of the Main Himalayan Thrust. Quatern Int 462:3–21

Biswas S, Grasemann B (2005) Quantitative morphotectonics of the southern Shillong Plateau (Bangladesh/India). Aust J Earth Sci 97:82–93

Biswas S, Coutand I, Grujic D, Hager C, Stockli D, Grasemann B (2007) Exhumation and uplift of the Shillong Plateau and its influence on the eastern Himalayas: new constraints from apatite and zircon (U–Th–[Sm])/He and apatite fission track analysis. Tectonics 26:TC6013

Cawood PA, Kröner A, Collins WJ, Kusky TM, Mooney WD, Windley BF (2009) Accretionary orogens through Earth history. Geol Soc Lond Spec Publ 318:1–36. https://doi.org/10.1144/SP318.1

CDMP II (2013) Report of active fault mapping in Bangladesh: paleo-seismological study of the Dauki Fault and the Indian-Burman plate boundary fault. Comprehensive Disaster Management Programme (CDMP II), Ministry of Disaster Management and Relief, Bangladesh, 67 p

Curray JR (1991) Possible green schist metamorphism at the base of a 22 km sedimentary section, Bay of Bengal. Geology 19:1097–1100

Curray JR, Munasinghe T (1991) Origin of the Rajmahal traps and the 85°E Ridge: preliminary reconstructions of the trace of the crozet hotspot. Geology 19(12):1237–1240

DeCelles PG (2012) Foreland basin systems revisited: variations in response to tectonic settings. In: Busby C, Pérez A (eds) Tectonics of sedimentary basins: recent advances. Wiley, Chichester, pp 405–426. https://doi.org/10.1002/9781444347166.ch20

DeCelles PG, Kapp P, Gehrels GE, Ding L (2014) Paleocene-Eocene foreland basin evolution in the Himalaya of southern Tibet and Nepal: implications for the age of initial India-Asia collision. Tectonics 33(5):824–849

Ferguson EK, Seeber L, Steckler MS, Akhter SH, Mondal D, Lenhart A (2012) The Dauki Thrust Fault and the Shillong Anticline: an incipient plate boundary in NE India? AGU Abstract, T51F-2665

Hiller K, Elahi M (1984) Structural development and hydrocarbon entrapment in the Surma Basin, Bangladesh (northwest Indo-Burman fold belt). In: Fifth Offshore Southwest conference, Singapore, pp 656–663. https://doi.org/10.2118/12398-MS

Hossain MS, Khan MSH, Chowdhury KR, Afrooz M (2014) Morpho-structural classification of the Indo-Burman ranges and the adjacent regions. In: National conference on rock deformation & structures (RDS-III), Assam, India. Abstract volume, p 31. https://doi.org/10.13140/rg.2.1.3769.6240

Hossain MS, Chowdhury KR, Khan MSH, Abdullah R (2016) Geotectonic settings of the Dauki Fault—a highly potential source for a significant seismic threat. In: Kruhl JH (ed) International conference Humboldt Kolleg on Living under Threat of Earthquake—Kathmandu, Nepal. Abstract volume, p 25

Hossain MS, Islam MM, Islam O, Khan MSH, Chowdhury KR (2018) Deformation characteristics of the Chittagong Tripura Fold Belt—an insight from the geometrical analysis of folded structures. In: Mamun et al (eds) International conference on recent advances in mathematical and physical sciences. Jahangirnagar University, Dhaka. Abstract volume, pp 9–10

Hossain MS, Khan MSH, Chowdhury KR, Abdullah R (2019) Synthesis of the tectonic and structural elements of the Bengal Basin and its surroundings. In: Mukherjee S (eds) Tectonics & structural geology: Indian context. Springer International Publishing AG, Cham, 84 p. https://doi.org/10.1007/978-3-319-99341-6

Hossain MS, Xiao W, Khan MSH, Chowdhury KR (2020) Geodynamic model and tectonic features of the Bengal Basin and its surroundings. J Maps (in review)

Ingersoll RN, Graham SA, Dickinson WR (1995) Remnant ocean basins. In: Busby CJ, Ingersoll RV (eds) Tectonics of sedimentary basins. Blackwell, Oxford, pp 363–391

Khan AA, Chouhan RKS (1996) The crustal dynamics and the tectonic trends in the Bengal Basin. J Geodyn 22(3–4):267–286

Khan AA, Rahman T (1992) An analysis of gravity field and tectonic evaluation of the northwestern part of Bangladesh. Tectonophysics 206(3–4):351–364

Khan MSH, Hossain MS, Chowdhury KR (2017) Geomorphic implication and active tectonics of the Sitapahar Anticline—CTFB Bangladesh. Bangladesh Geosci J 23:1–24

Khan MSH, Hossain MS, Uddin MA (2018) Geology and active tectonics of the Lalmai Hills, Bangladesh—an overview from Chittagong Tripura fold belt perspective. J Geol Soc India 92(6):713–720

Kumar S, Rino V, Hayasaka Y, Kimura K, Raju S, Terada K, Pathak M (2017) Contribution of Columbia and Gondwana supercontinent assembly- and growth-related magmatism in the evolution of the Meghalaya Plateau and the Mikir Hills, Northeast India: constraints from U-Pb SHRIMP zircon geochronology and geochemistry. Lithos 277:356–375

Li S, Wang Q, Yang S, Qiao X, Nie Z, Zou R, Ding K, He P, Chen G (2018) Geodetic imaging mega-thrust coupling beneath the Himalaya. Tectonophysics 747–748:225–238

Lindsey EO, Almeida R, Mallick R, Hubbard J, Bradley K, Tsang LLH et al (2018) Structural control on downdip locking extent of the Himalayan megathrust. J Geophys Res Solid Earth 123. https://doi.org/10.1029/2018JB015868

Maurin T, Rangin C (2009) Structure and kinematics of the Indo-Burmese Wedge: Recent and fast growth of the outer wedge. Tectonics 28:TC2010. https://doi.org/10.1029/2008tc002276

Mitra S, Fisher GW (1992) Structural geology of Fold and Thrust belts. Johns Hopkins Univ. Press, Baltimore, 254 p

Mitra S, Priestley KF, Borah K, Gaur VK (2018) Crustal structure and evolution of the Eastern Himalayan plate boundary system, Northeast India. J Geophys Res: Solid Earth 123. https://doi.org/10.1002/2017JB014714

Mohanty WK, Mohapatra AK, Verma AK, Tiampo KF, Kislay K (2014) Earthquake forecasting and its verification in northeast India. Geomat Nat Hazards Risk 7(1):194–214

Morgan JP, McIntire WG (1959) Quaternary geology of the Bengal Basin East Pakistan and India. Bull Geol Soc Am 70(3):319–341

Najman Y, Allen R, Willett EAF, Carter A, Barford D, Garzanti E, Wijbrans J, Bickle M, Vezzoli G, Ando S, Oliver G, Uddin M (2012) The record of Hi-malayan erosion preserved in the sedimentary rocks of the Hatia Trough of the Bengal Basin and the Chittagong Hill Tracts, Bangladesh. Basin Res 24:499–519

Najman Y, Bracciali L, Parrish RR, Chisty E, Copley A (2016) Evolving strain partitioning in the Eastern Himalaya: the growth of the Shillong Plateau. Earth Planet Sci Lett 433:1–9

Nandy DR (2001) Geodynamics of Northeast India and the adjoining region. ABC Publications, Calcutta, pp 1–209

Panda D, Kundu B, Santosh M (2018) Oblique convergence and strain partitioning in the outer deformation front of NE Himalaya. Nature 8:10564. https://doi.org/10.1038/s41598-018-28774-3

Pickering JL, Goodbred SL, Reitz MD, Hartzog TR, Mondal DR, Hossain MS (2014) Late Quaternary sedimentary record and Holocene channel avulsions of the Jamuna and Old Brahmaputra River valleys in the upper Bengal delta plain. Geomorphology 227:123–136

Price RA (1981) The Cordilleran foreland thrust and fold belt in the southern Canadian Rocky Mountains. Geol Soc Lond Spec Publ 9:427–448. https://doi.org/10.1144/GSL.SP.1981.009.01.39

Rahman MJJ, Xiao W, Hossain MS, Ao S, Yeasmin R, Sayem ASM, Abdullah R, Dina NT (2019) Geochemistry and detrital zircon U-Pb dating of the Plio-Pleistocene sandstones of the Chittagong Tripura Fold Belt (CTFB), South-eastern Bengal Basin: Implications for provenance and tectonic setting. Gondwana Res (accepted)

Reimann K-U (1993) Geology of Bangladesh. Gebrueder Borntraeger, Berlin, 160 p

Rodgers J (1991) Fold-and-thrust belts in sedimentary rocks; part 2, other examples, especially variants. Am J Sci 291:825–886. https://doi.org/10.2475/ajs.291.9.825

Roeder D, Gilbert OE, Witherspoon WD (1978) Evolution and macroscopic structure of valley and ridge thrust belt, Tennessee and Virginia. University of Tennessee, Dept of Geol Sci, 25 p

Sandwell DT, Muller RD, Smith WHF, Garcia E, Francis R (2014) New global marine gravity model from CryoSat-2 and Jason-1 reveals buried tectonic structures. Science 346:65–67

Sihua Y, Guitang P, Liquan W, Xinsheng J, Fuguang Y, Wanping Z, Jiewen Z (2009) Accretionary orogenesis in the active continental margins. Earth Sci Front 16(3):31–48

Sikder AM, Alam MM (2003) 2-D modelling of the anticlinal structures and structural development of the eastern fold belt of the Bengal Basin, Bangladesh. J Sediment Geol 155:209–226

Singh A, Bhushan K, Singh C, Steckler MS, Akhter SH, Seeber L, Kim W-Y, Tiwari AK, Biswas R (2016) Crustal structure and tectonics of Bangladesh: New constraints from inversion of receiver functions. Tectonophysics 680:99–112

Srivastava V, Mukul M, Barnes JB, Mukul M (2018) Geometry and kinematics of Main Frontal thrust-related fault propagation folding in the Mohand Range, northwest Himalaya. J Struct Geol 115:1–18

Steckler MS, Akhter SH, Seeber L (2008) Collision of the Ganges-Brahmaputra Delta with the Burma Arc. Earth Planet Sci Lett 273:367–378

Steckler MS, Mondal D, Akhter SH, Seeber L, Feng L, Gale J (2016) Locked and loading megathrust linked to active subduction beneath the Indo-Burman ranges. Nat Geosci 9:615–618

Steckler MS, Stein S, Akhter SH, Seeber L (2018) The wicked problem of earthquake hazard in developing countries. Earth Space Sci News. https://doi.org/10.1029/2018EO093625

Thakur VC (2004) Active tectonics of Himalayan Frontal Thrust and Seismic Hazard to Ganga Plain. Curr Sci 86(11):1554–1560

Uddin A, Lundberg N (2004) Miocene sedimentation and subsidence during continent–continent collision, Bengal basin, Bangladesh. Sediment Geol 164:131–146

Vernant P, Bilham R, Szeliga W, Drupka D, Kalita S, Bhattacharyya AK, Gaur VK, Pelgay P, Cattin R, Berthet T (2014) Clockwise rotation of the Brahmaputra Valley relative to India: tectonic convergence in the eastern Himalaya, Naga Hills, and Shillong Plateau. J Geophys Res Solid Earth 119(8):6558–6571

Wang Y, Sieh K, Tun ST, Lai K-Y, Myint T (2014) Active tectonics and earthquake potential of the Myanmar region. J Geophys Res 119:3767–3822

Wesnousky S, Kumar GS, Mohindra R, Thakur VC (1999) Uplift and convergence along the Himalayan frontal thrust of India. Tectonics 18:967–976

Yassaghi A (1998) Geometry, kinematics, microstructure, strain analysis, and P-T conditions of the shear zones and associated ductile thrusts in the southern Mt. Lofty Ranges/Adelaide Hills area, South Australia. Ph.D. thesis, University of Adelaide, Dept. of Geology and Geophysics, 351 p

Yin A, Dubey CS, Webb AAG, Kelty TK, Grove M, Gehrels GE, Burgess WP (2010) Geologic correlation of the Himalayan orogen and Indian craton: Part 1. Structural geology, U-Pb zircon geochronology, and tectonic evolution of the Shillong Plateau and its neighboring regions in NE India. Geol Soc Am Bull 122(3–4):336–359

Zhang P, Najman Y, Mei L, Millar I, Sobel ER, Carter A, Barfod D, Dhuime B, Garzanti E, Govin G, Vezzoli G, Hu X (2019) Palaeodrainage evolution of the large rivers of East Asia, and Himalayan-Tibet tectonics. Earth Sci Rev 192:601–630

Structural Imprints of Andaman Accretionary Prism and Its Tectonic Relation with Ophiolite Belt of Indo-Burma Ranges

Tapan Pal

Abstract In Andaman subduction complex Indian plate is subducting towards east below Burma plate. In this subduction complex an accretionary prism comprising of Upper Cretaceous ophiolite and Eocene sediments is exposed in Andaman Islands. Map pattern and the field distribution show dismembered ophiolites occurring in different N-S trending thrust slices where the western slices have low dip (8°–10°) and the eastern slices have steep dip (65°–70°) towards east. Besides, few N-S trending back thrust and E-W trending out of sequence thrust have affected the disposition of the litho units. Oligocene-Miocene forearc sediments are exposed on both side of the accretionary prism. Eocene sediments deposited in trench slope basin have very irregular fold geometry which is due to changes to the basin floor topography during its upliftment along with ophiolite. The forearc sediments showing proximal to distal fan facies have regular fold geometry with N-S striking axial plane and very low plunge (18°–25°) either north or south. Based on the field disposition it is suggested that on land emplacement of Andaman ophiolite took place after the deposition of Oligocene-Miocene sediments. The forearc sediment was deformed during on-land emplacement of the accretionary prism. Similarity in petrographic character, age of the ophiolite, occurrence in an accretionary prism and field disposition suggest that both the ophiolite bodies of Andaman and Naga Hills of Indo Burma Ranges (IBR) represent Neotethyan crust. The ophiolites with Neotethyan crust situated in western belt of IBR have also similarity with the ophiolites of Eastern belt of IBR and on-land emplacement of IBR ophiolites took place during India-Asia/Burma plate collision in Late Miocene time.

Keywords Andaman accretionary prism · Deformation · Ophiolite belt · Indi-Burma ranges

T. Pal (✉)
Geological Survey of India, North Eastern Region, Shillong 793003, India

T. K. Biswal et al. (eds.), *Structural Geometry of Mobile Belts of the Indian Subcontinent*, Society of Earth Scientists Series,
https://doi.org/10.1007/978-3-030-40593-9_5

1 Introduction

The Andaman Islands, lying in the central part of the 5000 km long Burma–Andaman–Java subduction complex, displays different tectonic elements of subduction zone. Here the Indian plate is subducting below Burma plate with easterly dipping subduction. The closure of Tethyan Sea was the outcome of India–Asia collision. The Indus–Yarlung Tsangpo suture zone (ITYSZ) display the remnants of the once extensive Mesozoic–age Tethyan Ocean. Along the northeastern margin of the Indian plate ophiolitic rocks and associated rocks of ITYSZ continue sweeps in a clock wise direction around the eastern Himalaya syntaxis and turns towards southward along Indo Burma ranges to the Andaman Sea. This suture zone is traceable through SE Tibet into Myanmar and is offset by the dextral strike slip fault known as Sagaing fault which shows Miocene to recent displacement (e.g., Morley and Searle 2017). The Neotethyan boundary follows a trend of ophiolite occurrences, lying east of present day plate boundary, running from Indo Burma ranges (IBR) to southwards in Andaman islands (Acharyya 2007; Searle et al. 2007; Baxter et al. 2011; Sloan et al. 2017). The IBR mountain ranges lies beyond the eastern extremity of Indian continental crust and thus did not develop due to continent-continent collision (e.g., Aitchison et al. 2019). Within India the IBR runs through SE Arunachal Pradesh, Nagaland, Manipur and Mizoram. The IBR is interpreted as a part of extensive accretionary prism beneath which subduction is still continuing, and the Burma arc situated to the western edge of Burma microplate collides with northeastern Indian passive margin (Steckler et al. 2008, 2016; Sloan et al. 2017). At present the trench around the Andaman islands, known as Java trench, is to the West of the main Andaman Islands and the Java trench has little expression in the gravity field (20 mgal) towards Burma but stronger signatures (100 mgal) towards south in Nicobar Islands (Mukhopadhyay and Dasgupta 1988). Seismic data from the Andaman Islands and Andaman Sea also indicates a Benioff zone dipping 40°–55° towards the east and record epicentres at 200 km focal depth (e.g., Mukhopadhyay and Dasgupta 1988).

The Andaman group of Islands represent mainly Cretaceous-Eocene accretionary prism and Oligocene–Miocene forearc sediments. In the east of Islands, Quaternary magmatic arcs, known as dormant Narcondam and active Barren volcanoes are present. In the accretionary prism besides Eocene trench sediments, dismembered ophiolite thrust slices are present in the central to eastern part of the islands.

Many works have been done on petrology and tectonics of Andaman ophiolite and depositional environment of the Tertiary outerarc and forearc sediments but the structural attributes of this belt have not been dealt comprehensively. In this review work, map pattern, field features, structural elements, petrological characters of ophiolite and associated Tertiary sediments have been dealt together to understand the Andaman accretionary prism in the background of subduction tectonics. In many literatures Andaman ophiolite belt has been correlated with ophiolite of IBR (e.g., Acharyya 2007, 2010). But detailed field and petrological attributes of those ophiolite belts were not discussed during correlation. Here detailed features of accretionary prism setting of Andaman ophiolite belt have been dealt for tectonic correlation of Andaman ophiolite belt with the ophiolite belts of Indo Burma Ranges.

2 Geological Setting

Andaman Islands being a part of an accretionary prism comprises of (a) numerous north–south-trending slices of upper Cretaceous ophiolites and (b) Tertiary sedimentary rocks (Fig. 1). The disposition of ophiolite along with sediments have been described by many (Karunakaran et al. 1967; Ray 1982; Pal et al. 2003; Pal 2011; Ghosh et al. 2013, 2017; Bhattacharya et al. 2013). Although slices are dismembered in nature, all the members ophiolite stratigraphy along with sole metamorphic are exposed in Andaman Islands. The thrust slices of ophiolite as a coherent bodies, run several kilometers in the eastern part and small, isolated klippes in the western part of the island chain. The different members of ophiolite bodies occur mainly in eastern part of the North-, Middle- and South Andaman Islands (Fig. 1a–c). The mantle section representing mainly peridotites along with few dunite pods and rare chromitite pods occurs in large volume in Middle and North Andaman and Rutland islands whereas cumulate, intrusives and extrusives dominate in South Andaman Island (Pal and Bhattacharya 2016). The cumulate section is represented mainly by dunite-lherzolite-wehrlite-pyroxenite-gabbro, intrusive unit is represented by plagiogranite-diorite-dolerite dykes, and extrusives are represented by massive to pillowed tholeiitic basalt and boninite-andesite-dacite of calc alkaline suite (Pal. 2011). The field disposition of the ophiolite along with sediments shows characters of accretionary prism (Pal et al. 2003; Bhattacharya et al. 2013). In South Andaman, largest ophiolite slice with 300–750 m thickness runs in kilometer scale. Radiolarian chert interlayered with basalt gives Cretaceous–Palaeocene an upper age limit of ophiolite (Roy et al. 1988) whereas the palgiogranite gives a radiometric age ~95 Ma (Pedersen et al. 2010; Sarma et al. 2010).

Tectonic slices of ophiolitic rocks are interleaved with Middle to Late Eocene trench slope sediments (c. 1.4 km thick) (Ray 1982; Chakraborty et al. 1999). Eocene sediments cover a large span ranging from conglomerate to sandstone to shale where the clasts are derived mainly from ophiolitic rocks. Pyroclastic sediments has also been reported from Eocene sediments (e.g., Bandopadhyay and Carter 2017). A thick pile (c. 3 km thick) of Late Eocene to Oligocene turbidite forearc sediments occur above the ophiolite–Eocene-sediments sequence (Ray 1982; Chakraborty and Pal 2001). Within forearc regime turbidite display proximal to mid fan to distal fan facies (Chakraborty and Pal 2001). Miocene forearc sediments occur mainly in the Archipelago Group of Islands and are not in direct contact with any of the ophiolite slices. These sediments have huge volcanic inputs and consist of subaqueous pyroclastic rocks and siliciclastic turbidites in the lower part and carbonate turbidites in the upper part (Pal et al. 2005). Besides these lithounits of the Andaman and Nicobar islands two quaternary volcanoes, named as Barren and Narcondam volcanos located within Andaman Sea represent sub-aerial volcano in the inner volcanic arc chain. The active Barren volcano throughout its evolutionary history is erupting basaltic magma (Pal et al. 2010). Oldest age of the Narcondam lava has been reported as 1-5-1.8 Ma (Ray et al. 2015). The dormant Narcondam Volcano dacite-andesite domal volcano (Pal et al. 2007; Pal and Bhattacharya 2011). Oldest age of this lava has been reported as 1.8 and 3.5 Ma (Bhutani et al. 2014).

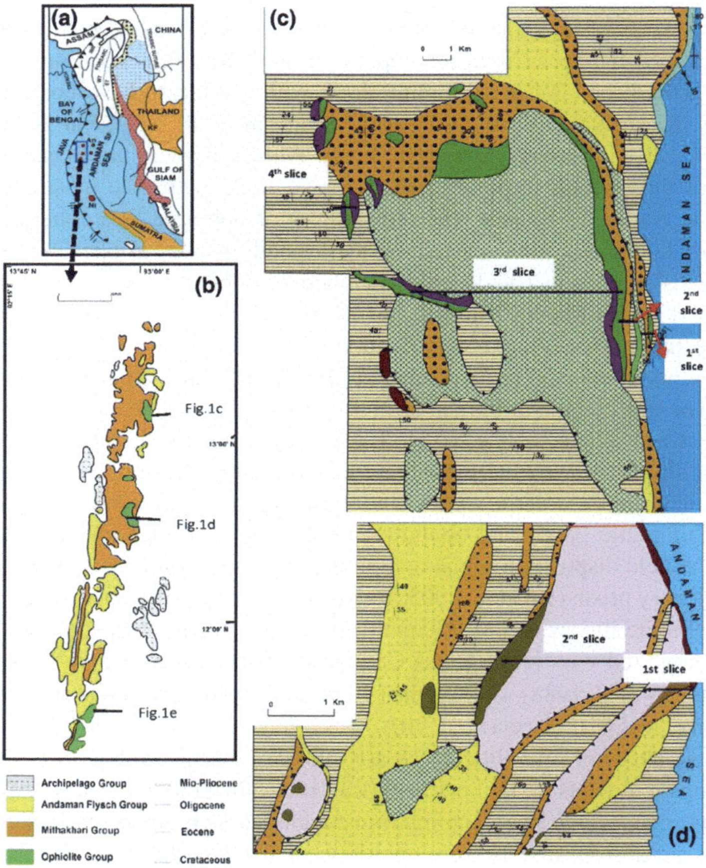

Fig. 1 **a** Regional tectonic framework of SE Asia after Mitchell (1986). **b** Generalized geological map of Andaman Islands showing distribution of the ophiolite and Tertiary sedimentary units with their stratigraphic relationship. (**c**, **d**, **e**) Geological map of part of North, Middle and South Andaman respectively (modified after Pal and Bhattacharya 2016) showing disposition of different ophiolite thrust slices and distribution of Tertiary sediments, four major thrust slices are demarcated in map. An, Andaman Island; ET, Eastern Trough; IBR, Indo Burma ranges; Ni, Nicobar Island; SF, Sagaing Fault; WT, western trough; open circles, Mogok Belt; crosses, Early Tertiary tin–granite belt

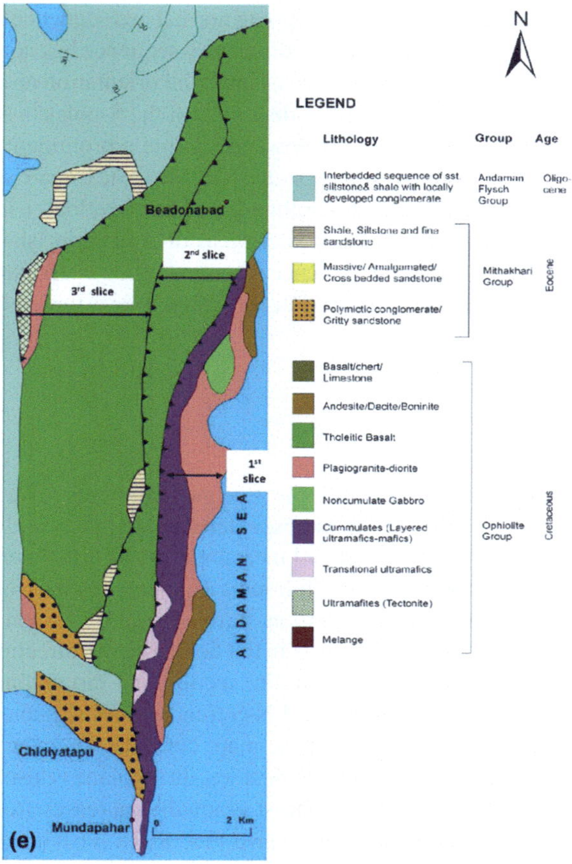

Fig. 1 (continued)

3 Deformation Structures of Andaman Accretionary Prism

3.1 Folds

Both Eocene trench slope basin and Oligocene–Mio-Pliocene forearc sediments show mesoscopic folds. In Eocene sediments folds are in general open cylindrical with low (29° towards 50°) to moderate (52° towards 240°) plunge with axial planes are inclined to vertical. The different generations could not be deciphered from the isolated fold exposures. On the other hand folds in Oligocene sediments are open with a low plunge either towards south or north (25° towards 224°), in Korang nala, Middle Andaman, 18° towards 185°, in Kalipur beach, North Andaman (22° towards 32°, in Shadipur, South Andaman). In general Oligocene sediments in eastern part of the islands record overturned sequence whereas the Oligocene sediments in the western side of the islands show steeply dipping normal bed sequence. Detailed study on the correlation of both the Oligocene sedimentary sequence has not been reported.

Eocene sediments do not show any consistent axial orientation and no cylindrical fold axis could be deciphered from the wide scatter of the S-poles in stereoeplot (Pal et al. 2003). The Oligocene sediments however, define a prominent girdle in E-W with a low plunge either towards north or south. The Miocene forearc sediments show very regular fold geometry with a low plunge, e.g., in Havelock island mesocopic folds show a 15° to the south. In the western part of Andaman Islands the forearc Oligocene sediments show normal sequence whereas in the eastern part of the Islands those forearc sediments in general show steeply dipping overturned beds.

3.2 Thrust

The Cretaceous ophiolite sitting over the Eocene sediments have thrust contact (Fig. 2a, b). Due to lack of good exposures, extensive alteration in the contact zone recognition of thrust features in field and measurement of thrust plane are difficult. In major cases the distribution of the lithounis helps to identify the thrust plane. Mylonitic rocks, silicified zone, brecciation, development of slicken side and dragging of beds also help to recognize thrusts in the field. The dip direction is postulated from the younging direction of the ophiolite section and dip amount is estimated from the geological cross section. Several N-S trending thrust planes with easterly dip have affected the area. Normally in the map, contact of Eocene sediments and ophiolite units are marked as thrust plane. Besides, thrust plane is also passing within ophiolitic members. In general dip of thrust gradually increases from west to east of the area and thrust in the western part have low dip (8°–10°) and in the eastern

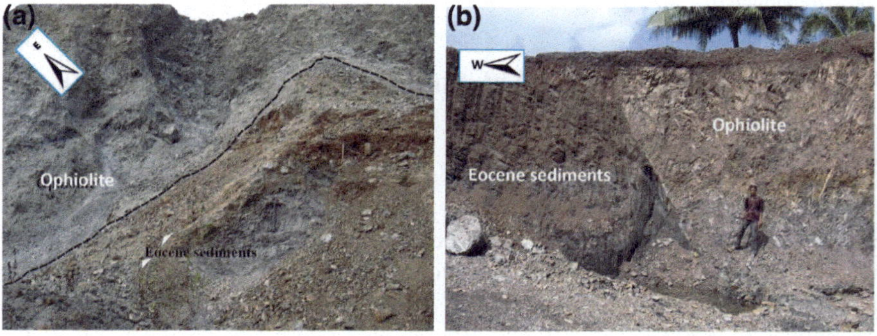

Fig. 2 Field photograph showing thrust contact of Eocene sediments and ophiolite exposed in South Andaman, **a** gently dipping thrust, exposed in Badmashpahar area, south Andaman, nature of the contact is thrust, dipping 25° towards East, here mantle lherzolite is occurring above Eocene sediments, gouge has also developed at the contact and **b** steeply dipping thrust, exposed in Bimlitan quarry section, South Andaman where basalt is in thrust contact with shale, bed attitude could not be measured, thrust is dipping 65° (modified) towards east

part of the ophiolite belt it has steep dip (65°–70°). Besides easterly dipping thrust few westerly dipping but N-S trending back thrust are also reported in Middle and North Andamans (Pal et al. 2003). A few 'out of sequence' thrust trending E-W is also recorded in North Andaman. The brittle and ductile nature of spatially different thrust is common. In general western thrust has ductile behavior whereas thrust in eastern part depicts brittle behavior. Foliations, wavy shear surfaces, anastomosing shearplanes, shear lenses in basalt is commonly found in thrust contact of the western part. The different ophiolite slices occur in different structural levels and are emplaced by thrust which are mainly imbricated. The geological maps of the north, middle and South Andaman Islands show in general four major thrust slices of ophiolites (Fig. 1a–c). In the eastern part of North Andaman repetition of ophiolite members may be due to thrust imbrications. Tectonic overlap also resulted omission of sediments in between ophiolite bodies.

3.3 Strike-Slip/Normal Faults

In many parts of the Andaman Islands N-S normal faults affect the disposition of litho units of Cretaceous ophiolites-Eocene sediments and Oligocene-Miocene sediments. A number of N-S trending normal and E-W trending strike slip faults have also affected the entire Andaman Islands.

4 Discussion

4.1 Evolution of Andaman Accretionary Prism

4.1.1 Thrust Emplacement of Ophiolite

The disposition of different thrusts and faults could be explained by accretionary prism model of Plat (1986). This is wedged shaped prism resting over the subducting slab. It is approximated that ophiolite slices are scrapped off from the slab. But the geochemistry of the Andaman ophiolite shows SSZ character formed in forearc condition (Pal and Bhattacharya 2016). A convincing mechanism is still not available to explain the formation of accretionary prism involving SSZ (forearc) ophiolites. The model of Whattam and Stern (2011) explains emplacement of forearc lithosphere over the subducting slab. During progressive subduction either buoyant crust or sea mount of the subducting slab jams the subduction and gradually forearc lithosphere emplaced over the subducting slab. In Andaman Islands thick sediments over the subducting slab perhaps behaved as buoyant crust. The emplacement of the ophiolite could take place through a series of thrust planes forming an accretionary prism (Fig. 3a, b). With continued eastward subduction a new thrust will generate at the back

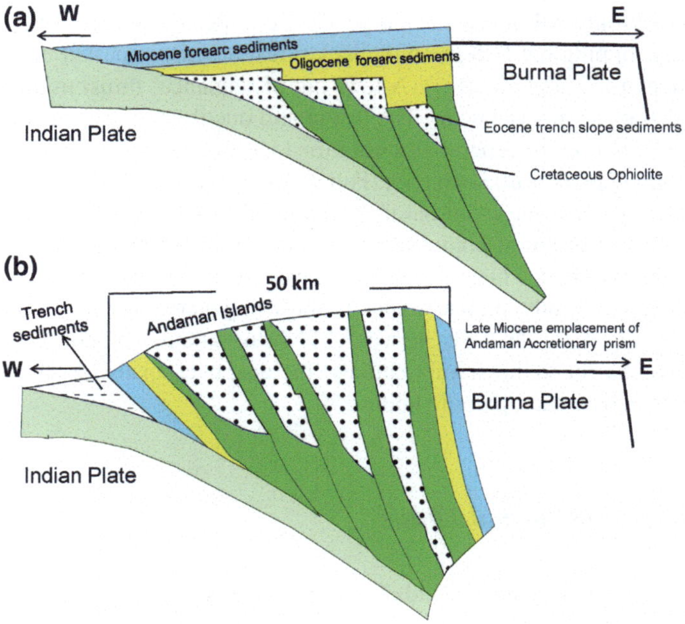

Fig. 3 A sketch diagram showing **a** the evolution of accretionary prism of ophiolite slices and Eocene trench sediments, and development of forearc basin and disposition of Oligocene-Miocene forearc sediments, **b** uplift of accretionary prism and on land emplacement of accretionary prism during Late Miocene. Forearc sediments are also deformed by thrust, disposition of forearc sediments on both side of accretionary prism is also reflected

of earlier thrust and the earlier thrust is modified and become steepened (e.g., Condie 1989). As a result of this process dip of thrust planes increased towards east. During steepening the reactivation of thrust perhaps took place in brittle regime. Accreted material lengthened the wedge resulting internal shortening. This led to "Out of sequence" and back thrusting across the earlier accretionary structures. To stabilize the wedge, sedimentation towards the rear side of wedge favoured extension and series of normal and strike slip faults developed and allowed further sedimentation in forearc basin of Andaman Island.

4.1.2 Evolution of Eocene Trench Slope Sediments

Polymictic conglomerate, sandstone, and shale comprises the Eocene sediments. This ~1.4 km thick sedimentary unit (Ray 1982) contains larger foraminifera and calcareous algae of Middle to Late Eocene age (Karunakaran et al. 1967; Roy et al. 1988). The sedimentary facies of these sediments suggest deposition of coarse detritus along with fine sediments in trench-slope basins where coarse detritus were derived from accreted ophiolite slices (Chakraborty et al. 1999). Individual small basins separated by structural highs with local sources of coarse detritus were uplifted and became part of the accretionary complex.

These sediments showing shearing effect in contact with ophiolites suggest that structural uplift during and after sedimentation and progressive uplift raised from below storm wave base to a shallow coal-depositing paralic setting (Chakraborty et al.1999). Lack of any definite fold geometry in Eocene sediments as well as the effect of superposition may be the effect of tectonically induced episodic changes to the basin floor topography during emplacement of ophiolite. Such changes may be correlated to development and growth of fault bend and fault propagation folds at the tip line of thrusts pushing up slices of ophiolite basement.

4.1.3 Evolution of Forearc Sediments

Oligocene-Miocene sediments displaying wide lateral extent of beds and development of a large-scale submarine fan and regular fold patterns deposited in forearc environment where quartz dominated turbidites were derived mainly from Irrawardy delta of Myanmar (Pal et al. 2003). During Miocene times, the inner arc probably supplied huge amounts of pyroclastic sediments in the form subaqueous pyroclastic flows and towards the end of forearc deposition, shallowing of the basin favoured carbonate deposition (Pal et al. 2005). The force resolved into margin normal and parallel components resulting strike slip movement in between accretionary complex and arc.

4.1.4 Emplacement of Ophiolite

The structural attributes show that different ophiolite slices along with Eocene sediments forms the accretionary prism where the isolated basins of Eocene sediments were uplifted along with ophiolite bodies. Again forearc sediments occur on both side of ophiolite-Eocene accretionary complex. Those sediments are steeply dipping in the eastern part of the islands and moderate to steeply dipping in the western part of the island (Fig. 3). Previously it was suggested that ophiolite were emplaced on land during late Eocene time (Pal et al. 2003; Pal and Bhattacharya 2016). In many cases the Eocene to Oligocene sediments have conformable boundary. The disposition of the forearc sediments on both side of the ophiolite-Eocene sedimentary complex, steep dip of the beds and fold pattern of the Oligocene-Miocene forearc sediments suggest that the accretionary prism uplifted to land position through thrust systems during late Miocene time (Fig. 3). Late Miocene emplacement of ophiolite could be linked with major collision time of Indian plate with Asian plate. Subduction in Andaman islands are intra oceanic and disposition of the ophiolite slices and Eocene sediments in accretionary complex and the field relationship with forearc sediments suggest autochthonous nature of the ophiolite-Eocene sediment.

4.2 *Tectonic Correlation Between Andaman Ophiolite with Other Ophiolites of IBR*

Detailed study has been done on Andaman ophiolites. But if we move further north in Indo Burma region there is diverse opinion on number of belt, age, nature of emplacement, and timing of collision of the ophiolite belt (e.g., Morley and Searle 2017). Towards South in Sumatra region continuous subduction related activity is reported from 90 to 45 Ma (Hall 2012). For Burma regions available literatures do not give a clear knowledge on the type, nature, age, mode of emplacement and the nature of underlying crust of the ophiolite. In Indo Burma region and Myanmar area two ophiolite belts are present viz. (1) Western belt ophiolite (Naga Hills- Chin Hills), (2) Eastern belt ophiolite (Fig. 4a–c). The Eastern belt ophiolite can further be divided into two: (a) Jade mines belt ophiolite (JMB ophiolite) and (b) Tagaung-Myktyina belt ophiolite (TMB ophiolite).

4.2.1 Andaman Ophiolite

It has a metamorphic sole of greenschist to amphibolites grade quartzite-phengite schist to chlorite schist to meta chert. Age of this metamorphic is still unknown. The metamorphic sole is overlain by mantle peridotites and which again is overlain by cumulates, intrusive and extrusive (Fig. 5a). Nature of ophiolite and the extensive melt-rock interaction features in the mantle rocks, geochemical characters of the mantle rocks and the geochemical behavior of the extrusives of the ophiolite suggest the suprasubduction zone character of the ophiolite (Pal 2011, Ghosh et al. 2017). Age of the Andaman ophiolite has been assigned to Upper Cretaceous (~95 Ma). On land emplacement of it took place in the form of accretionary prism possibly in late Miocene age.

4.2.2 The Naga Hills Ophiolite (NHO)

This 200 km long 15 km wide ophiolite belt runs in SE Nagaland to NE Manipur (Brunnschweiler 1966). The ophiolite units occur as thrust slices in an accretionary prism. Detailed geological map of the terrain displaying thrust slices is still not available. It has been reported that ophiolite slices are thrust upon Late Eocene trench deposit known as Disang Formation (Ghose et al. 2014). Although dismembered in nature all the units of ophiolite ranging from mantle peridotites to ultarmafic-mafic cumulate to dykes to pillowed basalt to pelagic sediments (Chert-limestone) have been reported (Acharyya 2007, 2010; Pal et al. 2014; Ghose et al. 2014; Fig. 5b). Isolated exposures of high P/T metamorphic rocks including eclogite and lawsonite bearing blueschist are also associated with ophiolites (Vidyadharan et al. 1989; Ghose et al. 2010, AO and Bhowmik 2014; Bhowmik and AO 2016). The mantle rocks show in general SSZ character (Pal et al. 2014; Ghose et al. 2014). The geochemistry of the

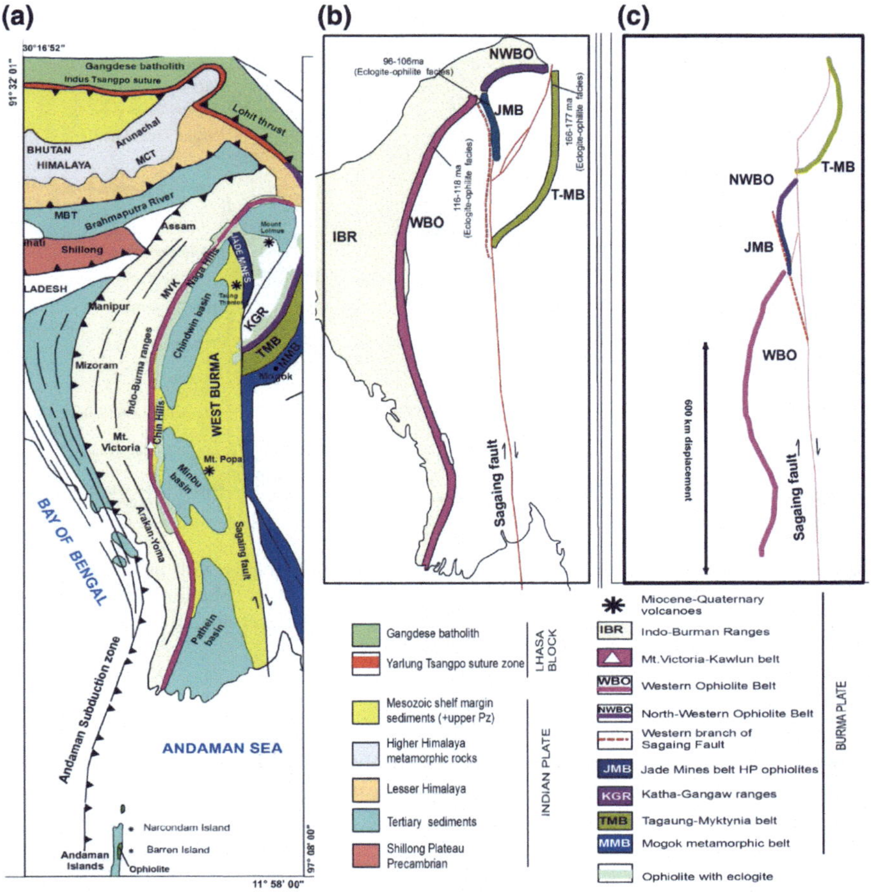

Fig. 4 **a** Geological map of SE Asia, Burma, and the Andaman Sea showing the major suture zones, faults, and terrain boundaries (after Searle et al. 2007) and distribution of eclogite ophiolite belt of Indo Burma Region (IBR) (after Morley and Searle 2017) is also reflected. **b** Present day distribution of the western belt and eastern belt ophiolites of IBR. **c** Restoration of western and eastern ophiolite belts in a single belt before the effect of Sagaing fault (**b**, **c** after Morley and Searle 2017). The reported age of the ophiolites is also shown

volcanics show a wide range from MORB to Ocean Island (OIB) and suprasubduction zone affinities (Singh et al. 2012, 2016; Kingson et al. 2017, 2019). Ophiolite records Middle to Late Jurassic fossil from chert (Baxter et al. 2011; Aitchison et al. 2019) and Early Cretaceous U/Pb date (116–118 Ma) from plagiogranite (Singh et al. 2017). Low grade metapelites and gniesses, known as Naga metamorphics, occurring in the eastern part of the ophiolite belt has Ordovician age (Aitchison et al. 2019) rather than previous report of Proterozoic age (Ghose et al. 2010). The prevalent idea is that Naga metamorphics thrust over the ophiolites during collision of Indian plate with Burma plate. But the recent field observation in Luthur-Thanemier section

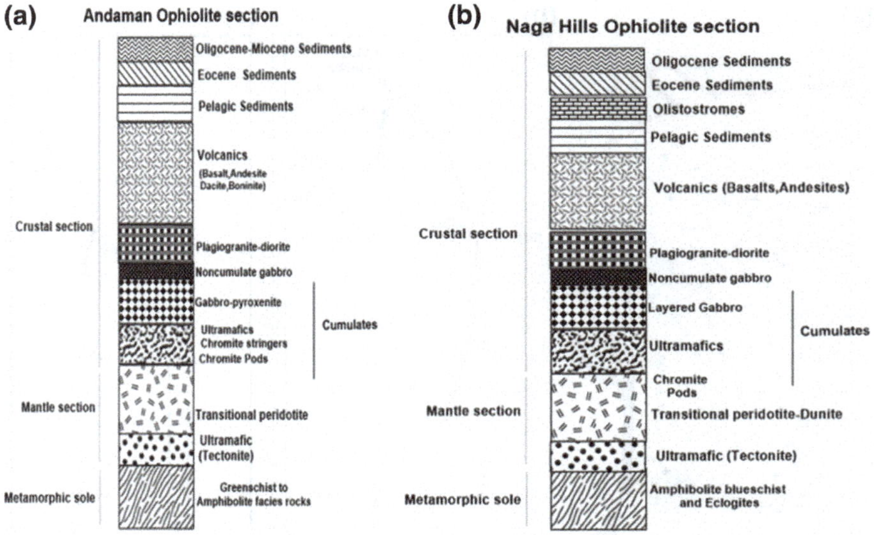

Fig. 5 A diagram showing different lithunits of ophiolite stratigraphy with metamorphic sole, **a** Andaman ophiolite (after Pal 2011; Pal et al. 2003); **b** Naga Hills ophiolite (section prepared based on the reports of different units by Acharyya 2010; Ghose et al. 2014; Pal et al. 2014; Bhowmik and Ao 2016; Singh et al. 2017 and present study)

of Nagaland by the author shows that in the eastern part of the ophiolite belt the ophiolite slices are sitting over the Naga metamorphic with westerly dipping thrust contact (back thrust) (Fig. 6). The ophiolites in the western part has thrust contact with Middle Eocene sediments with abundant olistoliths and olisotromal masses (Ghose et al. 2014). The late Eocene sediments, locally sitting directly over the ophiolites are known as Phokpur Formation records paralic facies to marine facies and these are equivalent to the coarser facies of Eocene sediments of Andaman Island (Ghose et al. 2014). The Oligocene sediments, known as Barail Group, occurring over Eocene sediments has shallow marine affinity (Ghose et al. 2014). Detailed sedimentological analysis of those sedimentary units is not available. Similar to Andaman ophiolite the Naga ophiolite has same Upper Cretaceous igneous age. It records eclogite facies metamorphic sole whereas Andaman shows low to medium P-T amphibolite facies metamorphic sole. Ghose et al. (2014), Imchen et al. (2014) suggest that Naga ophiolite is a part of accretionary prism and ophiolite shows thrust contacts with Eocene sediments. The ophiolite being a part of accretionary prism again is overlain by ophiolite derived Late Eocene–Oligocene sediments (Imchen et al. 2014). The Naga ophiolite perhaps emplaced over continental margin during Late Miocene time during major collision of Indian plate with Asian/Burma plate.

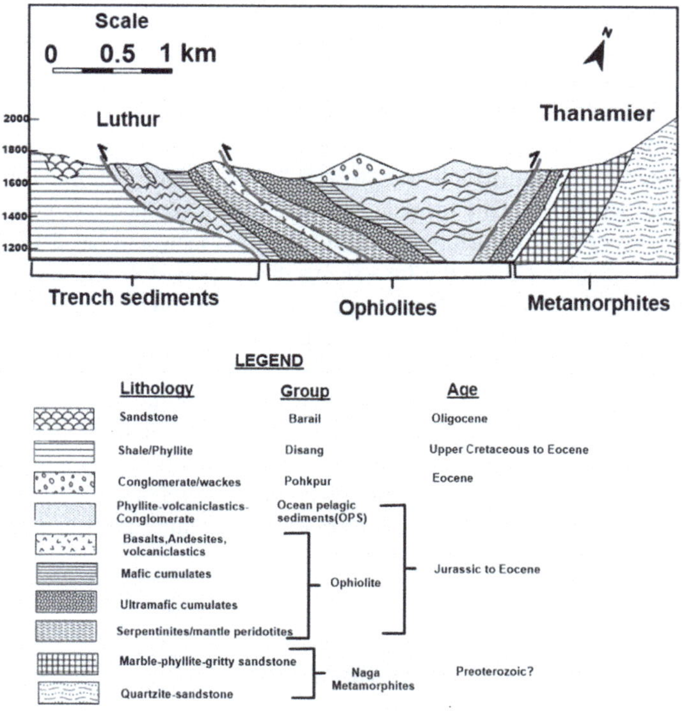

Fig. 6 A sketch showing geological section of the ophiolite exposed in Luthur (25.834° N, 94.883° E)-Thanamier (25.7773° N, 94.954° E) section of Nagaland showing easterly dipping thrust in the western part and westerly dipping back thrust where ophiolite resting over Naga metamorphics

4.2.3 Chin Hills (Mt Victoria) Ophiolite

Very little information is available for ophiolite but it comprises gabbros, dykes, pillow lavas and radiolarian chert (Socquet et al. 2002). This ophiolite belt also has eclogite facies metamorphics (e.g., Morley and Searle 2017).

4.2.4 JM Belt Ophiolite

This has high pressure eclogitic metamorphic sole represented by eclogite, and other units are poorly preserved. However, talc chlorite schist is overlain by basalt, pillowed andesite, plagiogranite and chert where the plagiogranite shows 96–106 Ma age. The metamorphic sole and age of the ophiolite are similar to the Naga ophiolite.

4.2.5 The TMB Ophiolite

It has also a metamorphic sole of chlorite and talc schists (Mitchell et al. 2007) and elcogite facies (e.g., Morley and Searle 2017). The metamorphic sole is overlain by serpentinised harzburgites, basalts and chert and the ophiolite is the host to famous Taung Ni deposit (Mitchell et al. 2007). U–Pb dating shows 166–177 Ma age (Yang and Xu 2012). This Jurassic ophiolite with SSZ character may represent either Neotethys or had a connection with Mesotethys (Yang and Xu 2012). Yui et al. (2013) however doubted about the Jurassic age and reported 77 Ma age.

The TM belt and TMB belt ophiolites are also correlated with Naga Hills ophiolite which are offset by strands of dextral Sagaing strike slip fault suggesting that it represent Neotethys rather than Mesotethys (Mitchell et al. 2012; Morley and Searle 2017). A model for a restoration of displacement by Morley and Searle (2017) suggest that the ophiolite belts in the region e.g., Naga Hills–Victoria trend, Jade Belt-, and Tagaung-Myitkyina may align into a single ophiolite belt (Fig. 4c). This single belt in that case marking the closure of Neotethys could be correlated with Indus Tsangpo suture zone of Tibet and displacement (600 km) of Sagaing fault is post-ophiolite emplacement (Morley and Searle 2017).

4.3 Nature of the Crust of Andaman and IBR

There is a diverse view on the nature of the crust below Andaman Sea. Much of the crust below East Andaman basin and Central Andaman basin is oceanic in nature (Curray 2005; Radhakrishna et al. 2008). Crustal thickness has been estimated as 30–40 km (Radhakrishna et al. 2008) which may match with the upper limit of island arc crust or accretionary prism. From regional gravity modeling, however, 25–35 km thick crust has been reported (Morley and Alvey 2015). To explain the ophiolite emplacement Acharyya (2007, 2010) suggested that ophiolite emplaced over a proto continental block. Currey, 2005 however, postulated an oceanic crust around the Andaman Islands. Based on modern seismic tomography and gravity modeling Singh et al. (2013) has reported an accretionary prism overlying lower plate oceanic crust around the islands.

Morley and Alvey (2015) suggested that Neogene oceanic crust may be restricted to Central Andaman basin whereas East Andaman basin and Alcock and Sewel rises may represent hyper-extended continental crust. Further west, the Invisible bank again represent continental crust (Roy and Chopra 1987). However Cretaceous oceanic crust in a suprasubduction zone setting exist to the west of Invisible bank (Morley 2012). Therefore a sliver of continental crust may be extended from the West Burma to further south in Andaman Sea representing Invisible Bank.

The crust of Shan plateau is being known as a part of the continental block of Sibumasu, which was derived from the Australia. Some workers also suggest that continental part of West Burma block could be part of Sibumasu (e.g., Morley and Searle 2017). West Burma was a part of SE Asia during Carboniferous to early

Permian, prior to Mesozoic and it was not a part of Indian plate origin (Sevastjanova and Hall 2016). The schists and meta sediments of the Mt Victoria is considered to represent a continental crust derived from Australia (Ridd, 2015; Rangin et al. 2013). Towards further west in the Indo Burma Ranges the ophiolite forming accretionary prism is sitting over the subducting Indian plate. The seismic signatures in northern Bay of Bengal, hoverer, is consistent with oceanic crust overlain by extremely thick and metamorphosed sediments (Curray 1991; Mitra et al. 2011). The seismically active slab extending to a depth of 160 km beneath Central Burman lowlands also confirm that continental crust cannot continue up to such great depth. Rangin et al. 2013 viewed that the Indo Burma ranges is underlain by Burma continental micro block but the interpretation of Sevastjanova and Hall (2016) is against that view. It is therefore suggested that from IBR to Andaman oceanic crust runs for the entire belt.

4.4 Volcanic Arc of Andaman and IBR Region

In Myanmar and further south in Andaman Sea a magmatic arc belt is present and in Myanmar area it is known as Wuntho-popa Arc (WPA) (Fig. 4a). The WPA shows volcanics of different age with main magmatic activity in 110–80 Ma and a subordinate stage of 70–40 Ma then in Oligocene-Miocene age (Wang et al. 2014; Mitchell et al. 2012). In the south Mt popa area the Miocene (16–13 Ma) volcanism is accompanied by Quaternary stage (<1 Ma) volcanism (Maury et al. 2004; Lee et al. 2010). This WPA continues southerly in offshore like in Barren and Narcondam Islands of Andaman Sea. The calc alkaline lavas of Narcondam volcano have a influence of the continental crust (Pal et al. 2007; Streck et al. 2011; Pal and Bhattacharya 2011). On contrary, Barren Volcano showing basaltic magmatism has no contribution from continental crust and lie on oceanic crust (Pal et al. 2010; Ray et al. 2015). Evidence of past igneous activity is also reported from Eocene sediments (Bandopadhyay 2005). The Miopliocene felsic volcanic has also been reported from the Andaman Islands (Pal et al. 2005). Subduction of Neotethys and later oceanic crust of Indian plate resulted magmatic arc and reduction of volcanic activity in WPA after 40 Ma could be linked to reduction of subduction and post collision decelarting the convergence rate (Morley and Searle 2017). The presence of the magmatic arc in the entire belt of IBR and Andaman Sea Mio Pliocene onwards with a same volcanic line suggest that the effect of subduction in IBR and Andaman area continued even after Miocene age.

5 Conclusions

The entire ophiolite belt running from Andaman to Indo Burma Ranges (IBR) has accretionary prism setup. As a part of accretionary prism the ophiolite belts running from Andaman to Chin Hills to Naga Hills, Jade mines belt have Cretaceous igneous age. Although dismembered in nature all the units of ophiolite succession are reported in Andaman and Naga hills ophiolite. Petrological attributes of the of Andaman and Naga Hills ophiolite suggest suprasubduction zone ophiolite representing forearc lithosphere. Eclogitic facies metamorphics occur as metamorphic sole in the IBR ophiolites whereas amphibolite grade metamorphic sole is present in Andaman.

Andaman ophiolite belt shows four major N-S trending east dipping thrust slices where western slices have low dip whereas eastern most slices have steep dip. In Andaman accretionary prism Eocene sediments deposited in trench slope basin and those isolated basins were also participated along the thrust uplift of the ophiolite. Subsequent to the Eocene sedimentation, Oligocene-Miocene forearc sediments deposited in the form of proximal to mid to distal fan deposits where the major source of clastics are presumed to be Irrawardy delta. The Eocene sediments display complex fold geometry whereas Oligocene-Miocene sediments show a regular fold geometry with a North-South axial plane and a low plunge. Occurrence of the forearc sediments on both side of the ophiolite-Eocene sediment complex and deformation features of forearc sediments suggest that on land emplacement of ophiolite belt took place in Late Miocene time. Field disposition of the different litho units of the Naga Hills ophiolite also suggest Late Miocene emplacement. We can guess similar situation for other ophiolite of IBR. Similarity in the age and litho logical characters of the Western belt and Eastern belt of IBR ophiolites suggest that initially there was a single ophiolite belt. Later on Sagaing strike slip fault has split that into two separate belts. Further one of the magmatic arc volcano (Mt Loimue) is situated near to Jade Mines belt which suggests that Jade Mines belt has moved down to arc zone through the Sagaing fault. The entire ophiolite belt marking the Neotethys suture probably emplaced on land during Late Miocene collision of Indian plate with Asian plate/Burma microplate.

Acknowledgements The author acknowledge helps of Annie D. Nelson, K. V. Thennuo, Rohit Gajbhiye and Ramesh Laishram of Geological Survey of India during preparation of maps, figures and finalization of the manuscript.

References

Acharyya SK (2007) Collisional emplacement history of the Naga-Andaman ophiolites and the position of the eastern Indian suture. J Asian Earth Sci 29:239–242

Acharyya SK (2010) Tectonic evolution of Indo-Burma range with special reference of Naga-Manipur Hills. Memoir Geol Soc India 75:25–43

Aitchison JC, Ao A, Bhowmik S, Clarke GL, Ireland TR, Kachovich S, Lokho K, Stojanovic D, Roeder T, Truscott N, Zhen Y, Zhou R (2019) Tectonic evolution of the western margin of the Burma microplate based on new fossil and radiometric age constraints. Tectonics (in press)

Ao A, Bhowmik (2014) Cold subduction of the Neothethys: the metamorphic record from finely banded lawsonite and epidote blueschists and associated metabasalts of the Nagaland Ophiolite in C. K. Morley & M. Searle. Downloaded from http://mem.lyellcollection.org/ at University of Colorado Boulder on February 3, 2017 Complex, India. J Metamorph Geol. https://doi.org/10.1111/jmg12096

Bandopadhyay PC (2005) Discovery of abundant pyroclasts in Namunagarh Grit, South Andaman; evidence for arc volcanism and active subduction during the Palaeogene in the Andaman area. J Asian Earth Sci 25:95–107

Bandopadhyay PC, Carter A (2017) Mithakhari deposits. In: Bandopadhyay PC, Carter A (eds) The Andaman-Nicobar accretionary ridge: geology, tectonics and hazards. Geological Society, London. Memoir 47:111–132, 235p

Baxter AT, Aitchison JC, Zyabrev SV, Ali JR (2011) Upper Jurassic radiolarians from the Naga Ophiolite, Nagaland, northeast India. Gondwana Res 20:638–644

Bhattacharya A, Pal T, Ghosh B (2013) Characterisation of the accreted ophiolite slices of Rutland Island, Andaman Sea: evolution in a suprasubduction zone setting. Ofioliti 38:121–142

Bhowmik SK, Ao A (2016) Subduction initiation in the Neo-Tethys: constraints from 687 counterclockwise P-T paths in amphibolite rocks of the Nagaland Ophiolite Complex, 688 India. J Metamorph Geol 34(1):17–44

Bhutani R, Pande K, Ray JS, Smitha RS, Awasthi N, Kumar A (2014) 40Ar-39Ar geochronology of Narcondam island volcano, Andaman-Indonesian island arc, Andaman Sea: voluminous andesitic eruption at 0.56 Ma and preservation of 1.8 Ma old plagioclase xenocrysts. Geological Society of America Abstract 46, 653

Brunnschweiler RO (1966) On the geology of the Indoburman ranges. J Geol Soc Aust 13:137–194

Chakraborty PP, Pal T (2001) Anatomy of a forearc submarine fan: Upper Eocene–Oligocene Andaman Flysch Group, Andaman Islands, India. Gondwana Res 4:477–487. https://doi.org/10.1016/s1342-937x(05)70347-6

Chakraborty PP, Pal T, Dutta Gupta T, Gupta KS (1999) Facies pattern and depositional motif in an immature trench-slope basin, Eocene Mithakhari Group, Middle Andaman, India. J Geol Soc India 53:271–284

Condie KC (1989) Plate tectonics and crustal evolution (3rd edn). Pergamon Press, 469 pp

Curray JR, Moore DG (1974) Sedimentary and tectonic processes in the Bengal deep sea fan and geosyncline. In: Burke CA, Drake CL (eds) The geology of continental margins. Springer, New York, pp 617–627

Curray JR (1991) Possible greenschist metamorphism at the base of a 22 km sediment section, Bay of Bengal. Geology 19:1097–1100

Curray JR (2005) Tectonics of the Andaman Sea region. J Asian Earth Sci 25:187–232

Ghose NC, Agrawal OP, Chatterjee N (2010) Geological and mineralogical study of 762 eclogite and glaucophane schists in the Naga Hills Ophiolite, Northeast India. Island 763 Arc 19(2):336–356

Ghose NC, Chatterjee N, Fareeduddin (2014) A Petrographic Atlas of Ophiolite, an example from the Eastern India-Asia collision zone. Springer, India, https://doi.org/10.1007/97-81-322-1569-1

Ghosh B, Bandyopadhyay D, Morishita T (2017) Andaman-Nicobar ophiolites, India: origin, evolution and emplacement. In: Bandopadhyay PC, Carter A (eds) The Andaman-Nicobar accretionary ridge: geology, tectonics and hazards. Geological Society, London, pp 95–110. Memoir 47, 235p

Ghosh B, Morishita T, Bhatta K (2013) Significance of chromian spinels from the mantle sequence of the Andaman ophiolite, India: paleogeodynamic implications. Lithos 164–167:86–96

Hall R (2012) Late Jurassic-Cenozoic reconstructions of the Indonesian region and the Indian Ocean. Tectonophysics 570–571:1–41

Imchen W, Thong GT, Pongen T (2014) Provenance, tectonic setting and age of the sediments of the Upper Disang Formation in the Phek District, Nagaland. J Asian Earth Sci 88:11–27

Karunakaran C, Ray KK, Saha SS (1967) A revision of the stratigraphy of Andaman and Nicobar islands, India. Bulletin of the National Institute of Science, India 38:436–441

Kingson O, Bhutani R, Balakrishnan S, Dash JK, Shukla AD (2019) Subduction-related Manipur Ophiolite Complex, Indo-Myanmar Ranges: elemental and isotopic record of mantle metasomatism. Geological Society, London, Special Publications, 481, doi:10.1144/SP481.9

Kingson O, Bhutani R, Dash JK, Sebastian S, Balakrishnan S (2017) Resolving the conundrum in origin of the Manipur Ophiolite Complex, Indo-Myanmar range: constraints from Nd isotopic ratios and elemental concentrations in serpentinized peridotite. Chem Geol 460:117–129

Lee H, Chung S, Yang H, Chu C, Lo C, Mitchell A (2010) Cenozoic volcanic rocks from central Myanmar: age, geochemical characteristics and geodynamic significance. AGU, Fall Meeting, San Francisco, 13–17th December, 2010, abstract

Maury RC, Pubellier M et al (2004) Quaternary calc-alkaline and alkaline volcanism in an upper-oblique convergence setting, central Myanmar and western Yunnan. Bull *Soc Géol France* 175:461–472

Mitchell AHG (1986) Ophiolite and associated rocks in four settings: relationships to subduction and collision. Tectonophysics 125:269–285

Mitchell AHG, Chung S-L, Oo Thura, Lin T-H, Hung C-H (2012) Zircon U-Pb ages in Myanmar: magmaticmetamorphic events and the closure of a neo-Tethys ocean? J Asian Earth Sci 56:1–23

Mitchell AHG, Htay MT, Htun KM, Win MN, Oo T, Hlaing T (2007) Rock relationships in the Mogok Metamorphic belt, Tatkon to Mandalay, central Myanmar. J Asian Earth Sci 29:891–910

Mitra S, Priestley K, Acton C, Gaur VK (2011) Anomalous surface wave dispersion and the enigma of 'continental-like' structure for the Bay of Bengal. J Asian Earth Sci 42:1243–1255

Morley CK, Searle M (2017) Ch-5: Regional tectonics, structure and evolution of the Andaman–Nicobar Islands from ophiolite formation and obduction to collision and back-arc spreading. In: Bandopadhyay PC, Carter A (eds) The Andaman-Nicobar accretionary ridge: geology, tectonics and hazards. Geological Society, London, Memoir 47:51–74 (235p)

Morley CK, Alvey A (2015) Is spreading prolonged, episodic or incipient in the Andaman Sea? Evidence from deepwater sedimentation. J Asian Earth Sci 98:446–456

Morley CK (2012) Late Cretaceous-Early Palaeogene tectonic development of SE Asia. Earth-Sci Rev 115:37–75

Mukhopadhyay M, Dasgupta S (1988) Deep structure and tectonics of the Burmese arc: constraints from earthquake and gravity data. Tectonophysics 149:299–322

Pal T (2011) Petrology and geochemistry of the Andaman ophiolite: melt–rock interaction in a suprasubduction-zone setting. J Geol Soc London 168:1031–1045. https://doi.org/10.1144/0016-76492009-152

Pal T, Bhattacharya A (2011) Block-and-ash flow deposit of the Narcondam Volcano: product of dacite–andesite dome collapse in the Burma-Java subduction complex. Island Arcs 20:520–534

Pal T, Bhattacharya A, Nagendran G, Yanthan NM, Singh R, Raghumani N (2014) Petrogenesis of chromites from the Manipur ophiolite belt, NE India: evidence for a supra-subduction setting prior to Indo-Myanmar collision. Mineral Petrol 108:713–726

Pal T, Bhattacharya A (2016) Evolution of Andaman ophiolite in the backdrop of Andaman-Java subduction—a comprehensive approach integrating field and Laboratory data Geological Survey of India Sp. Pub. No 105, pp 137

Pal T, Chakraborty PP, Duttagupta T, Singh CD (2003) Geodynamic evolution of an outer arc in convergent margin of active Burma-Java subduction complex, a document from Andaman islands, Bay of Bengal. Geol Mag 140:289–307

Pal T, Duttagupta T, Chakraborty PP, Dasgupta SC (2005) Pyroclastic deposits of Mio-Pliocene age in the Arakan Yoma–Andaman–Java subduction complex, Andaman Islands, Bay of Bengal, India. Geochem J 39:69–82

Pal T, Mitra SK, Sengupta S, Katari A, Bandopadhyay PC, Bhattacharya AK (2007) Dacite-andesites of Narcondam volcano in the Andaman Sea—an imprint of magma mixing in the inner arc of the Andaman-Java subduction system. J Volcanol Geoth Res 168:93–113

Pal T, Raghav SR, Bhattacharya A, Bandopadhyay PC, Mitra SK, Renjit ML, Shankar MS, Ghosh B (2010) The 2005-06 eruption of Barren Volcano, Andaman Sea: evolution of basaltic magmatism in island arc setting of Andaman-Java subduction complex. J Asian Earth Sci 39:12–23

Pedersen RB, Searle MP, Carter A, Bandopadhyay PC (2010) U-Pb zircon age of the Andaman ophiolite: implications for the beginning of subduction beneath the Andaman-Sumatra arc. J Geol Soc London 167:1105–1112. https://doi.org/10.1144/0016-76492009-151

Radhakrishna M, Lasitha S, Mukhopadhyay M (2008) Seismicity, gravity anomalies and lithospheric structure of the Andaman arc, NE Indian Ocean. Tectonophysics 460:248–262

Rangin C, Maurin T, Masson F (2013) Combined effects of Eurasia/Sunda oblique convergence and East-Tibetan crustal flow on the active tectonics of Burma. J Asian Earth Sci 76:185–194
Ray JS, Pande K, Bhutani (2015) 40Ar/39Ar geochronology of subaerial lava flows of Barren Island volcano and the deep crust beneath the Andaman Island Arc, Burma Microplate. Bull Volcanol 77:57. https://doi.org/10.1007/s00445-015-0944-9
Ray KK (1982) A review of the geology of Andaman and Nicobar islands, vol 42. Geological Survey of India Miscellaneous Publications, pp 110–125
Ridd MF (2015) Should Sibumasu be renamed Sibuma? The case for a discrete Gondwana-derived block embracing western Myanmar, Upper Peninsular Thailand and NE Sumatra. J Geol Soc London 173(2):249. https://doi.org/10.1144/jgs2015-065
Roy DK, Acharyya SK, Ray KK, Lahiri TC, Sen MK (1988) Nature of occurrence and depositional environment of the oceanic pelagic sediments associated with the ophiolite assemblage, South Andaman Island. Indian Miner 42:31–56
Roy TK, Chopra NN (1987) Wrench faulting in Andaman forearc basin, India. Proc Offshore Technol Conf 19:393–404
Sarma DS, Jafri SH, Fletcher IR, McNaughton NJ (2010) Constraints on the tectonic setting of the Andaman ophiolites, Bay of Bengal, India, from SHRIMP U–Pb zircon geochronology of plagiogranites. J Geol 118:691–697
Searle MP, Noble SR, Cottle JM, Waters DJ, Mitchell AHG, Hlaing T, Horstwood MS (2007) A tectonic evolution of the Mogok metamorphic belt, Burma (Myanmar) constrained by U-Th-Pb dating of metamorphic and magmatic rocks. Tectonics 26:1–24
Sevastjanova I, Hall R (2016) Myanmar and Asia united, Australia left behind long ago. Gondwana Res 32:24–40
Singh AK, Chung S-L, Bikramaditya RK, Lee HY (2017) New U-Pb zircon ages of plagiogranites from the Nagaland-Manipur Ophiolites, Indo-Myanmar Orogenic Belt, NE India. J Geol Soc 174(1):170–179
Singh AK, Khogenkumar S, Singh LR, Bikramaditya RK, Khuman CM, Thakur SS (2016) Evidence of Mid-ocean ridge and shallow subduction forearc magmatism in the Nagaland-Manipur ophiolites, northeast India: constraints from mineralogy and geochemistry of gabbros and associated mafic dykes. Chem Erde 76(4):605–620
Singh AK, Singh NI, Devi LD, Singh RKB (2012) Geochemistry of Mid-Ocean Ridge Mafic intrusives from the Manipur Ophiolitic Complex, Indo-Myanmar Orogenic Belt, NE India. J Geol Soc India 80(2):231–240
Sloan RA, Elliott JR, Searle MP, Morley CK (2017) In: Barber AJ, Zaw K, Crow MJ (eds) Chapter 2. Active tectonics of Myanmar and the Andaman Sea. Geological Society, London. Memoirs, pp 19–52
Socquet A, Goffé B, Pubellier M, Rangin C (2002) Le métamorphisme Tardi-Crétacé à Éocène des zones internes de la chaîne Indo-Birmane (Myanmar occidental): implications géodynamiques. Compte Rendus Geosci 334:573–580
Steckler MS, Akhter SH, Seeber L (2008) Collision of the Ganges–Brahmaputra Delta with the Burma Arc: implications for earthquake hazard. Earth Planet Sci Lett 273(3):367–378
Steckler MS, Mondal DR, Akhter SH, Seeber L, Feng L, Gale J, Howe M (2016) Locked and loading megathrust linked to active subduction beneath the Indo-Burman Ranges. Nat Geosci 9(8):615–618
Streck MJ, Ramos F, Gillam A, Haldar D, Duncan RA (2011) Thei intra-oceanic Barren Island and Narcondam arc volcanoes, Andaman Sea: implications for subduction inputs and crustal overprint of depleted mantle source. In: Ray J, Sen G, Ghosh B (eds) Topics in igneous petrology. Springer, Dordrecht, Heidelberg, pp 241–273
Vidyadharan KT, Joshi A, Ghose S, Gaur MP, Sukla R (1989) Manipur Ophiolites. Its geology, tectonic setting and metallogeny. In: Ghose NC (ed) Phanerozoic Ophiolites of India and associated monerial resources. Sumna Publishers and Distributors, Patna, pp 197–212
Wang J-G, Wu F-Y, Tan X-C, Liu C-Z (2014) Magmatic evolution of the Western Myanmar Arc documented by U-Pb and Hf isotopes in detrital zircons. Tectonophysics 612–613:97–105

Whattam SA, Stern RJ (2011) The 'subduction-initiation rule': a key for linking ophiolites, intra-oceanic forearcs and subduction initiation. Contributions to Mineral Petrol 162:1031–1045. https://doi.org/10.1007/s00410-011-0638-z

Yang JS, Xu ZQ (2012) Discovery of a Jurassic SSZ ophiolite in the Tagaung-Myitkyina region of Myanmar. Acta Petrol Sin 28:1710–1730

Yui T-F, Fukoyama M, Iizuka Y, Wu C-M, Wu T-W, Liou JG, Grove M (2013) Is Myanmar jadeitite of Jurassic age? A result from incompletely recrystallized inherited zircon. Lithos 160–161:268–282

Two and Three-Dimensional Structural Modelling of Central Kohat Plateau, Northwestern Himalaya, Pakistan

Nawaz Ikram, Syed Ahsan Hussain Gardezi, Sajjad Ahmad, Gohar Rehman and Adnan Khalid

Abstract The Kohat Fold and Thrust Belt (*KFTB*) lies in the north western margin of Himalayan arc in the Pakistan. It covers a part of outer Himalayan Foreland Basin where 800 m thick Eocene package is predominantly exposed at the surface forming different structures. Current research describes the structural geometry and kinematics of the central *KFTB* and identifies the role of mechanical heterogeneity in defining the surface structuration as well as its impacts on the deeper horizons. The east west trending, open to tight and overturned folds having curvilinear axes associated with thrust fault system dominates the exposed structural geometry of the area. Three mechanically different stratigraphic packages have been identified in the research area. The thick middle (Late Paleocene to Middle Eocene) clayey and mechanically soft package is sandwiched between upper and lower mechanically competent packages. The upper package comprises of Middle Eocene to Late Miocene rocks while the lower package consists of Paleozoic up to Paleocene rocks. Our structural modelling suggests that the deformation within the exposed upper competent stratigraphic package is disharmonic in structural relation to the lower stratigraphic package, which is comprised of south verging, ramp-related fore-folds. The middle, soft and plastic stratigraphic package is acting as surface of fault-fold decoupling between the lower and upper competent packages. The structural balancing reveals

N. Ikram (✉)
Institute of Geology, University of Azad Jammu and Kashmir, Muzaffarabad 13100, Pakistan
e-mail: nawazikram@hotmail.com

S. A. H. Gardezi
Geological Survey of Pakistan, AJK Directorate, Muzaffarabad 13100, Pakistan
e-mail: ahsangardezi2504@gmail.com

S. Ahmad · G. Rehman · A. Khalid
Department of Geology, University of Peshawar, Peshawar 25120, Pakistan
e-mail: sajjadahmad@uop.edu.pk

G. Rehman
e-mail: goharrehman@uop.edu.pk

A. Khalid
e-mail: adnanmarwat@yahoo.com

T. K. Biswal et al. (eds.), *Structural Geometry of Mobile Belts of the Indian Subcontinent*, Society of Earth Scientists Series,
https://doi.org/10.1007/978-3-030-40593-9_6

that the area has undergone 42.35% shortening. The mode of deformation is thin-skinned, while the facing of structures show southward propagation of deformation as a result of the Indo-Eurasian plates collision.

Keywords Structural geology · 2D and 3D structural modelling · Stress analysis · Central Kohat Plateau · KP Pakistan

1 Introduction

The study area lies in the central part of Kohat Fold and Thrust Belt (*KFTB*). The *KFTB* represents the western segment of the Upper Indus Basin of Pakistan and lies in the north western margin of Himalayan arc (Fig. 1). The Himalayas formed about 57 Ma ago due to Indian and Eurasian landmass collision (Gansser 1964; LeFort 1975; Molnar and Tapponnier 1975; Fraser et al. 2001). The *KFTB* is an east west tending foreland basin comprised of Eocene to Plio-Pleistocene sediments (Gee 1945; Fatmi 1973; Meissner et al. 1974; Sameeni et al. 2009; Shah 2009). The lack of igneous and metamorphic rocks makes this region a part of the external zone of Himalayas (Coward et al. 1986).

The *KFTB* is characterized by tens of kilometers extended anticlines and synclines (Meissner et al. 1974) and the limbs of many of these structures shows thrust/reverse faulted relation with younger rocks (Khan et al. 1986; Gardezi et al. 2017). During the orogeny of Himalayas, the *KFTB* was the major depositional center of eastwestern Himalayan sediments in the Miocene age (Dutta et al., submitted). Gradually, it uplifted tectonically and became folded and faulted, recording the southward propagation of Himalayan deformation as a result of ongoing collision between Indian and Eurasian plates (Wells 1984; Pivnik 1992).

The stratigraphy of the western Himalayan collision zone is predominantly preserved in the Kohat region. The stratigraphy of this region and adjoining areas reveals a broad spectrum of oceanic and continental environment facies ranging from Triassic to Recent (Fig. 2). The continental rock units of the Miocene to Plio-Pleistocene mark a regional unconformity with the underlying marine rock units of the Eocene age. Jatta Gypsum of the Eocene is the oldest exposed rock unit in the study area. The northern and north western border of *KFTB* is demarcated by the Main Boundary Thrust (*MBT*) (Khan et al. 1986), the Bannu Basin and the Surghar Range mark its southern boundary. The axial belt of Kurram accretionary wedge, which is transpressional left lateral fault, delimits its western boundary while the Indus River marks its eastern frontier and separates it from the Potwar Plateau. The present research focuses on the detailed study of structural geology of the central *KFTB*, which lies in the north west of outer Himalayas in the north Pakistan. The structuration in this area is the result of southward propagating deformation due to Indo-Eurasian collision. These structures are mainly anticlines, synclines, thrust faults and their associated back thrusts. The overall structural trend is east west with local exceptions (Fig. 1).

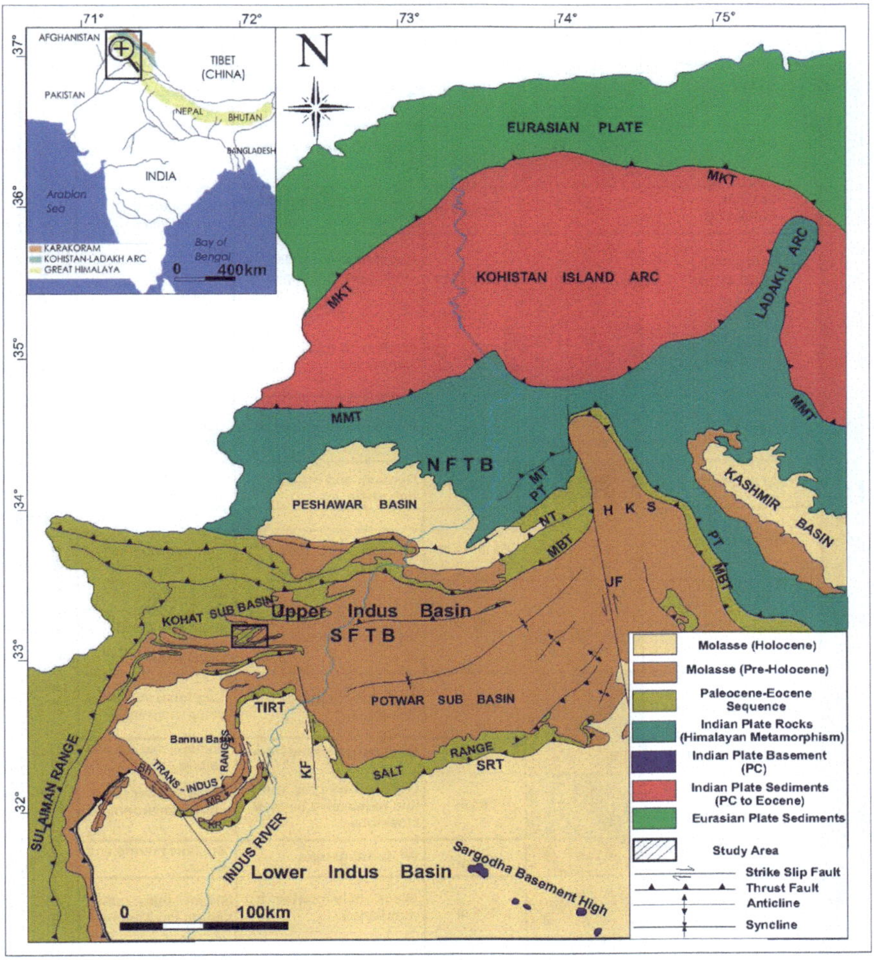

Fig. 1 Generalized geologic map of the NW Himalayan foreland fold and thrust belt, modified after Kazmi and Rana (1986). Inset shows the location of study area. *MKT* Main Karakorum Thrust, *MMT* Main Mantle Thrust, *MBT* Main Boundary Thrust, *NFTB* Northern Fold and Thrust Belt, *SFTB* Southern Fold and Thrust Belt, *MT* Mansehra Thrust, *PT* Panjal Thrust, *NT* Nathiagali Thrust, *HKS* Hazara Kashmir Syntaxis, *JF* Jhelum Fault, *KF* Kalabagh Fault, *SRT* Salt Range Thrust, *TIRT* Trans Indus Range Thrust, *MR* Marwat Range and *KR* Khissor Range

The *KFTB* has been explored by many researchers mainly because of its extensive salt, gypsum, oil and gas occurrences (Burnes 1832; Davies 1926; Gee 1945; Pinfold 1918; Raza and Khattak 1972). Fatmi (1973) carried out a detailed work to establish the lithostratigraphy of this geological province. In 1974, Meissner et al., mapped this region on a scale of 1: 50,000. Wells (1984) published his worked on the depositional environments of the Early Eocene strata of the *KFTB*. Yeats and Hussain (1987) worked on the timing of structural events in the Himalayan foothill of *NW* Pakistan.

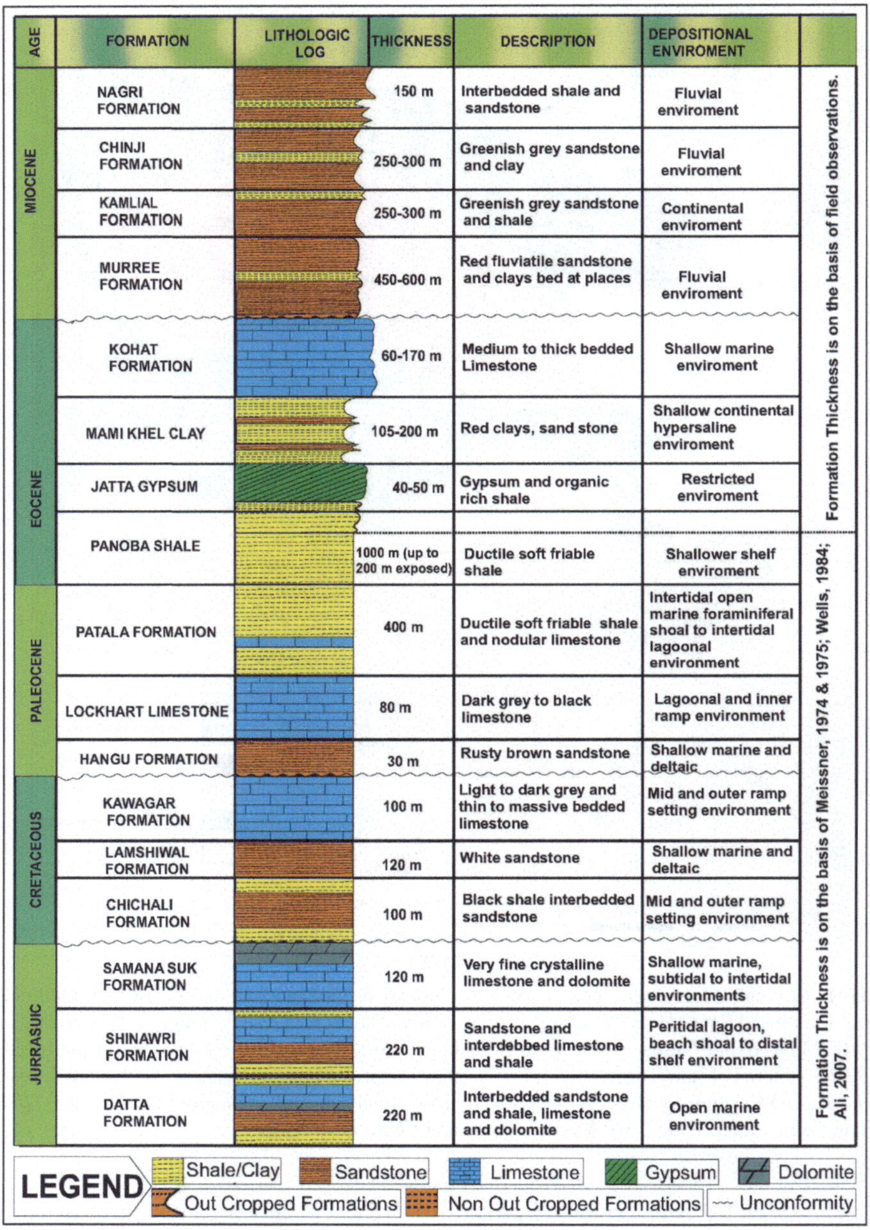

Fig. 2 Generalized Stratigraphic column of the *KFTB*, modified after Meissner (1974, 1975), Wells 1984

Abbasi and McElroy (1991) conducted the structural modelling of some parts of the Kohat region. McDougall and Hussain (1991) discussed the thrust kinematics of the region. Other workers described the *KFTB* as a complex, hybrid terrain consisting of wrench faults and compressional structures (Pivnik 1992; Pivnik and Sercombe 1993; Sercombe et al. 1994a, b). Saleemi and Ahmed (2000) have worked on the chemical and mineralogical composition of mudstone of the Karak area. Hussain et al. (2004) work on the geological map of the *NWFP*. Paracha (2004) published a research paper on the *KFTB* with reference to the Himalayan Tectonics. Gardezi et al. (2017) worked on the structural styles of western *KFTB*. They concluded that the area has undergone thin-skinned style of deformation which is the result of compressional stress generated form on going collision between Indian and Eurasian plates.

2 Tectonic Setting and Stratigraphy of the Study Area

The collision between the Indian and Eurasian plates have produced ~2500 km long Himalaya having *NW* to *SE* in India which switches to *NE* to *SW* orientation in Pakistan (Fig. 1). The Indo-Eurasian convergence is still sustained since Eocene at the rate of 5 mm per year due to which, uprising of Himalaya and establishment of a chain of foreland fold-and-thrust belts thrust over the Indian craton (Kemal 1991). The Himalayas in the north Pakistan are divided into five major tectono-stratigraphic terrains, separated by thrust system. The northern most fault of the thrust system is the Main Karakorum Thrust (*MKT*), which separates the Karakorum Block in the north from the Kohistan Island Arc in the south (Tahirkheli 1983). Coward et al. (1986) named it as the Northern Suture Zone that formed during Late Cretaceous. Due to collision and subsequent subduction of the Indian Plate underneath the Kohistan Island Arc (*KIA*), the Main Mantle Thrust (*MMT*) or the Indus Suture Zone formed (Tahirkheli et al. 1979; Tahirkheli 1982, 1983). The *MMT* and the *MKT* combinedly forms the Indus-Sangpo Suture (*ITS*) of the central Himalaya in the India and the Tibet.

The continent-arc-continent convergence that deformed the Indian crust remained continued with the formation of the *MMT* (Fig. 1). This convergence developed a regional foreland fold-and-thrust belt situated in south of the *MMT*. This foreland fold-and-thrust belt is further sub divided into a northern fold-and-thrust belt (*NFTB*) which is 260 km wide along *NS* and extends around 500 km along *EW*. This northern belt is comprised of severely deformed rocks of sedimentary, meta-sedimentary and igneous origin displaying festoon geometry. The *NFTB* is bounded by Kashmir Basin in the east and the Kurram area in the west. The *MBT* demarcates its southern boundary and separate it from the southern fold-and-thrust belt (*SFTB*). The *SFTB* is underlain by sedimentary rocks and because of the presence of petroleum system the area is very important for oil and gas exploration. The study area is located within the western premise of the *SFTB* i.e. in the central *KFTB*. This external part of the foreland basin is thrust southward over the undeformed sediments of the Punjab Foreland Basin along the *SRT*.

The *KFTB* and the adjacent areas represent a complete depositional record from Mesozoic up to Pleistocene (Fig. 2). The stratigraphic framework of the central *KFTB* demonstrates that until Middle Eocene in was a restricted sedimentary basin followed by the Late Eocene to Oligocene exposure marked by a regional unconformity. Since Miocene, it has become a syn-orogenic foreland basin filled with fluvial sediments mainly derived from north. The Eocene, Miocene and Pliocene rock units are exposed, which are mapped during the current field work (Fig. 4). The Pliocene rocks are not deposited throughout the entire mapped area but their presence in the extreme south shows that the foreland basin migrated southward due to the Miocene onward slip associated with the *MBT* (Burbank 1983).

The rock sequence exposed in the study area include Eocene age platform rocks unconformably overlain by Miocene to Mid Pleistocene molasse sequence (Fig. 2). The Early Eocene Panoba Shale is the oldest formation that is exposed in the core of the anticlinal folds. The study area shows a complete stratigraphic record from Eocene to Early Jurassic Datta Formation (Fig. 2). The stratigraphic description of the exposed Eocene to Miocene rock sequence is based on the field observations whereas the description of the non outcropping Paleocene to Jurassic rock sequence is based on the literature review (Meissner et al. 1974, 1975; Wells 1984).

3 Methodology

3.1 Reconnaissance and Travers Selection

Google Earth imagery (30 m resolution), already published literature and maps as well as Digital Elevation Model (DEM) (30 m resolution) were analyzed to understand structure, geology and drainage pattern of the area. Different systematic traverses were marked in the Google Earth through out the study area by following road network and drainage in the study area for detail geological mapping.

3.2 Field Data Collection

Different field trips were planned to collect geological and structural field data. This field data included determination of attitude of bedding planes and different structures (folds and faults) and marking different contacts along systematic traverses. Sketches of different faults and folds were drawn during field in order to visualize the true pictures of the outcrop. Photographs were taken of different faults, folds and formations during field work.

3.3 *Remote Sensing*

The field data was plotted on Google Earth to pick color tone of different geological contacts along the systematic traverses. That is how the normal as well as faulted contacts were extended between the systematic traverses.

3.4 *Surface Geological Map Construction*

Geographic Information System (*GIS*) software was used to construct a comprehensive geological map at scale 1:50,000 of the mapped coordinates by incorporating field data and Google Earth work.

3.5 *Structural Study*

Two different structural models were proposed by previous author to explain the structural style of the *KFTB*. One model considers the structures to be part of a passive thin-skinned fashion roof thrust translating southwards. It overlies a passive roof duplex formed by an active wedge of thrust slices directed southwards deforming Pre-Tertiary stratigraphy (McDougall and Hussain 1991; Abbasi and McElroy 1991). The second model suggests that the thick-skinned deformation has greatly influenced the structural evolution of the *KFTB*. This thick-skinned deformation is

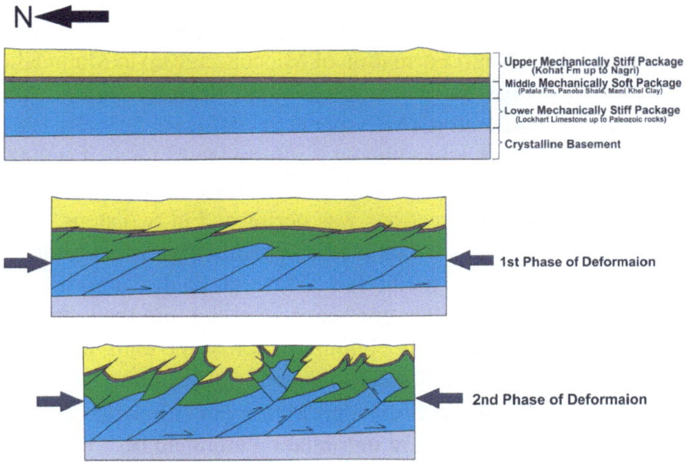

Fig. 3 Model showing structural evolution concept and mechanical stratigraphy of the *KFTB*

demonstrated in strike slip faulting recognizable only in the subsurface and exposing themselves in the Eocene cover as anticlinal trends (Pivnik 1992; Pivnik and Sercombe 1993; Sercombe et al. 1998). The proposed structural evolution model of the study area (Fig. 3) deduces that the subsurface structures do not mimic the surface structures. The stratigraphy of the study area is presumed in three different stratigraphic packages on the basis of stiffness i.e. the upper mechanically stiff package, the middle mechanically soft package and the lower mechanically stiff package (Fig. 3). This mechanical stratigraphy has a vital role in the structural architecture of the region.

3.6 Two and Three Dimensional Structural Modelling

Surface geology is always helpful for understanding the subsurface structural geometries especially in the areas where the surface structuration systematically passes down into subsurface. In the study area the stress vectors do not penetrate homogeneously into the depth (Fig. 3) which makes it tricky to interpret the subsurface using the surface data. But the relationship does exist between the surface and the subsurface structures, which once figure out, becomes the key while making admissible subsurface structural geometries. For two dimensional structural modelling, *2D* module of MOVE software was used. This modelling consisted of preparation of *2D* balanced geological cross sections (*XX′*, *YY′* and *ZZ′*) and their restored sections. For the construction of geological cross sections *DEM* (30 m resolution) and surface geological map were used.

The *DEM* is used to generate the topographic profiles along cross section lines. By utilizing apparent dips the contacts and faults along profile lines interpolated in subsurface using "construct horizons from template objects" tool of MOVE software. Horizon thicknesses from Nagri Formation of Miocene age up to Mami Khel Clay of Eocene age are according to field observations. While the rest of horizon thicknesses are according to previous literature (Meissner et al. 1974, 1975; Wells 1984). On the basis of previous data, the basal detachment is inferred to be located at 9000 m depth dipping towards north having slope of 1.6° in the study area (Sercombe et al. 1998). Restoration of geological cross sections was performed in two steps. The first step includes restoration along faults by sequentially removing the movements on each fault using "Fault Parallel Flow" method of "Move on Faults" algorithm. In "Move on Faults" algorithm different values for heave were fed into the software until the horizons were fully restored and were continuous across the fault. In the second step the beds are unfolded to their undeformed state using "Unfolding" method of "Flexural Slip" algorithm. In "Flexural Slip" algorithm a pin is inserted at the relatively less deformed area and horizons were unfolded on template bed. Restored sections were used to calculate the magnitude and percentage of shortening strains. For *3D* structural modelling, *3D* module of MOVE software was used. The *3D* structural modelling consist of *3D* fault and horizon models. By utilizing same horizon or fault lines in different *2D* sections using "Spline curves" method different *3D* surfaces were generated.

3.7 Stress Analysis

The "Geo-Mechanical" module of MOVE software was use to perform the stress analysis of the Sumari Payyan Anticline. This module works on the theory of the Mohr's stress circle diagram which is overlapped on Coulomb Failure Envelope. Plunge, plunge azimuth and stress amount (MPa) of different principal stress axis (σ_n) i.e. the maximum stress (σ_1), the minimum stress (σ_2) and the intermediate stress (σ_3) were calculated form *3D* surface of Sumari Payyan Anticline of existing model. These values along with *3D* surface used by the software to generated the final stress model. The software convert whole *3D* surface into large number of planes on the basis of different dip direction domains. Each separate plane is represented by a black dot which is mark on the centre of the plane. The software calculate the stress analysis of each plane using the input values. The stress data of each point was plotted on the Mohr's stress circle diagram in the software. The Coulomb Failure Envelope was plotted over the Mohr's stress circle diagram and they intersect each other. The intersected region is marked by the red color. The points which fall within the red color region of Mohr's stress circle diagram are the plane where the failure can occur.

4 Results

4.1 Surface Geology

It has been found that the structuration in the study area is mainly controlled by east west trending open to tight normal and overturned folds. Major folds in the area having curvilinear axis. The faults is the area are thrust and back thrust in nature. The vergence of major structures is towards south (Fig. 4). Dominated dip directions in the area is in north and south.

4.2 Two-Dimensional Structural Modelling

On the basis of structural evolution model of the study area (Fig. 3) three structural cross sections along lines *XX′*, *YY′* and *ZZ′* (Figs. 6, 7, 8) are selected which are approximately perpendicular to the major structural trend in the study area (Fig. 4).

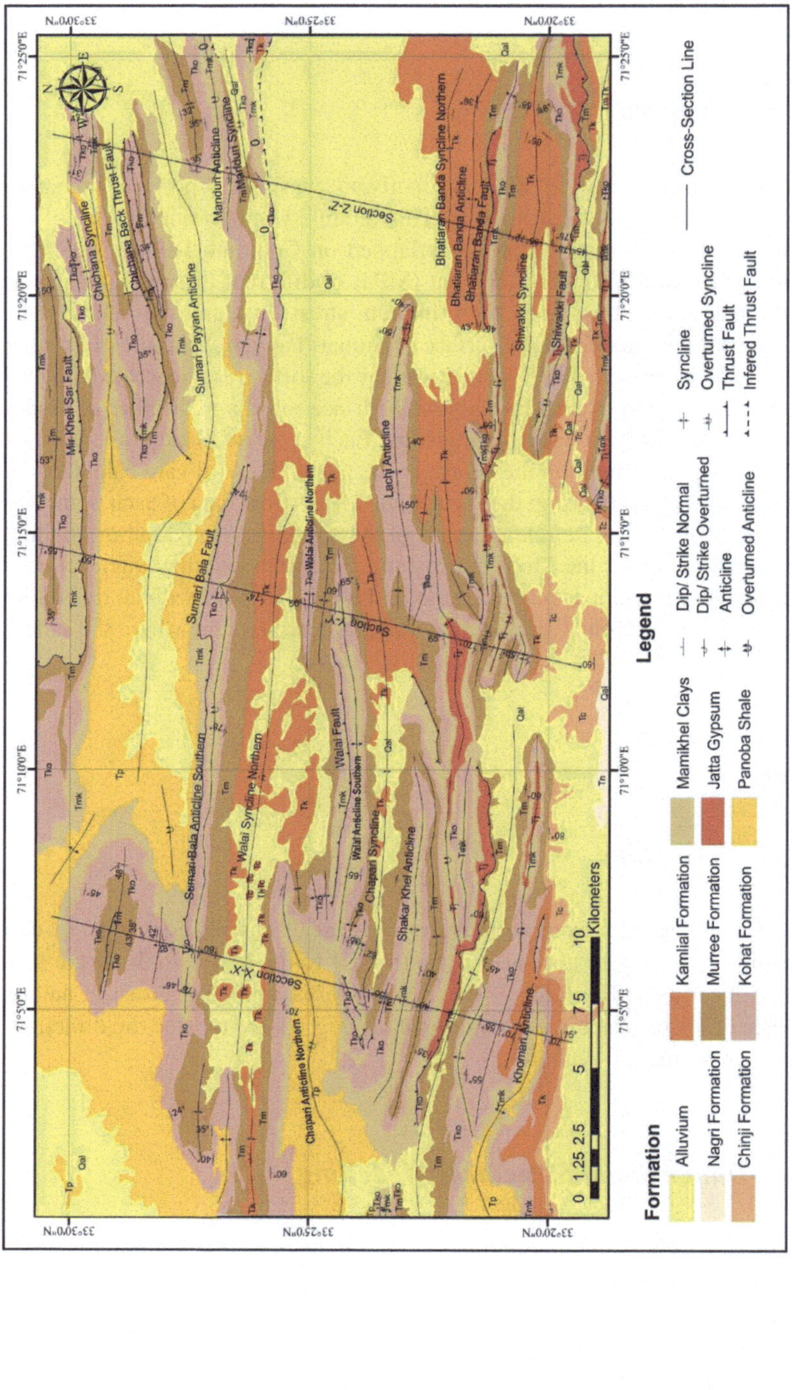

Fig. 4 Geological map of the study area

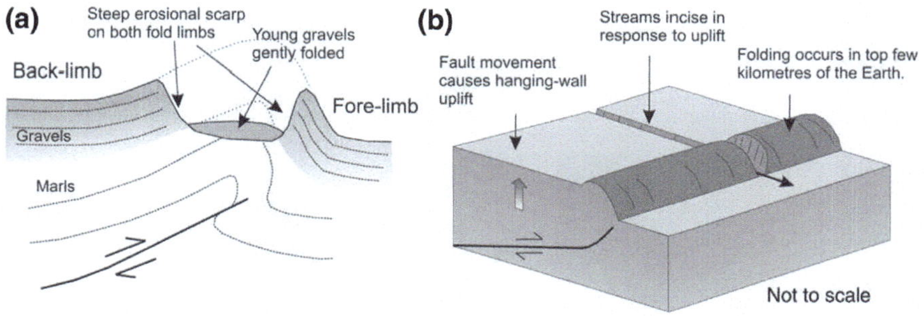

Fig. 5 **a** Schematic cross-section and **b** block diagram of a 'blind' thrust fault (Stein and Yeats 1989)

The east west trending flat-topped ridges and valleys are commonly interpreted to be alternating low-amplitude anticlines and synclines in the study area. Anticlines are mostly uplifted by sub-cropping blind faults, the slip on which is in most of the cases accommodated by the tight folding in the upper stiffer package (Stein and Yeats 1989) (Fig. 5).

Three cross sections (Figs. 6, 7 and 8) were constructed for the better understanding of subsurface structural geometries, kinematics, and behavior of the various stratigraphic units across different structures of the study area. The structural model constructed along the lines XX', YY' and ZZ' suggests that within the Eocene carapace the disharmonic folds of varying amplitudes dominate the subsurface structural geometry of the area (Figs. 6, 7 and 8). As the Panoba Shale occupies the base of the Eocene, its plastic nature has resulted in thickening of the anticlinal cores. Maximum folds are overturned and south verging. The cross sections suggest that the dominant structures style within the area is comprised of a series of south verging alternate anticlines and synclines.

Restoration and balancing of cross sections estimate the amount and percentage of shortening along the lines XX', YY' and ZZ' of lengths 20,482.3 m, 21,321.4 m and 21,810.8 m, respectively (Figs. 6b, 7b, 8b). The total amount of shortening occurred along folds and faults of the cross section XX' is 18,383.6 m so the percentage of shortening is 47.3%. The shortening amount occurred along folds and faults of the cross section YY' is 15,473.1 m with calculated percentage of shortening is 42.05%. The total amount of shortening occurred along folds and faults of the cross section ZZ' is 13,197.5 m with calculated percentage of shortening is 37.69%. Average amount and percentage of shortening along the cross sections XX', YY' and ZZ' are 15,685.07 and 42.35%, respectively. The percentage of shorting along cross sections XX', YY' and ZZ' indicated that in that area the rate of deformation increases from east to west in the study area.

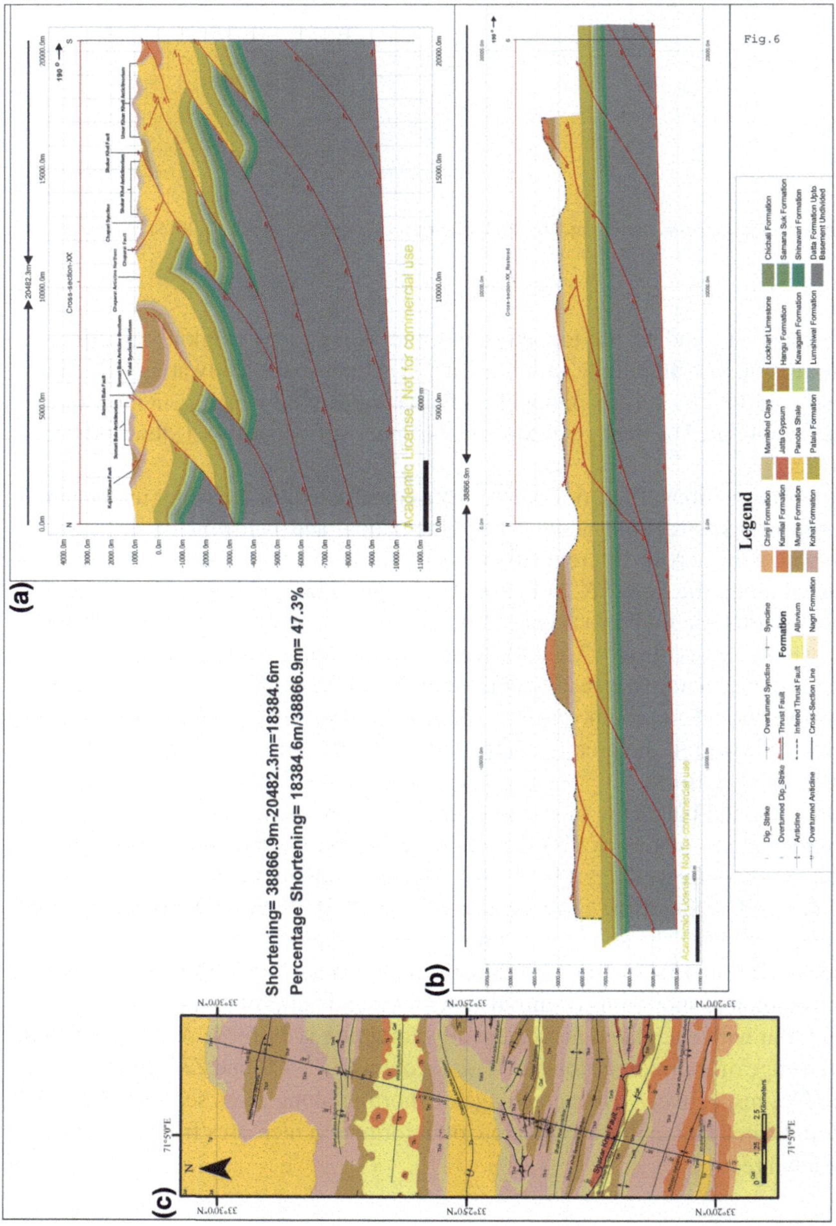

Fig. 6 **a** Structural interpretation before restoration, **b** final stage of restoration and **c** base map of cross-section XX′

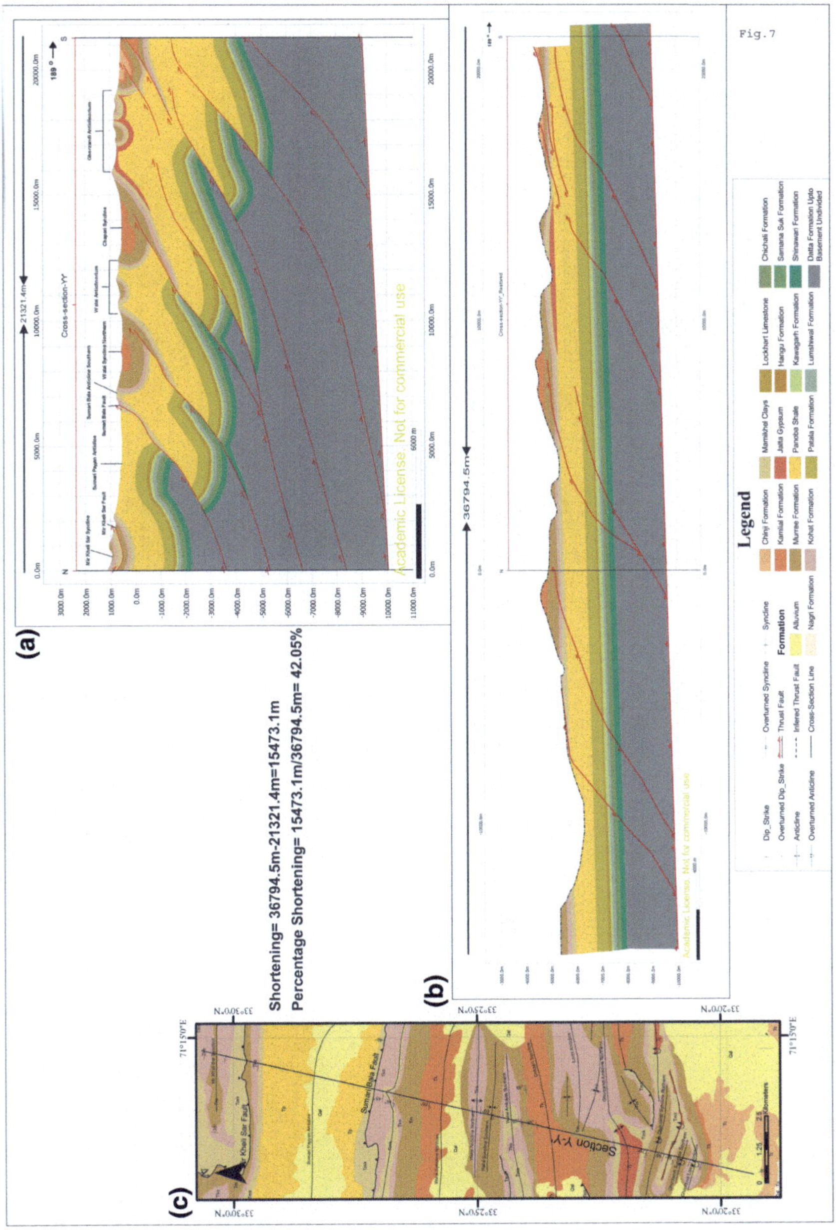

Fig. 7 **a** Structural interpretation before restoration, **b** final stage of restoration and **c** base map of cross-section YY′

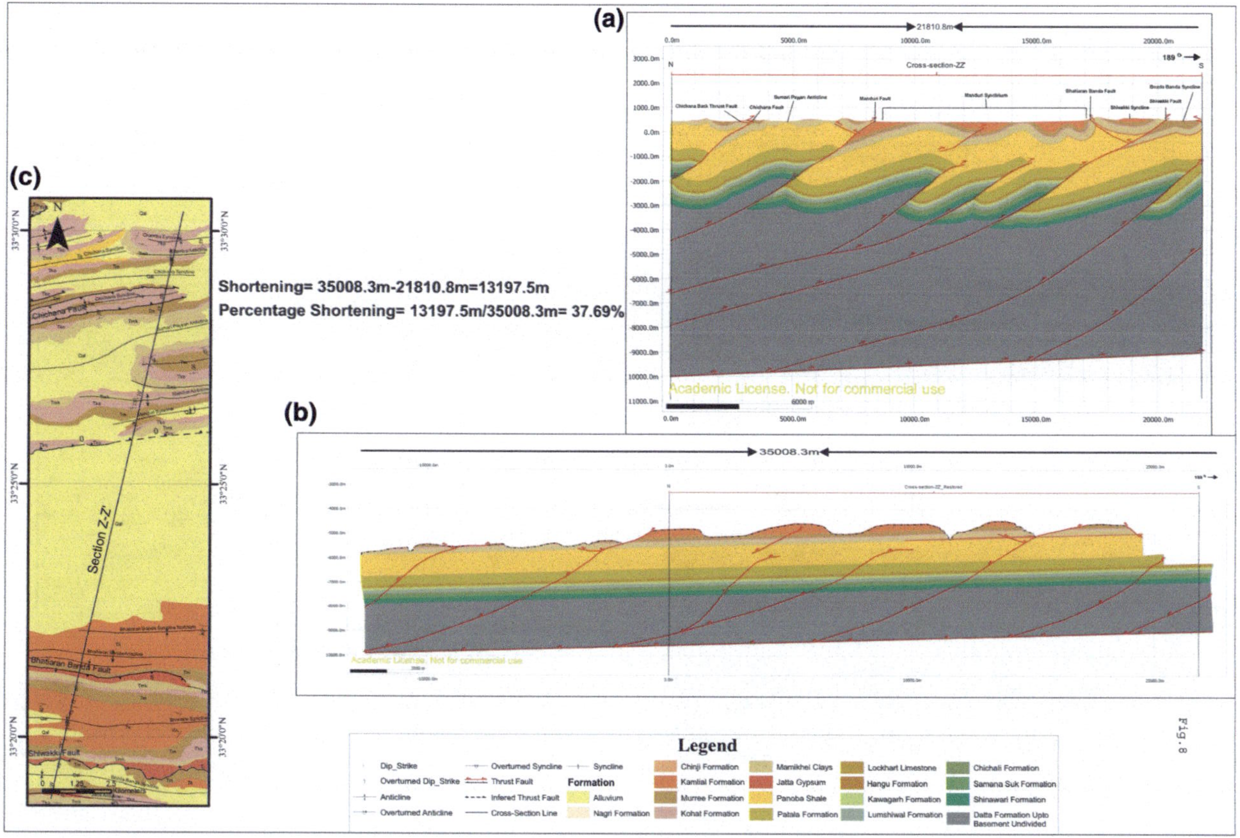

Fig. 8 **a** Structural interpretation before restoration, **b** final stage of restoration and **c** base map of cross-section ZZ′

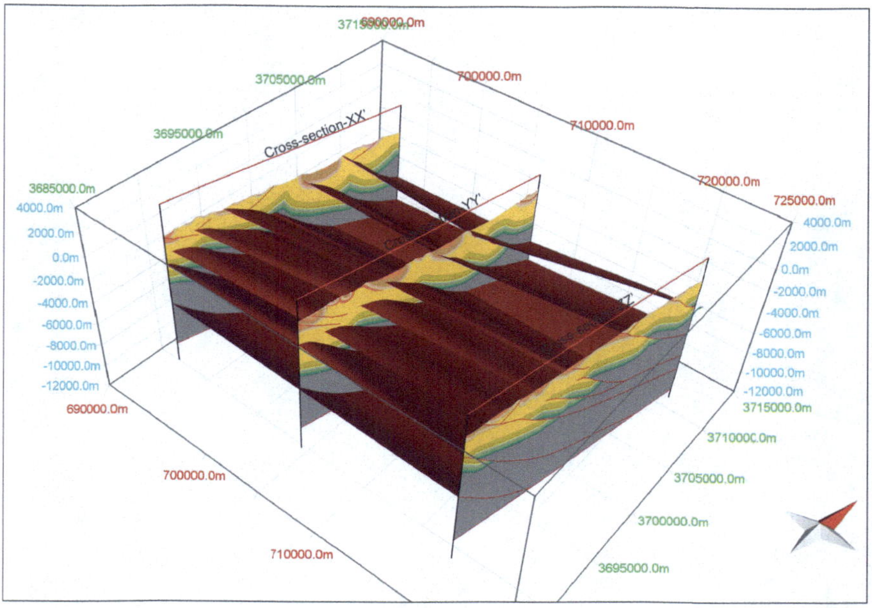

Fig. 9 Image showing south eastern view of *3D* fault model of the study area

4.3 Three-Dimensional Structural Modelling

Total, 21 fault planes from *2D* sections were used to generate *3D* fault model of the area (Fig. 9). The model consist of seven fault surfaces which shows trend and behaviour of different faults in subsurface. This shows that the major faults in the area are east west trending and south verging. The *3D* fault model shows that the fault system in the study area is the thrust related faults. The strike direction of the fault system is ENE (N80°E) with the dip angle varying 35°–40°. The majority of the faults have seen as "blind faults" and only a few faults cuts up section at the surface in the entire model.

The *3D* horizon model of top of the Lockhart Limestone (Paleocene) was generated according to the developed fault model (Fig. 10). The model shows the relation between four major anticlinal folds with intervening depressions and fault system. These anticlines can be suitable traps for hydrocarbon accumulation. The *3D* horizon model also describes the petroleum play concept of the area where fault plane can act as the path for primary migration of hydrocarbon from source rock to reservoir rock.

The *3D* model shows majority of structural trend is *EW* and some with minor tilt. This structuration is the result of southward propagation of deformation as a result of *NS* compressions and localize lateral forces due to Indo-Eurasian plates collision.

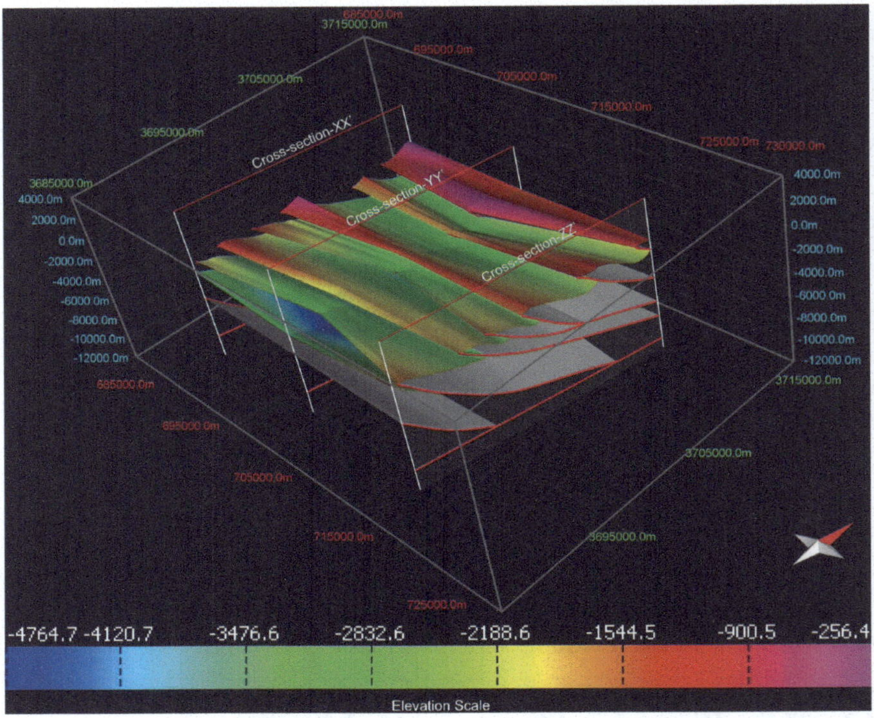

Fig. 10 Image showing south eastern view of *3D* horizon model at the top of the Lockhart Limestone

4.4 Stress Analysis of the Area

Stress analysis of a *3D* surface of the Sumari Payyan Anticline (Fig. 11) at a level of the Lockhart Limestone was performed (Fig. 12). This structure covers an area of about 30 km^2 and it rest on the Sumari Bala Fault at the depth of 1700 m. There are

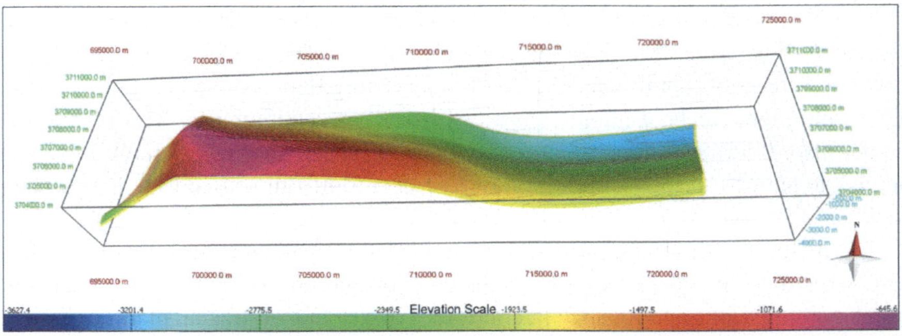

Fig. 11 Image showing *3D* surface of Sumari Payyan Anticline at the top of the Lockhart Limestone

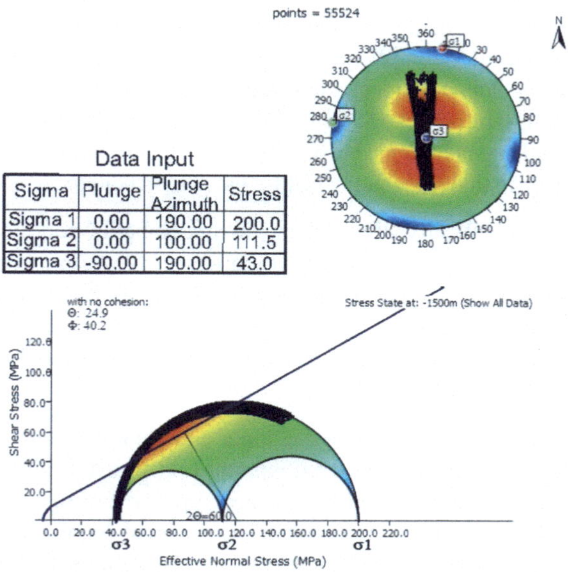

Fig. 12 Stress analysis of the Sumari Payyan Anticline at the top of Lockhart Limestone

55,524 different points selected by the software on the basis of different dip direction domain of planes of *3D* surface. Total 22,209 points lies in the intersected red color region of the Mohr's stress circle diagram and the Coulomb Failure Envelope. This means that the stress analysis of the Sumari Payyan Anticline at a level of the Lockhart Limestone shows that total 40% of plane are susceptible to failure.

5 Discussion

In this study, field and remote sensing data sets are utilized for detailed structural analysis of the central *KFTB*. Different softwares e.g. Google Earth, *GIS* and MOVE (*2D* and *3D)* for the construction of geological map and structural cross sections, *3D* models as well as to perform stress analysis of the area. The study area lies in the western part i.e. *KFTB* of the Upper Indus Basin. The *KFTB* lies of the north western margin of Himalayan arc in the Pakistan. The rock sequence that is exposed in the study area include Eocene age platform rocks unconformably overlain by molasses sediments of Miocene to Plio-Pleistocene age. The structural geometry is well developed in the area as a result of southward propagating deformation due to Indo-Eurasian collision.

The structural fabric is dominated by east west trending open to tight folds and southward verging thrust faults system as well as back thrust faults. Our *2D* and *3D* structural modelling deduced that there is disharmonic relation between surface and subsurface structures. Thrusting related to decollement tips within Paleocene and older rocks. This folding is complicated further by shale diapirism. Based on the relative stiffness and mechanical aspect, the stratigraphy of the study area is presumed in three different stratigraphic packages (Fig. 3). This mechanical stratigraphic arrangement and thick shale sequence have a vital role in the structural architecture of the region. These three distinct mechanical lithic packages include:

(i) Uppermost mechanically stiff package having Kohat Formation and overlying molasse sediments up to the Nagri Formation.
(ii) Middle mechanically soft package consisting of Patala Formation, Panoba Shale, and Mami Khel Clay, and
(iii) Lower mechanically stiff package comprising of Lockhart Limestone up to Paleozoic rocks.

The structural disharmony between the upper and the lower units is mostly credited to the middle soft package. The predominant fold-thrust system verging southward at the surface is characterized by the structures within the upper unit. Shale diapirism has modified this fold thrust system. These local decollements provided by the soft mechanical strata at shallow depth are greatly responsible for the structural disharmony between the underlying and overlying strata. Although, overall structural trend is east west which is the result of regional north south stresses but because of en echelon spread of the folds (Fig. 13) it can be inferred that letteral component of stress also present in the area. The Sumari Payyan Anticline, Walai Syncline Northern, Walai Anticline Northern, Walai Anticline Southern, Chapari Anticline, Chapari Syncline, Shakar Khel Anticline, Shakar Khel Syncline, Khomari Syncline and Khomari Anticline plunged eastward in an en echelon fashion. This shows that right handed letteral movement acting over here along line AA'. Same type of movement can be inferred along line BB′ which changed the trend of Chapari Anticline toward *NE-SW*. This proves that the structuration in the *KFTB* is the result of both north south compression stresses and the letteral stresses components.

Many authors worked on the structure geology of the *KFTB* but their work was on the regional scale also they not incorporate the role of the mechanical stratigraphy in their analysis (Meissner et al. 1974; Yeats and Hussain 1987; Abbasi and McElroy 1991; McDougall and Hussain 1991; Hussain et al. 2004; Paracha 2004). Previous authors did not use advance softwares for structural modelling and no one perform the *3D* structural modelling of the area.

6 Conclusions

- Three different stratigraphic packages based on their mechanical response to deformation are presumed to control the structural disharmony within the central *KFTB*. The uppermost package is mechanically stiff and is comprised of Kohat Formation (Eocene) up to Siwalik Group (Miocene) which is underlain by a middle, mechanically soft mobile package of Mami Khel Clay, Jatta Gypsum and Panoba

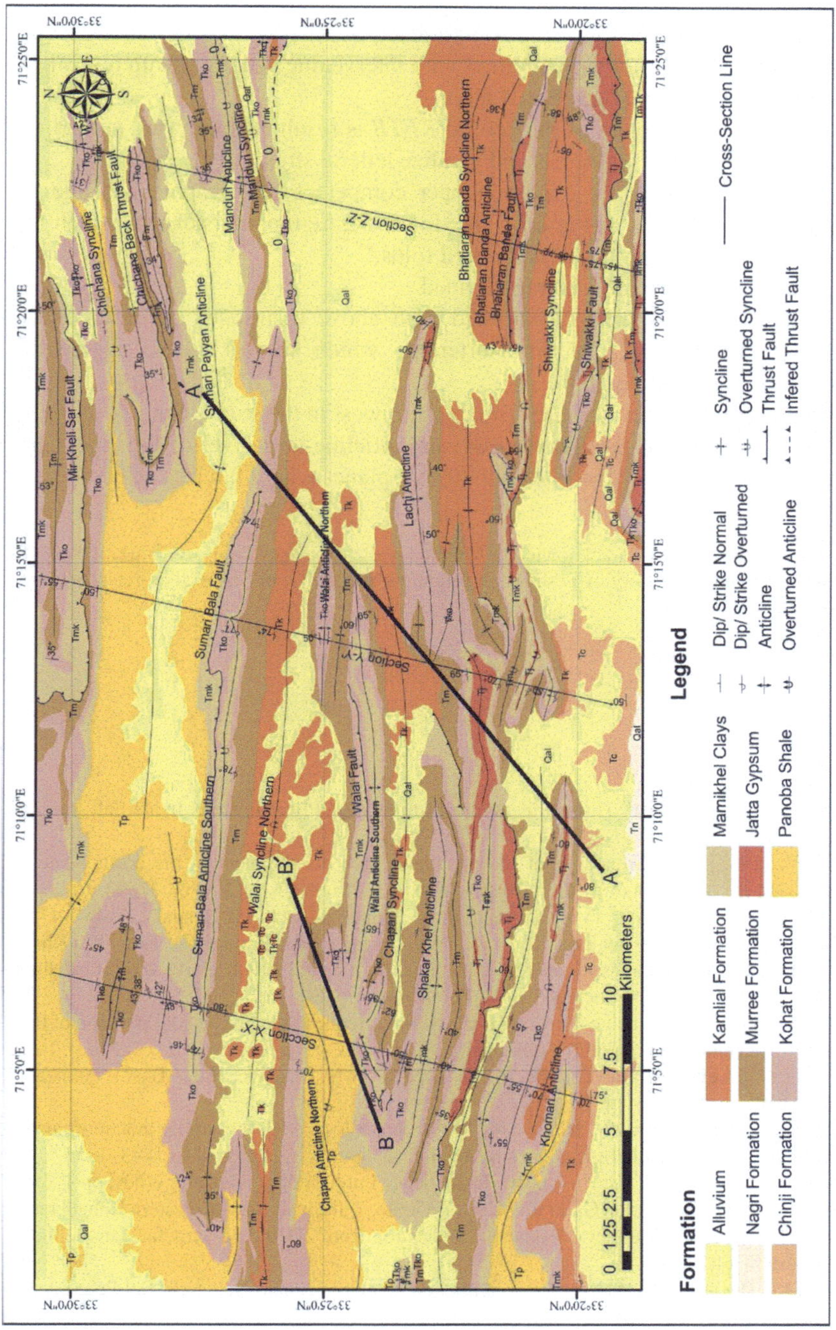

Fig. 13 Geological map of the study area showing concept of letteral stresses acting along lines AA′ and BB′

Shale (Eocene) and is underlain by a lower stiff package of Lockhart Limestone (Paleocene) and older rock units.

- The structural geology of the area is dominated by east west (with minor exceptions) trending, south verging anticlines with intervening synclines with few blind and emergent thrust and back thrust faults.
- The structural architecture of the central *KFTB* is result of southward spread of north south compressional stress and localize letteral stress.
- Structural style within the exposed upper competent stratigraphic package is disharmonic and does not mimic the lower stratigraphic package which is comprised of south verging, ramp-related folds.
- The deformation in the area is thin-skinned.
- Calculated average shorting associated with above mentioned deformation is 15,685.07 km with 42.35% of shortening, which shows intense degree of deformation and orogeny in the area.
- The rate of deformation increases from east to west in the study area.
- The stress analysis of the Sumari Payyan Anticline at a level of the Lockhart Limestone demonstrates that the total 40% of planes where failure can occur.

Acknowledgements I wish to acknowledge Midland Valley for providing academic license of MOVE software and University of Peshawar for funding the field work as well as for providing great research environment.

References

Abbasi IA, McElroy R (1991) Thrust kinematics of the Kohat Plateau, Trans Indus Salt Range, Pakistan. J Struc Geol 13:319–327

Burbank DW (1983) The chronology of intermontane basin development in the north-western Himalaya and the evolution of the north-west syntaxis. Earth Planet Sci Lett 64:77–92

Burnes A (1832) Some account of the salt mines of the Punjab. Asiatic Soc Bengal J 1:145–147

Coward MP, Windley BF, Broughton RD, Luff IW, Petterson MG, Pudsey CJ, Rex DC, Khan MA (1986) Collision tectonics in the NW Himalaya. In: Coward MP, Ries AC (eds) Collision tectonics, vol 19. Geological Society Special Publication, pp 201–219

Davies LM (1926) Notes on the geology of Kohat, with reference to the homotaxial position of the salt marl at Bahadur Khel. Asiatic Soc Bengal J 20:207–224

Fatmi AN (1973) Lithostratigraphic units of the Kohat-Potwar Province, Indus Basin, Pakistan. Pakistan Geological Survey, pp 10–80

Fraser JE, Searle MP, Parrish RR, Noble SR (2001) Chronology of deformation, metamorphism and magmatism in the southern Karakorum Mountains. Geol Soc America 113:1443–1455

Gansser A (1964) Geology of the Himalayas. Interscience Publishers London etc. Wiley, pp 1–289

Gardezi SAH, Ahmad S, Rehman G, Ikram N (2017) Geological constraints on western Kohat Foreland Basin, Khyber Pakhtunkhwa, Pakistan: implication from 2D and 3D structural modelling. Arab J Earth Sci 4(2):95–116

Gee ER (1945) The age of the Saline Series of the Punjab and of Kohat. Natl Acad Sci Proc Sec B 14(6):269–310

Hussain A, Dipietro JA, Pogue KR (2004) Geological map of the 43/8-degree sheet, NWFP, Pakistan Geological Survey of Pakistan. Geological map No. 11

Kazmi AH, Rana RA (1886) Tectonic map of Pakistan, 1:1,000,000. Geological Survey of Pakistan Quetta, Pakistan

Kemal A (1991) Geology and new trends for hydrocarbon exploration in Pakistan. In: Ahmed G, Kemal A, Zaman ASH, Humayon M (eds) New directions and strategies for accelerating petroleum exploration and production in Pakistan: Proceedings, international petroleum seminar, November 22-24: Islamabad. Ministry of Petroleum and Natural Resources, Pakistan, pp 16–57

Khan MA, Ahmed A, Raza HA, Kemal A (1986) Geology of petroleum in Kohat-Potwar Depression. Pakistan. American Association of Petroleum Geoscientists 70(4):396–414

LeFort P (1975) Himalaya: the collided range, Present knowledge of the continental arc. Am J Sci 275A:1–44

McDougall, Hussain A (1991) Fold and Thrust propagation in the western Himalaya based on a balanced cross-section of the Surghar Range and Kohat Plateau, Pakistan. AAPG Bull 75:462–476

Meissner CR, Master JM, Rashid MA, Hussain M (1974) Stratigraphy of the Kohat Quadrangle, Pakistan. United States Geological Survey, Professional Paper: vol 716 D, pp 30–36

Meissner CR, Hussain M, Rashid MA, Sethi UB (1975) Geology of the Parachinar quadrangle, Pakistan. US Geol Surv Prof Pap 76F:19–26

Molnar P, Tapponnier P (1975) Cenozoic tectonics of Asia: Effects of a continental collision, New York. Science 189:419–426

Paracha W (2004) Kohat Plateau with Reference to Himalayan Tectonic General Study. Can Soc Expl Geophysicist Conf, pp 46–51

Pinfold ES (1918) Notes on structure and stratigraphy in the north-west Punjab: India. Geol Survey Recs 49:137–160

Pivnik DA (1992) Depositional response to encroachment of Himalayan compressional and transpressional deformation on the north Pakistan foreland (Ph.D. thesis). Hanover, New Hampshire, Dartmouth College, pp 1–258

Pivnik DA, Sercombe WT (1993) Compression and transpression-related deformation in the Kohat Plateau, NWFP, Pakistan. In: Treloar PT, Searle, MP (eds) Himalayan tectonics, vol 74. Geol Soc Lond, Special Publ., pp 559–580

Raza SQ, Khattak AK (1972) Gypsum deposits of Kohat District N.W.F.P, West Pakistan

Saleemi AA, Ahmed Z (2000) Mineral and chemical composition of Karak mudstone, Kohat Plateau, Pakistan: implications for smectite-illitization and provenance. Sed Geol 130:229–247

Sameeni SJ, Haneef M, Rehman O, Lipps JH (2009) Paleogene biostratigraphy of Kohat area, Northern Pakistan. Geol Bull Punjab Univ 44:27–42

Sercombe WJ, Pivnik DA, Stratton MA, Albertin M, Beck RA, Wilson WP, Roth BL (1994a) Wrench faulting in the northern Pakistan foreland region. The Leading Edge 13:1107–1110

Sercombe WJ, Pivinik DA, Stratton MA, Albertin M, Beck RA, Wilson WP, Roth BL, Nieuwenhuise RE (1994b) Wrench faulting in the northern Pakistan foreland region: rethinking a deformed belt (abs.). Am Assoc Petrol Geol Bull 3:256

Sercombe WJ, Pivnik DA, Wilson WP, Albertin ML, Beck RA, Stratton MA (1998) Wrench Faulting in the northern Pakistan foreland. AAPG Bull 82:2003–2030

Shah SMI (2009) Stratigraphy of Pakistan. Geol Surv Memoir 22:381

Stein RS, Yeats RS (1989) Hidden earthquakes. Sci Am, pp 48–57

Tahirkheli RAK, Mattaure M, Proust F, Tapponnier P (1979) The India-Eurasia suture in northern Pakistan: synthesis and interpretation of data on plate scale. In: Farah A, DeJong KA (eds) Geodynamics of Pakistan. Geol Surv Pakistan, pp. 125–130

Tahirkheli RAK (1982) Geology of Himalaya, Karakorum and Hindukush in Pakistan. Geol Bull Univ Peshawar 15:1–15

Tahirkheli RAK (1983) Geological evolution of Kohistan Island arc on the southern flank of Karakorum-Hindukush in Pakistan. Bull Geofis Teorica Appl 25(99–100):351–364

Wells NA (1984) Marine and continental sedimentation in the early Cenozoic Kohat basin and adjacent north-western Indo-Pakistan. Unpublished Ph.D. Dissertation. University of Michigan

Yeats RS, Hussain A (1987) Timing of structural events in the Himalayan foothills of north-western Pakistan. Geol Soc Am 99:177–198

The Northern Margin of the Eastern Ghats Mobile Belt: Evidence for Strike-Slip Tectonics Along a Craton-Mobile Belt Boundary

Saibal Gupta, Ritabrata Dobe, Amol Dayanand Sawant, Surajit Misra and William Kumar Mohanty

Abstract The Eastern Ghats Mobile Belt (*EGMB*) lies to the east of the Archaean Bastar and Dharwar Cratons, and to the southeast of the Singbhum Craton in the Indian shield. Along its western boundary, the *EGMB* granulites have been thrust westward as a nappe over the Bastar Craton along a mylonitic contact zone. Earlier studies considered the northern boundary of the *EGMB* with the Singhbhum Craton to be a thrust, although this interface is geometrically parallel to the west-directed transport direction of the granulitic nappe. Detailed geological studies along this northern margin reveal that the c. 1.0 Ga granulites of the *EGMB* do not share a direct contact with the Archaean granite-greenstone terrane of Singhbhum, but are actually juxtaposed against a Late Archaean (2.8–2.5 Ga) high grade terrane referred to as the Rengali Province. Structural studies reveal that the *EGMB*-Rengali Province contact has a WNW-ESE strike with sub-vertical dip, with prominent asymmetric markers indicating dextral strike-slip shearing along a horizontal transport vector. Microstructural studies indicate that fabric formation during strike-slip deformation is controlled by plastic deformation of quartz, while other minerals remained passive or deformed in a brittle manner. Electron Back-Scatter Diffraction (*EBSD*) studies on selected samples from the contact zone indicate that quartz deformed mostly by prism <a>, rhomb <a> and basal <a> slip, with asymmetry indicating dextral simple shearing. The Rengali Province samples show that earlier shortening (pure shear) microstructures and quartz CPO patterns were also sheared dextrally by this later deformation. These results confirm that the northern boundary of the *EGMB* is not a thrust, but a strike-slip shear zone that operated well after granulite metamorphism, and at lower temperatures characteristic of the greenschist facies. Shortening structures in the Rengali Province and the Singhbhum Craton are related to an older deformation event unrelated to emplacement of the *EGMB*. Gravity studies across the contact confirm that the shear zone continues vertically to a depth of at least 25 km. Thus, integrated geological and geophysical studies confirm the strike-slip nature of the craton-mobile belt boundary.

S. Gupta (✉) · R. Dobe · A. D. Sawant · W. K. Mohanty
Department of Geology and Geophysics, Indian Institute of Technology Kharagpur, Kharagpur, West Bengal 721302, India

S. Misra
Department of Geological Sciences, Gauhati University, Guwahati, Assam 781014, India

T. K. Biswal et al. (eds.), *Structural Geometry of Mobile Belts of the Indian Subcontinent*, Society of Earth Scientists Series,
https://doi.org/10.1007/978-3-030-40593-9_7

Keywords Eastern Ghats Mobile Belt · Rengali Province · Northern boundary of *EGMB* · Strike-slip deformation · Field and microstructural studies · Electron back-scatter diffraction (*EBSD*)

1 Introduction

Craton-mobile belt boundaries are intrinsic components of Precambrian shields in different parts of the world and have been the subject of significant research (e.g. Katz 1985; Khoza et al. 2013). A typical shield comprises of an Archaean nucleus bordered by a Proterozoic high-strain mobile belt (Black and Liegeois 1993). The mobile belt is either assumed to be a reworked part of the adjoining craton (e.g. Rivers et al. 1989), or part of a separate craton/continent that was amalgamated following an ancient collision event. In peninsular India, the Eastern Ghats Mobile Belt (*EGMB*) lies along the southern and eastern boundaries of the Singhbhum and Bastar-Dharwar Cratons, respectively. The *EGMB* was part of an East Antarctica continent that collided with the amalgamated cratons of peninusular India during assembly of the supercontinent Rodinia, around ~1000 Ma (see Gupta 2012; Dasgupta et al. 2013 for reviews). The contact of the *EGMB* with the bounding cratons is therefore of great interest, as it should represent a Precambrian suture zone. Moreover, the tectonics associated with this contact would potentially give an insight into the nature of Precambrian plate tectonics on continents, and the processes of terrane amalgamation in the past.

The EGMB extends inland from the northeastern coastal region of the Indian peninsula (Fig. 1). North of the Godavari rift, the EGMB is flanked to its west by the Bastar Craton; south of the rift, the Dharwar Craton defines its western boundary. For the most part, the boundary trends NE-SW to N-S, but along the northern boundary, the trend swings into a WNW-ESE orientation. Geological studies are in general agreement that along the western boundary, granulites of the *EGMB* have been thrust westward over the craton along a mylonitized zone, variously referred to as the Terrane Boundary Shear Zone or Lakhna Shear Zone, in a style resembling a modern fold-thrust belt (Biswal et al. 2000; Biswal and Sinha 2003; Bhadra et al. 2004; Biswal et al. 2007). The northern boundary of the *EGMB* was considered to be the Sukinda Thrust, along which granulites were thrust northward onto the Singhbhum Craton; this contact was also initially interpreted as a suture zone (Banerjee et al. 1987; Banerjee 1997). However, structural evidence for the presence of a thrust along the northern boundary was equivocal (e.g. Bhattacharya 1997; Ghosh et al. 2010), in part owing to the poor outcrop in much of the area. All along the contact zone, however, there initially appeared to be a general assumption by most workers that the *EGMB* had been thrust onto the Dharwar, Bastar and Singhbhum Cratons.

With further structural, metamorphic and geochronological studies along the better exposed western boundary of the EGMB with the Bastar Craton (e.g. Gupta et al. 2000; Bhadra et al. 2004; Biswal et al. 2007), it appeared clear that a thrust nappe of granulites had been thrust westward over the Bastar Craton, in all probability around 500 Ma. However, this implied a major contradiction about the nature of the

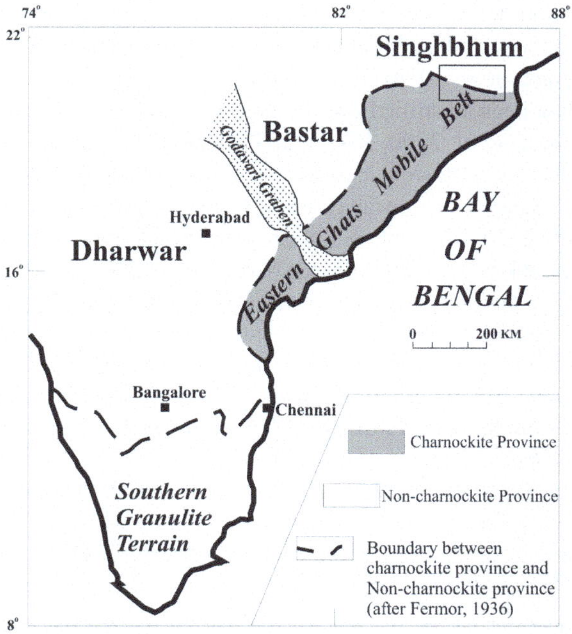

Fig. 1 Geological map of Peninsular India, showing the location of the *EGMB* with respect the Dharwar, Bastar and Singhbhum Cratons. The Southern Granulite Terrain is another granulite terrane distinct from the *EGMB* that lies to the south of the Dharwar Craton. The part discussed in the present study is marked with a rectangle

northern boundary of the *EGMB*—if the thrust was directed westward, the nature of the WNW-ESE trending northern boundary would be geometrically incompatible with a thrust, and is more likely to be strike-slip in nature. However, none of the earlier studies on the northern boundary domain considered this possibility. In the latter part of the last century, Nash et al. (1996) first suggested that the northern boundary could be strike-slip in nature based on satellite imagery data, although the idea received little support. Subsequently, Misra and Gupta (2014) and Sawant et al. (2017) presented field and microstructural data that corroborated the satellite observations. In this study, we integrate the evidence acquired by our group through a series of structural, metamorphic, geochronological and geophysical (primarily gravity) studies, and together with additional information based on ongoing studies, to demonstrate that the northern boundary of the *EGMB* is primarily strike-slip in nature, overprinting earlier, unrelated episodes of shortening.

2 Geological Setting

The tectonic history of the *EGMB* is complex and was affected by episodes of supercontinent assembly and dispersal (Upadhyay and Raith 2006; Chetty 2010; Mahapatro et al. 2012; Dasgupta et al. 2013; Pant and Dasgupta 2017; Sawant et al. 2017). The EGMB was contiguous with the Rayner Province of East Antarctica, and collided with proto-India in a late Mesoproterozoic to Neoproterozoic (~1000 Ma) collisional event during which the Bastar and Dharwar Cratons acted as the collisional fronts to the west and the Singhbhum Craton to the north (Biswal et al. 2007; Gupta 2012; Dasgupta et al. 2013). While the precise location of the western margin of the EGMB has been identified and its nature relatively well constrained (Biswal et al. 2000; Gupta et al. 2000; Biswal and Sinha 2003; Bhadra et al. 2004; Biswal et al. 2007), there is confusion about both the location and the nature of the northern margin. Part of the confusion was caused by the poorly exposed nature of the rocks to the north of the EGMB, but also because of the identification of a separate geological entity called the Rengali Province (Crowe et al. 2003) sandwiched between the Singhbhum Craton in the north and the *EGMB* in the south (Fig. 2). The Rengali Province has a distinct lithological set-up, predominantly metamorphosed in the amphibolite facies, distinct from the greenschist facies metamorphism in the Singhbhum Craton to the north and the granulites of the EGMB to the south, and is of late Archaean age (see Dobmeier and Raith 2003). This implies that along its northern margin, the *EGMB* is in direct contact with the Rengali Province, while the Rengali Province may have a separate contact further to the north with the Singhbhum Craton sensu stricto. This study focuses on the location and nature of the EGMB-Rengali Province contact, and does not discuss the contact between the Rengali Province and the Singhbhum

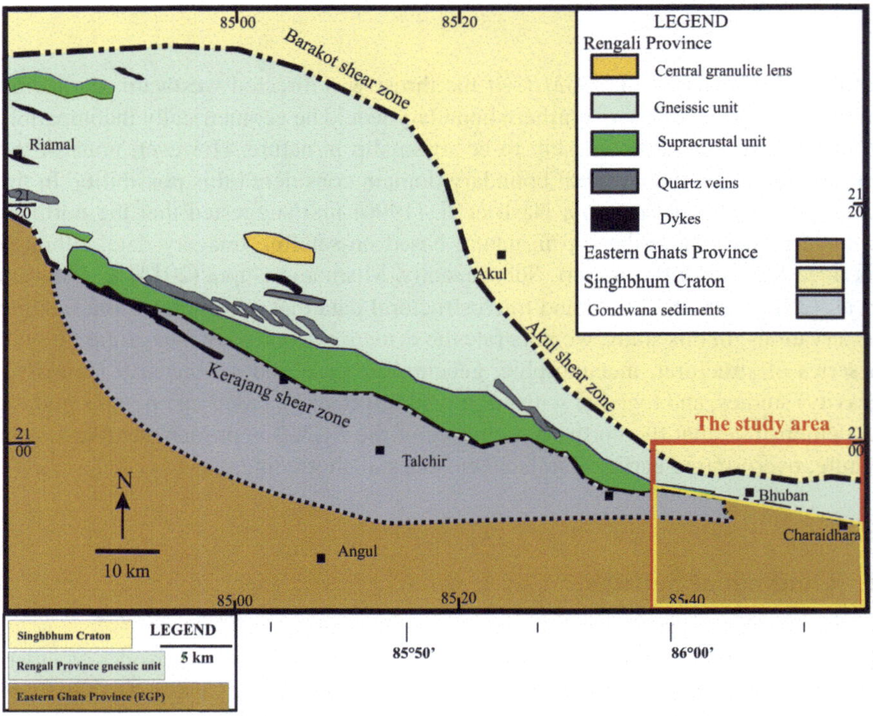

Fig. 2 Geological map of the northern margin of the *EGMB*, modified after Misra and Gupta (2014). The study area around the contact zone has been delineated using a red rectangle

Craton. Geological information on the EGMB-and the Rengali Province, and parts of the Singhbhum Craton relevant to this study, are summarized below.

2.1 Eastern Ghats Mobile Belt

The *EGMB* played a prominent role in the assembly and dispersal of the supercontinents Columbia and Rodinia (Zhao et al. 2004; Li et al. 2008; Dasgupta et al. 2013; Harley et al. 2013; Sawant et al. 2017; Ranjan et al. 2018). Based on the differences in geological history and evolution, Dobmeier and Raith (2003) classified the *EGMB* into a number of provinces and domains. The part of the *EGMB* north of the Godavari rift, referred to as the Eastern Ghats Province, records ultrahigh temperature metamorphism (M1) of dominantly Neoproterozoic age (~1.0 Ga) which is correlated with collisional orogeny during Rodinia formation. Strain associated with this orogeny was penetrative resulting in planar and linear fabric formation throughout the terrane (Gupta 2012), whereas strain associated with subsequent events at 900–650 Ma and 550–500 Ma were restricted to terrane boundaries (Biswal et al. 2007; Gupta 2012; Ranjan et al. 2018). This change in the nature of strain localization was attributed to a change in the rheological character of the EGMB following the onset of the first episode of 'hot orogenesis'. The 'M2' (980–930 Ma) metamorphic event was the most pervasive record of an orogenic event within the EGMB (Gupta 2012) and lithologies near the northern margin of the EGMB also record this 'Grenvillian signature'. This phase of metamorphism involved reworking of older granulites at 850 °C and 8–8.5 kbar pressure. Emplacement of porphyritic granitoids and anorthosite complexes, which suffered syn-emplacement deformation is also associated with this orogenic episode. Concomitant compressive deformation led to penetrative fabric formation, which is variably oriented, from N-S to NE-SW in the interior of the mobile belt (EGMB), to WNW-ESE and E-W near the northern margin. The change in orientation was attributed to the Kerajang shear zone at the boundary of the EGMB and the Rengali Province (Misra and Gupta 2014; see Fig. 2), which is a zone of intense strain localization in the vicinity of which the regional fabric trend undergoes a significant change (Sawant et al. 2017). The dominant lithologies along the northern margin of the EGMB include migmatitic quartzofeldspathic gneisses, metapelitic gneisses (or khondalites) and charnockite suite rocks. Away from the contact zone, khondalites, K-feldspar augen gneiss, deformed alkaline intrusives and calc-silicate gneisses are the dominant lithologies along with localised occurrences of mafic granulite. The structural imprints on these lithologies have been discussed by Sawant et al. (2017).

2.2 Rengali Province

The existence of the Rengali Province as a unique geological entity with an evolutionary history different from that of the Singhbhum Craton to the north and the

EGMB to the south was first explicitly postulated by Crowe et al. (2003). The major segment of the Rengali Province trends WNW-ESE, mirroring the orientation of the Kerajang fault which separates it from the EGMB further south. This segment is composed of a sequence of basement gneisses and a meta-volcanosedimentary sequence (Crowe et al. 2003). The basement gneisses of the Rengali Province are divided into two distinct complexes; the Kansal complex in the eastern and central parts of the province with hornblende orthogneiss, migmatitic quartzofeldspathic gneiss and irregular patches of charnockite, and the Badarama complex further to the West which is composed of leucogneiss grading into charnockite along with hornblende granulite and mafic granulite (Crowe et al. 2003). The Pal-Lahara complex in the north-western part of the province contains a meta-volcanosedimentary sequence defined by quartzite, amphibolite, chert and meta-conglomerate (Crowe et al. 2003). The overall grade of metamorphism in the Rengali Province generally varies from the middle to upper amphibolite facies, with granulite facies enclaves defined by patchy and massive charnockites and mafic granulite enclaves (Misra and Gupta 2014; Bose et al. 2015; Bhattacharya et al. 2016). The deformation signatures encountered in the Rengali Province were interpreted as effects of the existence of a strike-slip dilational step-over zone (Misra and Gupta 2014). The zone of most intense deformation is bound by three shear zones (Kerajang in the south, Barakot in the north and Akul in the east; Fig. 2) and displays prominent evidence of strike-slip deformation. Two earlier deformation events, D_1 and D_2 were operative under a shortening regime and amphibolite facies conditions (Misra and Gupta 2014). A third deformation event D_3, variable in magnitude in different parts of the terrane, involving prominent mylonitization, dynamic recrystallization and development of dextral asymmetry, operated under greenschist facies conditions (Misra and Gupta 2014) and is best developed near the terrane boundary shear zones. The presence of shortening structures as well as sub-vertical shear zones with prominent dextral asymmetry has led some workers to classify the deformation of this terrane as transpressive (Nash et al. 1996; Bhattacharya 1997; Ghosh et al. 2016) but concomitant operation of shortening as well as strike-slip deformation is contentious based on field evidence. Recent geochronologic data indicate thermal imprints around 3.0, 2.8 and 2.5 Ga (Bose et al. 2016; Sawant et al. 2017); Chattopadhyay et al. (2015) inferred a common evolutionary history of the EGMB and proto-India since the Neoproterozoic based on similarity of ages obtained from the Malaygiri supracrustal rocks and adjacent parts of the EGMB.

2.3 Singhbhum Craton

The Singhbhum Craton is an example of an archetypal Archaean granite-greenstone terrane (Saha 1994; Mahalik 1996; Bose 2009; Mukhopadhyay et al. 2012; Chaudhuri et al. 2018; Olierook et al. 2019) flanked by the Proterozoic Eastern Ghats Mobile Belt (EGMB) along its southeastern flank. Before the Rengali Province was recognized as a distinct geological entity, the high grade rocks of the EGMB were considered to have been thrust over the low grade rocks of the Singhbhum Craton along a southerly dipping thrust known as the Sukinda thrust (Saha 1994; Ghosh et al. 2010). Subsequent detailed structural and petrological studies on the northern margin of the EGMB established the strike-slip Kerajang Fault as the major structural feature, along which amalgamation of the EGMB with the Rengali Province was effected (Misra and Gupta 2014). From the available geochronological data from the three terranes (Mezger and Cosca 1999; Mishra et al. 1999; Dasgupta et al. 2013; Bose et al. 2016; Sawant et al. 2017; Chaudhuri et al. 2018), it is now considered that the Rengali Province and the Singhbhum Craton shared a common history of Neoarchaean deformation and reworking, implying that the Rengali Province and the Singhbhum Craton had been juxtaposed and deformed much before their final amalgamation with the EGMB. A number of pervasive deformation episodes and multiple reworking along crustal scale shear zones involving supercontinent assembly and fragmentation (Bhadra et al. 2004; Biswal et al. 2007; Mahapatro et al. 2012; Misra and Gupta 2014; Sawant et al. 2017) characterise the contact zone between the three terranes. This study incorporates information across this zone, although this study does not discuss the nature of the original contact between the Rengali Province and the Singhbhum Craton.

3 Methodology

Extensive fieldwork has been carried out along the eastern segment of the EGMB-Rengali Province contact along a north-south transect from Duburi to Charaidhara, as well as within the step-over zone within the Rengali Province (Fig. 3). Thin sections from lithologies of the contact zone between the EGMB and the Rengali Province have been studied using a Leica DM 4500P microscope. Photomicrographs from which shear sense has been inferred (Fig. 4e–h) have been obtained from X-Z sections, which are always horizontal in lithologies within the study area. Vertical photomicrographs (Fig. 4a, c) are perpendicular to X-Z. Electron Backscatter Diffraction (EBSD) studies have been conducted on quartz grains within a metapelite sample from near the contact zone and a quartzite sample from the supracrustal unit within the Rengali Province. Deformation mechanisms and asymmetry in distribution of poles to <c> axes in quartz have been used to interpret the temperature and sense of shear during deformation mechanisms that are correlated with deformation temperature. EBSD analyses were carried out at the Central Research Facility at the Indian Institute of Technology Kharagpur. The EBSD system consists of a Zeiss-Auriga Compact system with a Gemini column Schottky type field emission filament, and an Oxford Nordlys detector for EBSD pattern detection. Step sizes of 5 and 10 μm were used to collect the Electron Backscatter Patterns (EBSPs). Post-processing of raw EBSD data was carried out using the HKL Channel 5 software.

4 Results

Results of the study conducted using the above methodologies have been are described in this section.

4.1 Structure

4.1.1 Eastern Ghats Mobile Belt

Field relations and penetrative fabric orientation data have been collected from the easternmost part of the northern boundary domain of the EGMB, near its contact with the Rengali Province. Major lithologies in this area include charnockites associated with K-feldspar augen gneiss (Fig. 3a, b), khondalite and local patches of amphibolite. Near the contact with the Rengali Province, the EGMB comprises a zone of khondalite-charnockite intercalations trending E-W to ESE-WNW, roughly parallel to the trend of the Kerajang Shear Zone. South of this intercalated zone, charnockites are scarce and lateritised khondalites are the dominant lithology. Foliations are indistinct in the massive charnockites, but ubiquitous in the K-feldspar augen gneiss hosted within the charnockites (Fig. 3b). Sub-vertical mylonite fabrics and pseudotachylite veins within the charnockite strike E-W to ESE-WNW and dip sub-vertically (Fig. 3c). The EGMB lithologies in the proximity of the contact have experienced at least three phases of deformation D_{EG1}, D_{EG2} and D_{EG3}. D_{EG1} led to the formation of a metamorphic segregation banding (S_{EG1}) in the charnockites and khondalites. Tight to isoclinal folding of S_{EG1} led to the development of an axial planar fabric S_{EG2}. This axial planar fabric is the most prominent fabric near the northern margin of the EGMB; S_{EG1} can only be distinguished from S_{EG2} in the hinges of D_{EG2} folds. The intersection of S_{EG1} with S_{EG2} defines an intersection lineation defined by elongate silimanite needles that invariably plunge down the dip of the foliation. $S_{EG1/2}$ strikes NE-SW and dips to the southwest to the south of the Kerajang Shear Zone. Near the shear zone, $S_{EG1/2}$ has an E-W to ESE-WNW, subvertical orientation. The D_{EG3} deformation represents a dextral, strike-slip shearing event along the Brahmani Shear Zone (BSZ) that juxtaposed the Rengali Province against the EGMB (Misra and Gupta 2014; Sawant et al. 2017); Sawant et al. (2017) suggested that this event passively rotated $S_{EG1/2}$ into an E-W/ESE-WNW striking, sub-vertical orientation. Near the contact with the Rengali Province, $S_{EG1/2}$ is drawn into open folds with sub-vertically plunging fold axes showing a definite dextral sense of drag. Asymmetric markers are much less visible in the khondalites and charnockites compared to the K-feldspar augen gneisses.

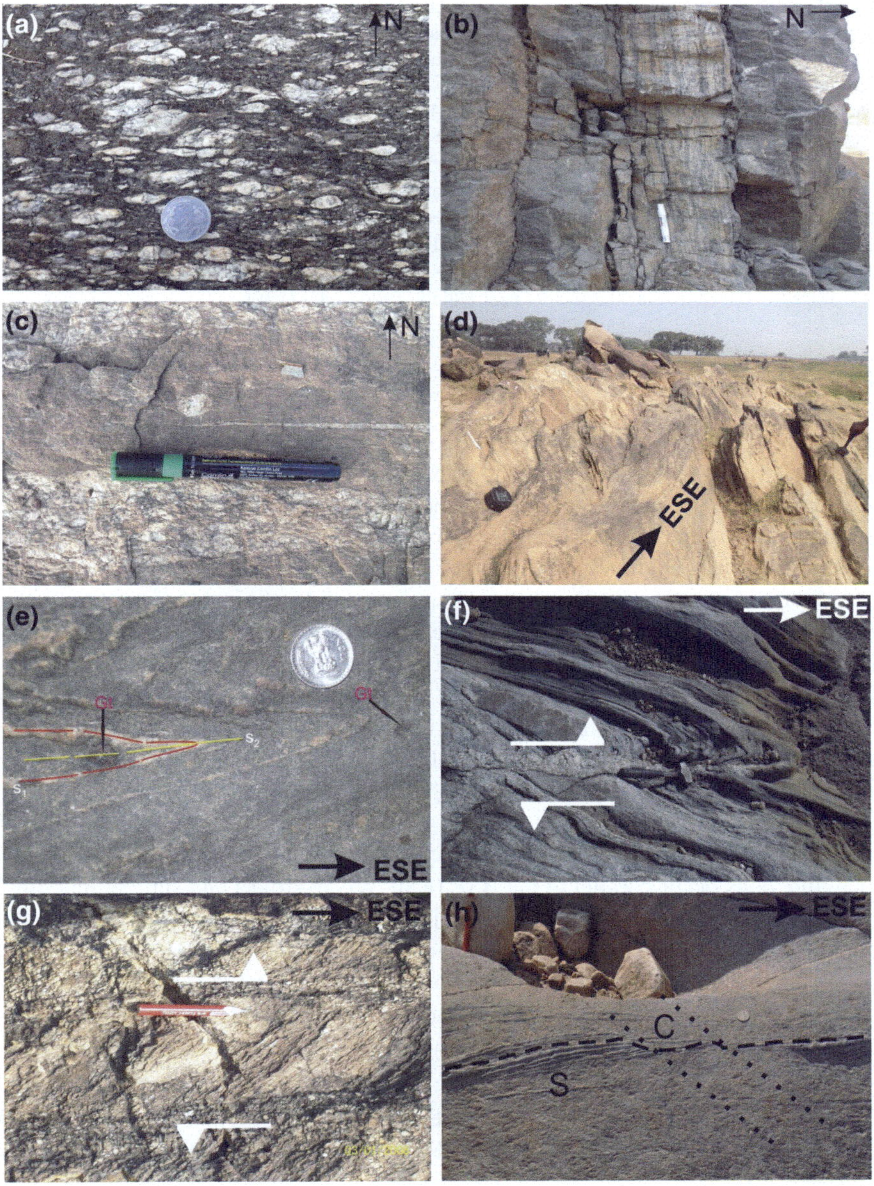

◂**Fig. 3** Mesoscopic structures observed in outcrop in the vicinity of the Brahmani Shear Zone and the dilational step-over zone to the north. **a** *EGMB*, location 20° 50′ 24.8″ N, 85° 51′ 7.6″ E: K-feldspar augen in a mylonitised matrix near Raipur (horizontal view); **b** *EGMB*, location 20° 47′ 43.4″ N, 86° 01′ 21.7″ E: augen gneiss domain within charnockite with a prominent sub-vertical fabric near Nityanandpur (vertical view); **c** *EGMB*, location 20° 48′ 16.9″ N, 85° 53′ 54.1″ E: mylonite with asymmetric delta clasts hosted within charnockite near Kashipur (horizontal view); **d** Rengali Province, location 20° 52′ 50.7″ N, 85° 57′ 10.0″ E: steeply dipping fabric in a quartzofeldspathic gneiss, near the Brahmani Shear Zone, on the northern bank of the Brahmani river near Hatibari (vertical view); **e** Rengali Province, location 21° 18.18′ N, 84° 55.71′ E: isoclinal folding of S_1 layers and development of an axial planar S_2 fabric near Rengali dam (horizontal view); **f** Rengali Province, location 21° 11.75′ N, 85° 04.66′ E: dextral drag of previously folded S_1 layers along a sub-vertical NW-SE trending shear fabric near Rengali dam (horizontal view); **g** Rengali Province, location 20° 55.20′ N, 85° 40.40′ E: S-C fabric in Rengali quartzofeldspathic gneiss with prominent dextral asymmetry (horizontal view); **h** Rengali Province, location 21° 12.08′ N, 84° 53.01′ E: S-C-C′ fabric developed in charnockite near Kulabir (horizontal view)

4.1.2 Rengali Province

The southeastern segment of the Rengali Province, north of the BSZ bears a striking structural similarity to the EGMB. The dominant lithologies in this area are migmatitic quartzofeldspathic gneiss, charnockite, metapelite, hornblende orthogneiss, granite porphyry and metapelite boudins within leucogneiss which locally grade into charnockite. Three penetrative fabric elements correlatable with deformation events D_{RNG1}, D_{RNG2} and D_{RNG3} are visible within the migmatitic quartzofeldspathic gneisses. A metamorphic segregation layering S_{RNG1} defines the earliest fabric, which is accentuated by melt segregations in the migmatitic gneisses. D_{RNG2} involved isoclinal folding of S_{RNG1} and development of an axial planar fabric S_{RNG2} which steepens with increasing proximity to the EGMB (Fig. 3d). S_{RNG2} is subsequently warped by a subvertical shear fabric (S_{RNG3}) trending WNW-ESE. Further south, in metapelite boudins within leucogneisses, the shear fabric shows a slightly different orientation. In these boudins, $S_{RNG1/2}$ is extensively warped by a sub-vertical shear fabric trending NE-SW. In the southern part of the boudins, prominent pseudotachylite veins occur parallel to the BSZ (ESE-WNW strikes and sub-vertical dips) and are parallel to $S_{RNG1/2}$. The difference in orientation of the shear fabric within the boudins may be related to passive rotation of the boudin within the rheologically less competent matrix leucogneisses during an earlier deformation event, while pseudotachylite generation occurred when both lithologies had cooled into the brittle regime. Structures within the high grade zone of the Rengali Province have been extensively documented by Misra and Gupta (2014). The central and western segments of the Rengali Province, bound by the Kerajang, Barakot and Akul shear zones, have a dominant strike-slip deformation signature and this zone has been characterised as a dilational step-over zone. Within the charnockites in this zone, the earliest fabric is a segregation layering S_{GR} and a primary bedding S_0 in the supracrustals. The first deformation event D_1 involved tight to isoclinal folding of S_{GR} as well as S_0 and created an axial planar fabric S_1 parallel to S_{GR} and S_0 (Fig. 3e). A second deformation event D_2 led to refolding of the earlier tight, isoclinal folds along subvertical axial planes. D_2 was followed by a dextral strike-slip shearing event D_3 (Fig. 3f) which variably affected the lithologies of this segment. Evidence of D_3 deformation in the granulites is limited to narrow, WNW-ESE trending sub-vertical shear planes with dextral shear sense indicators which do not reorient pre-existing structures in the granulites (Fig. 3f, g). The effect of D_3 is most pronounced in the supracrustals, with S_3 shear bands that preserve unambiguous dextral sigmoidal shear sense indicators. D_3 deformation also creates S-C-C′ fabrics in the charnockite (Fig. 3h). The intensity of deformation increases with increasing proximity to the terrane boundary shear zones, with extensively mylonitised granites near the Akul shear zone developing an almost schistose appearance. Shortening during D_3 is manifested as buckle folds on S_1; extension during D_3 is observable as quartz-filled tension gashes oriented at 45° to the ESE-WNW subvertical shear planes. Detailed structural documentation of these features is provided in Misra and Gupta (2014).

4.2 Petrography and Microstructure

The major observations are textural and microstructural observations in both units is described in detail below.

4.2.1 Eastern Ghats Mobile Belt

Khondalites of the EGMB unit consist of garnet, sillimanite, quartz, K-feldspar and minor rutile. The initial segregation layering, defined by layers rich in garnet and sillimanite alternating with quartzofeldspathic layers, is largely preserved even during D_{EG3} shearing, giving the impression that $S_{EG1/2}$ has been 'passively' rotated during this later deformation. Indeed, most of the strain related to this apparently 'passive' rotation was accommodated by plastic deformation of quartz, while other minerals underwent little internal deformation, or deformed in a brittle to brittle-ductile manner. Microstructurally, this strain is manifested in the form of elongate needles of sillimanite and flattened grains of quartz wrapping around garnet. With increasing strain, quartz grains in the matrix are drawn into ribbons whereas those that remain as inclusions within garnet preserve evidence of static recrystallization. In horizontal sections of the khondalites, pressure shadows around rigid garnet grains suggest that flow was sub-horizontal and directed E-W. Quartz veins intruding the khondalites along an E-W to ESE-WNW trend have also partitioned a significant degree of D_{EG3} strain, and their wavy grain boundaries with prominent bulges suggest recrystallization occurred under regime I of Hirth and Tullis (1992) (Stipp et al. 2002). The degree of elongation of quartz grains is significantly greater in horizontal sections compared to vertical sections, indicating that the X-direction was oriented E-W during D_{EG3} deformation. Since initial deformation during D_{EG1-2} was concomitant with granulite facies metamorphism, and D_{EG3} operated under much lower temperature conditions, no new mineral growth occurs during formation of the S_{EG3} fabric. Charnockite-suite

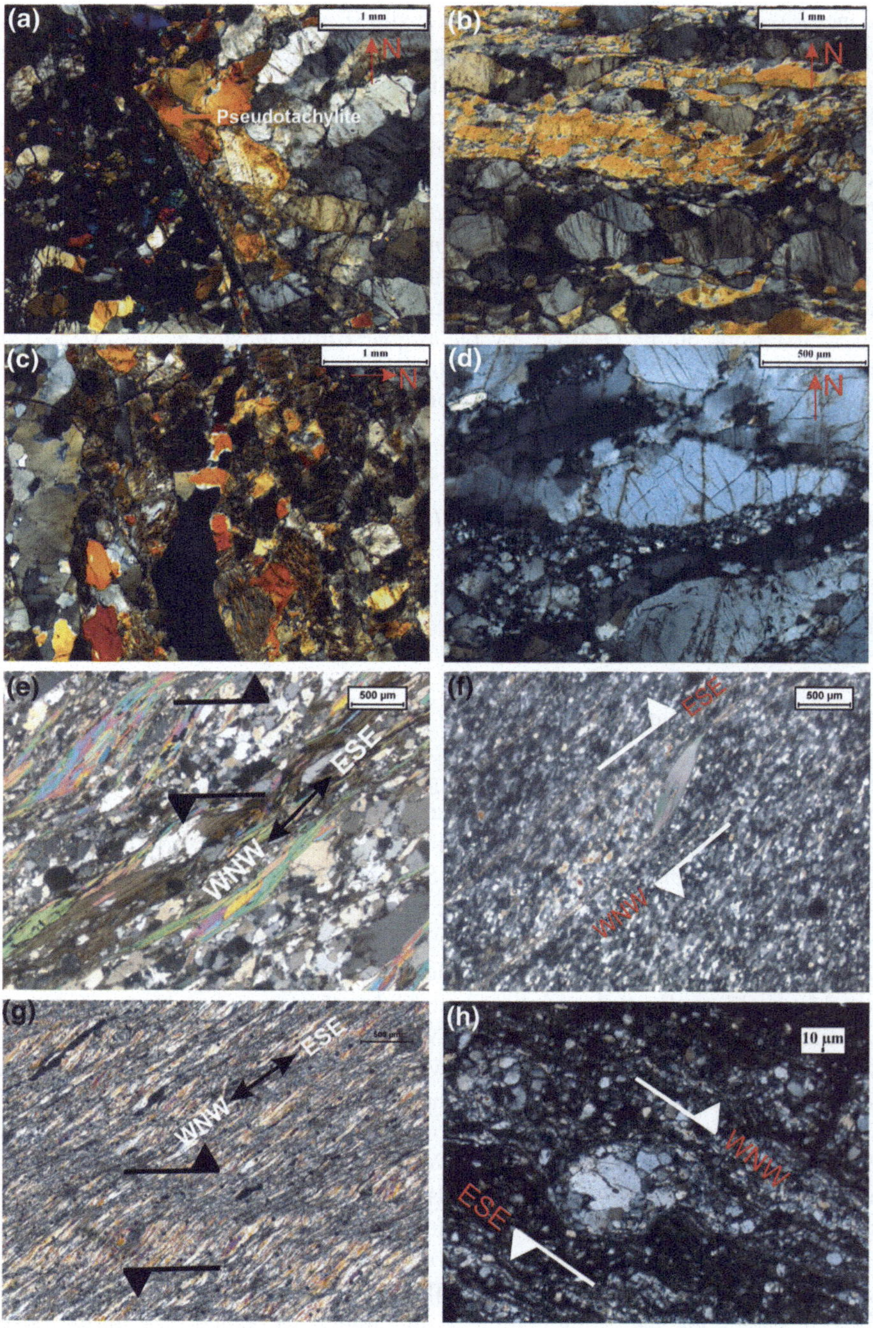
(a)
1 mm
N
Pseudotachylite
(b)
1 mm
N
(c)
1 mm
N
(d)
500 μm
N
(e)
500 μm
ESE
WNW
(f)
500 μm
ESE
WNW
(g)
ESE
WNW
(h)
10 μm
WNW
ESE

◂**Fig. 4** Photomicrographs of representative samples from lithologies of the *EGMB* and the Rengali Province in the vicinity of the Brahmani Shear Zone, and from the dilational step-over zone within the Rengali Province. Prefix AM, RN and RD are used for sample identification purposes **a** *EGMB* unit, location 20° 46′ 44″ N, 86° 02′ 57.4″ E; pseudotachylite associated with cataclasite, hosted within charnockites from the intercalated lithologies of the *EGMB* within the Brahmani Shear Zone, near Dankari (vertical section); **b** *EGMB* unit, location 20° 47′ 43.4″ N, 86° 01′ 21.7″ E: K-feldspar augen, without any prominent asymmetry, wrapped by layers of quartz near Nityanandpur (horizontal section); **c** *EGMB* unit, location 20° 50′ 58.3″ N, 85° 49′ 50.4″ E: connected quartz layer defining a subvertical fabric parallel to the trend of the Brahmani Shear Zone, on the southern bank of the Brahmani river near Dalar (vertical section); **d** *EGMB* unit, location 20° 50′ 24.8″ N, 85° 51′ 7.6″ E) mylonitised quartz grains wrapping around a K-feldspar porphyroclast, near Raipur (horizontal section); **e** Rengali Province, location 21° 12.08′ N, 84° 53.01′ E: dextral drag of a biotite defined fabric within quartzite, near Rengali dam (horizontal section); **f** Rengali Province, location 21° 12.08′ N, 84° 53.01′ E; muscovite fish within a quartzite displaying prominent dextral asymmetry, near Budhapal (horizontal section); **g** Rengali Province, 21° 03.79′ N, 85° 22.59′ E: S-C fabrics within quartzite in the vicinity of the Akul shear zone (horizontal section); **h** Rengali Province, location 20° 53′ 2.08″ N, 85° 52′ 57.49″ E: delta porphyroclast composed of recrystallized grains of quartz within metapelite boudins north of the Brahmani river, displaying prominent dextral asymmetry (horizontal section)

rocks in the contact zone are composed of quartz, clinopyroxene, garnet, K-feldspar, plagioclase, biotite, rutile, ilmenite, pyrite and rare orthopyroxene. K-feldspar augen gneiss associated with the charnockite is composed of K-feldspar, quartz, plagioclase, biotite, rutile, ilmenite and pyrite. $S_{EG1/2}$ in this rock is defined by layers of garnet-clinopyroxene-orthopyroxene alternating with quartz-K-feldspar layers. $S_{EG1/2}$ is oriented E-W to ESE-WNW within the charnockites, indicating rotation into parallelism with the boundary shear zone and a prominent D_{EG3} overprint. However, evidence of D_{EG3} dynamic recrystallization is limited and most grains have straight boundaries indicative of static recrystallization at high temperatures (Vernon 2004) and strain associated with D_{EG3} passively reoriented existing $S_{EG1/2}$ fabrics into their present trend (Fig. 4a). Both high D_{EG3} strain and low D_{EG3} strain domains are preserved in the K-feldspar augen gneiss hosted within the charnockite (Fig. 4b). Within the low strain domains, quartz grains are statically recrystallized without any low temperature overprint. The high strain domains contain quartz ribbons displaying wavy extinction patterns and these appear to have recrystallized under regime I of Hirth and Tullis (1992). The foliation within the high strain domains is sub-vertical and trends E-W to ESE-WNW. In these domains, mylonite bands within the K-feldspar augen gneiss contain recrystallized quartz mantles that form bands oriented E-W to ESE-WNW (Fig. 4d). Pseudotachylites within the charnockites contain crushed fragments of garnet, clinopyroxene and orthopyroxene and are associated with brittle micro-faults and altered cataclasites (Fig. 4a). Localized mineral alterations near the pseudotachylites suggest co-seismic enhanced fluid activity concomitant with D_{EG3} shearing.

4.2.2 Rengali Province

In the immediate vicinity of the Brahmani Shear Zone, the lithologies of the Rengali Province include hornblende orthogneiss, metapelite enclaves in leucogneiss locally grading into charnockite, migmatitic quartzofeldspathic gneiss, porphyry granite, and charnockite at its margin with the Singhbhum Craton. The northernmost charnockite contains a segregation layering which trends NW-SE and dips southeasterly. This segregation layering is overprinted by a prominent subvertical shear fabric oriented E-W. The shearing event operated under greenschist facies conditions as is evidenced by the abundance of chlorite and epidote. Chlorite rims on orthopyroxene are indicative of retrogression following high temperature metamorphism. The granite porphyry contains abundant phenocrysts of K-feldspar and numerous xenoliths of different lithologies such as chlorite schists and amphibolite. Microstructurally, quartz grains within the granite preserve prominently developed core-mantle structures. South of the granite, microstructures in the migmatitc quartzofeldspathic gneiss preserve evidence of high strain, manifested by elongate ribbons of quartz that have undergone substantial dynamic recrystallization by grain boundary migration (GBM) recrystallization. The gneissic fabric is defined by layers of biotite and hornblende with titanite rims on ilmenite occurring within these layers. The predominant fabric $S_{RNG1/2}$ steepens near the contact with the EGMB, and becomes subvertical near the Brahmani Shear Zone. The metapelite boudins within leucogneiss have a gneissic fabric defined by a garnet-biotite segregations alternating with a cordierite-quartz-orthopyroxene bearing layer. The metapelite hosts pseudotachylite veins with crushed fragments of garnet and orthopyroxene within the opaque matrix. Mylonitic bands within the metapelite, which are subvertical and oriented ESE-WNW, contain delta porphyroclasts with prominent dextral asymmetry (Fig. 4h). Further to the west, within the dilational step-over zone, evidence of dextral asymmetry is noticeable in the rocks of the supracrustal unit as well as the basement gneisses (Figs. 3f, g and 4e–g). Muscovite 'fish' within the quartzites (Fig. 3f) have undergone a dextral drag. Shear bands traversing through the quartzites define prominent S-C fabrics with a dextral shear sense (Fig. 4g). Away from the terrane boundary shear zones, quartz grains are flattened and drawn into ribbons and appear to have dynamically recrystallized at high temperatures. The quartzite units that are proximal to the granulites in this zone have largely escaped the strain associated with the regional shearing event, and consist of elongate quartz ribbons which show chessboard extinction inherited from the earlier, higher temperature deformation event. From the microstructural evidence available, it is abundantly clear that the degree of strain partitioned by the different lithologies of the Rengali Province varies based on proximity to the terrane boundary shear zones described previously. In rocks that have experienced relatively high strain associated with the D_{RNG3} shearing event, evidence of dextral asymmetry is abundant.

4.3 *Electron Backscatter Diffraction (EBSD) Studies*

Active slip systems in quartz have been inferred from <c> axes pole figure plots and asymmetry in the distribution of poles have been used to determine shear sense. RN 202, a quartzite from the supracrustal sequence within the step-over zone, deformed by a combination of rhomb <a> and prism <a> slip (Fig. 5a) under conditions of moderate temperature and strain rate. The entire pole figure pattern shows a dextral shift. Near the Brahmani Shear Zone, recrystallized delta porphyroclasts of quartz within a metapelite sample were also analysed using EBSD. The pole distribution resembles a type I crossed girdle with prominent dextral asymmetry (Fig. 5b), indicating preservation of an earlier pure shear shortening deformation signature, which was later modified by dextral shearing. The pole distributions obtained from the EBSPs from Rengali Province lithologies indicate the last phase of deformation that influenced the quartz CPO was strike-slip in nature, while shortening deformation signatures are preserved from an earlier event.

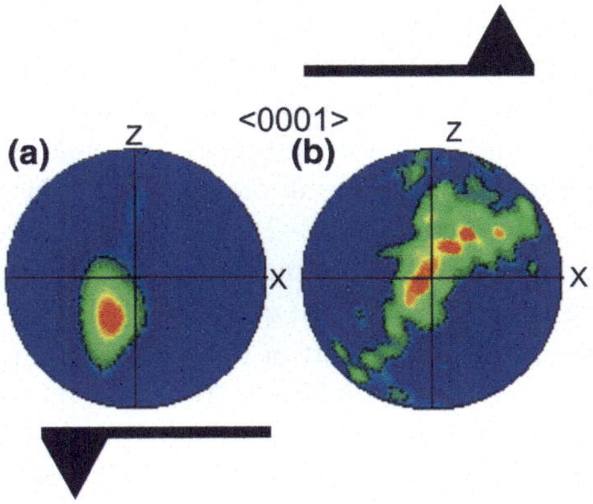

Fig. 5 <0001> pole figures of quartz grains from within **a** a quartzite sample within the Rengali Province; **b** metapelite near the Brahmani Shear Zone. Active slip systems have been estimated from the distribution of the poles to <c> axes. In **a**, a combination of rhomb <a> and prism <a> slip facilitated deformation, with a visible dextral shift in the pole distribution. The pole distribution in **b** approximates that of a type I crossed girdle that results from pure shear shortening. However, the distribution also shows a significant dextral asymmetry, indicating that a later shearing event was superposed on an initial shortening deformation episode

5 Discussion

5.1 *Evolution of Ideas About the Northern Margin of the EGMB*

While the western margin of the EGMB has been reasonably well constrained as a sequence of stacked thrust sheets (Biswal and Sinha 2003; Bhadra et al. 2004; Biswal et al. 2007), the nature and extent of deformation along the northern margin of the EGMB continues to be a subject of debate. The earliest interpretations, based largely on GSI studies, are summarized by Banerjee (1997), who considered the ultramafic rocks hosting chromitites to represent oceanic lithosphere that separated 'granulites' of the EGMB in the south from 'greenstones' of Singhbhum to the north. The involvement of oceanic lithosphere was used to interpret the contact as a suture zone representing continent-continent collision, along which oceanic lithosphere at the leading edge of the Singhbhum continent was initially subducted southward below granulites of the EGMB, before the passive margin of the subducting plate, with its shelf sediments, were involved in a 'hard collision' that followed oceanic subduction. However, the basis for the model was almost exclusively the spatial disposition of lithologies, with little structural or geochronological support.

As geochronological constraints were progressively obtained, it was understood that the high grade terrane south of the ultramafic rocks, that included granulites and amphibolites, were late Archaean (2.8–2.5 Ga) in age, unlike the Late Mesoproterozoic to Neoproterozoic (c. 1.0 Ga) granulites further south, mostly south of the River Brahmani (Bose et al. 2016; Sawant et al. 2017). The ~1.0 Ga signature was characteristic of granulites that were part of the Indo-Antarctica continent formed during Rodinia assembly (Dobmeier and Raith 2003), and therefore, these were distinct from the late Archaean amphibolites and granulites that were juxtaposed against them along the northern margin. This late Archaean high grade terrane represented the Rengali Province. Thus, the northern boundary zone of the EGMB essentially includes two terrane boundaries—a southern boundary between the EGMB and the Rengali Province, and a separate contact, further north, between the Rengali Province and the Singhbhum Craton. This study collates the information on the former.

5.2 *The EGMB-Rengali Province Boundary*

The precise location of the EGMB-Rengali Province boundary was unequivocally described by Misra and Gupta (2014) in the west, and Sawant et al. (2017) in the east. The boundary approximately coincides with the course of the River Brahmani, and is defined by the Kerajang Shear Zone in the west and the Brahmani Shear Zone in the east. While some workers make a case for the boundary between the EGMB and the Rengali Province to be transpressive in nature (e.g. Bhattacharya 1997; Ghosh et al. 2016), others have inferred the boundary to be dominantly strike-slip in character

(Misra and Gupta 2014; Sawant et al. 2017). While there are shortening structures present in the Rengali Province lithologies in the vicinity of the contact, EBSD studies show that these shortening structures were the result of an earlier pure shear deformation on which a later dextral strike-slip shearing event was subsequently superposed. Therefore, the shortening and strike-slip deformation observed in the Rengali Province rocks did not operate synchronously, but are discrete, temporally separated events. While the lithological set-up of the EGMB is different from the Rengali Province in the vicinity of the contact, the deformation signatures in both the terranes are similar, involving isoclinal folding of the initial segregation layering followed by mylonitisation and pseudotachylite formation associated with strike-slip shearing. Therefore, it is entirely possible that the Rengali Province and the EGMB shared at least a part of their deformation history. Similarities between the EGMB and the Rayner province have been documented previously (Harley 1987; Mezger and Cosca 1999; Dobmeier and Raith 2003; Sawant et al. 2017) and recent correlations between the Rauer group of rocks in Antarctica and the Rengali Province further reinforces models in which east Antarctica was juxtaposed against proto-India. The amalgamation of the EGMB with the Rengali Province may have occurred at 1 Ga or 500 Ma, based on the available geochronological evidence (Upadhyay and Raith 2006; Biswal et al. 2007), but irrespective of the timing of the event, the evidence overwhelmingly suggests that the boundary was strike-slip in nature. The shortening structures in the Rengali Province can be correlated with higher temperature deformation, while features related to strike-slip shearing operated under greenschist facies conditions. Thus, the temperature and timing of the two events are clearly different.

The sub-vertical foliations, mylonite bands and pseudotachylite veins in the vicinity of and parallel to the Brahmani Shear Zone within the EGMB lithologies, as well as the Rengali Province lithologies are clear indicators that the shear zone affected both the terranes. Prominent shear sense indicators both on mesoscopic and microscopic scale prove that the deformation event involved prominent dextral shearing, with maximum asymmetry observed in horizontal sections, implying the movement vector was subhorizontal during shearing. Microstructural features of this deformation, along with quartz deformation mechanisms confirm that the temperature of deformation is characteristic of the greenschist facies. Additionally, the steeply dipping WNW-ESE fabric that develops as a result of the shearing event is pervasive not only within the Rengali Province and the EGMB, but also extends into the granites of the Singhbhum Craton (Dobe and Gupta 2018). These features show uncanny similarities with the observations in the Rauer Group in East Antarctica (Sawant et al. 2017).

Bouguer anomaly studies across the Rengali Province-EGMB contact have been carried out by Mandal et al. (2015). Craton-mobile Belt contacts are typically characterised by a large positive Bouguer anomaly over the Mobile Belt, transitioning to a steep negative anomaly over the cratonic interior (Fountain and Salisbury 1981; Gibb 1983; Gibb et al. 1983; Tesha et al. 1997). A similar Bouguer gravity signature is observed along the EGMB-Rengali Province contact zone where lower and middle crustal layers have been exposed at shallower levels due to post-juxtaposition extensional tectonics (Mandal et al. 2015). Results of the Bouguer gravity anomaly study indicate the high density granulites of the EGMB extend till only a few kilometres

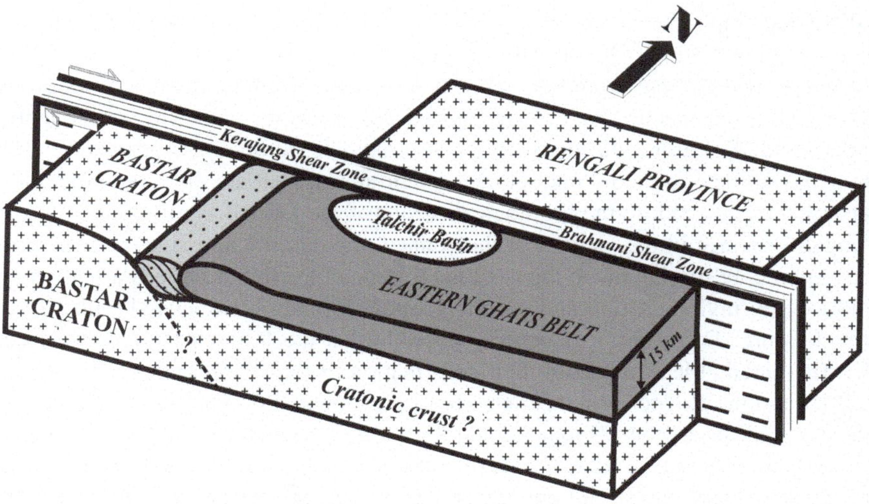

Fig. 6 Model showing the disposition of the Eastern Ghats Mobile Belt and the Rengali Province on either side of the subvertical Kerajang/Brahmani Shear Zone, as estimated from gravity models (modified after Mandal et al. 2015). Note the differences in lateral and vertical extent of the two terranes

below the surface and are underlain by less dense upper crustal material and conventional models which explain gravity anomaly signatures across Precambrian suture zones cannot explain this phenomenon. Using data collected from closely spaced gravity stations to determine depths up to which surface lithologies extend, Mandal et al. (2015) demonstrated that there is a steep negative anomaly to the north of the Talcher basin (the Rengali Province) and a steep positive anomaly to the south (the EGMB). Modelling of the Bouguer anomalies reveals a steeply dipping fault below the sediment cover at the northern margin of the Talcher basin (Fig. 6) separating crustal domains with different densities (Fig. 6). The sub-vertical contact can be traced to a depth of at least 15 km, further reinforcing the geological model that interprets the contact as a crustal scale strike-slip shear zone.

5.3 Implications for Further Studies on the Northern Margin of the EGMB

Results of recent studies, summarized and collated with the observations in this study, indicate clearly that unlike the western boundary of the EGMB, the northern boundary is essentially strike-slip in character. This boundary actually separates two high grade terranes—c. 1.0 Ga granulites of the EGMB from c. 2.8–2.5 Ga granulites and amphibolites of the Rengali Province. The implication is that the Sukinda Thrust, if it actually exists as a structural entity, must separate the Rengali

Province from the granite-greenstone terrane of the Singhbhum Craton. This contact therefore needs to be investigated in this context. As brought out in this compilation, much of the interpretation of a thrust along this contact is based on petrological arguments; unambiguous structural information is not yet available. Thus, the nature of this boundary, and its importance in the context of the construction of the Indian shield remains to be investigated.

The nature of the EGMB-Rengali Province contact is clearly strike-slip, which is consistent with the thrust-nappe model suggested by Biswal and Jena (1999) and Biswal et al. (2000) for the western boundary. However, a factor of concern here is the grade of metamorphism associated with the western boundary thrust, which was considered by Gupta et al. (2000) to be in the granulite facies, which is inconsistent with the greenschist facies metamorphism associated with the northern boundary shear zone. A model which can reconcile this apparent conflict may finally resolve the vexed problem of the northern boundary of the EGMB.

6 Conclusions

The EGMB formed during an Indo-Antarctic collision event during Rodinia supercontinent assembly. While the timing of the amalgamation of the EGMB with proto-India has been a matter of debate, the structural imprints along the northern margin of the EGMB and the southern margin of the Rengali Province (the part of proto-India that formed the amalgamation front) suggest that amalgamation occurred along an E-W to NW-SE trending shear zone. While shortening structures are observed in both the EGMB and the Rengali Province, these were inherited from earlier deformation events in the respective units, and are unrelated to the final assembly of the terranes. Field and microstructural evidence suggests that large scale movement occurred along a sub-vertical plane with a sub-horizontal transport vector, and shear sense indicators on both the mesoscopic and microscopic level suggest extensive dextral shearing brought the EGMB and the Rengali Province into their present configuration. Additionally, gravity anomaly studies clearly demonstrate a difference in crustal densities between the EGMB and the Rengali Province and their prominent separation along a steeply dipping shear fabric which coincides with the Kerajang shear zone. The contact, therefore, can be definitively characterized as a strike-slip shear zone.

Acknowledgements SG thanks Prof. T. K. Biswal for the invitation to contribute to the Special Volume on "Structural Geology of Mobile Belts of Indian Subcontinent", in commemoration of the IGC 2020, to be held at New Delhi. SG and WKM thank the Dept. of Geology & Geophysics, IIT Kharagpur, for providing facilities over the years that have culminated in this study. ADS and RD are grateful to the Council of Scientific and Industrial Research for the award of the S.P. Mukherjee Fellowship (No. SPM-06/081(0176)/2013-EMR-I), and Research Fellowship (No. 09/081(1242)/2015-EMR-I), respectively, which helped sponsor some of their fieldwork and analytical costs.

References

Banerjee PK (1997) Geodynamic implications of chrome-poor enstatitite bodies in the Sukinda-Boula-Nousahi ultramafic suites of Orissa. Proc Indian Acad Sci (Earth Planet Sci) 106(4):357–360

Banerjee PK, Mahakud SP, Bhattacharya AK, Mohanty AK (1987) On the northern margin of the Eastern Ghats in Orissa. Rec Geol Surv India 118(2):1–8

Bhadra S, Gupta S, Banerjee M (2004) Structural evolution across the Eastern Ghats Mobile Belt-Bastar craton boundary, India: hot over cold thrusting in an ancient collision zone. J Struct Geol 26:233–245. https://doi.org/10.1016/S0191-8141(03)00112-3

Bhattacharya A, Das HH, Bell E, Bhattacharya A, Chatterjee N, Saha L, Dutt A (2016) Restoration of Late Neoarchean—Early Cambrian tectonics in the Rengali orogen and its environs (eastern India): the Antarctic connection. Lithos 263:190–212. https://doi.org/10.1016/j.lithos.2016.06.006

Bhattacharya S (1997) Evolution of Eastern Ghats granulite belt of India in a compressional tectonic regime and juxtaposition against Iron Ore Craton of Singhbhum by oblique collision—transpression. Proc Indian Acad Sci Earth Planet Sci 106:65–75. https://doi.org/10.1007/BF02839281

Biswal TK, Jena SK (1999) Large lateral ramp in the fold-thrust belts of Mesoproterozoic Eastern Ghats Mobile Belt, Eastern India. Gondwana Res 2:657–660

Biswal TK, Jena SK, Datta S, Das R, Khan K (2000) Deformation of the Terrane Boundary Shear Zone (Lakhna Shear Zone) between the Eastern Ghats Mobile Belt and the Bastar craton, in Balangir and Kalahandi districts of Orissa. J Geol Soc Ind 55:367–380

Biswal TK, Sinha S (2003) Deformation history of the NW salient of the Eastern Ghats Mobile Belt, India. J Asian Earth Sci 22:157–169

Biswal TK, De Waele B, Ahuja H (2007) Timing and dynamics of the juxtaposition of the Eastern Ghats Mobile Belt against the Bhandara craton, India: a structural and zircon U-Pb SHRIMP study of the fold-thrust belt and associated nepheline syenite plutons. Tectonics 26. https://doi.org/10.1029/2006TC002005

Black R, Liegeois J-P (1993) Cratons, mobile belts, alkaline rocks and continental lithospheric mantle: the Pan-African testimony. J Geol Soc 150:89–98. https://doi.org/10.1144/gsjgs.150.1.0088

Bose MK (2009) Precambrian mafic magmatism in the Singhbhum craton, Eastern India. J Geol Soc India 73:13–35. https://doi.org/10.1007/s12594-009-0002-3

Bose S, Das K, Kimura K, Hidaka H, Dasgupta A, Ghosh G, Mukhopadhyay J (2016) Neoarchean tectonothermal imprints in the Rengali Province, eastern India and their implication on the growth of Singhbhum craton: evidence from zircon U-Pb SHRIMP data. J Metamorph Geol 34:743–764. https://doi.org/10.1111/jmg.12201

Bose S, Guha S, Ghosh G, Das K, Mukhopadhyay J (2015) Tectonic juxtaposition of crust and continental growth during orogenesis: example from the Rengali Province, eastern India. Geosci Front 6:537–555. https://doi.org/10.1016/j.gsf.2014.09.002

Chattopadhyay S, Upadhyay D, Nanda JK, Mezger K, Pruseth KL, Berndt J (2015) Proto-India was a part of rodinia: evidence from Grenville-age suturing of the Eastern Ghats Province with the Paleoarchean Singhbhum craton. Precambr Res 266:506–529. https://doi.org/10.1016/j.precamres.2015.05.030

Chaudhuri T, Wan Y, Mazumder R, Ma M, Liu D (2010) Evidence of enriched, hadean mantle reservoir from 4.2–4.0 Ga zircon xenocrysts from Paleoarchean TTGs of the Singhbhum craton, Eastern India. Sci Rep 8:7069. https://doi.org/10.1038/s41598-018-25494-6

Chetty TRK (2010) Structural architecture of the northern composite terrane, the Eastern Ghats Mobile Belt, India: implications for Gondwana tectonics. Gondwana Res 18:565–582. https://doi.org/10.1016/j.gr.2010.02.006

Crowe WA, Nash CR, Harris LB, Leeming PM, Rankin LR (2003) The geology of the Rengali Province: implications for the tectonic development of northern Orissa, India. J Asian Earth Sci 21:697–710. https://doi.org/10.1016/S1367-9120(02)00034-2

Dasgupta S, Bose S, Das K (2013) Tectonic evolution of the Eastern Ghats Belt, India. Precambr Res 227:247–258. https://doi.org/10.1016/j.precamres.2012.04.005

Dobe R, Gupta S (2018) Discriminating tectonic and magmatic fabrics in the Remal Granite Gneiss: implications for terrane amalgamation processes in Southeastern Singhbhum, India. J Geol Soc India 92:657–660. https://doi.org/10.1007/s12594-018-1083-7

Dobmeier CJ, Raith MM (2003) Crustal architecture and evolution of the Eastern Ghats Belt and adjacent regions of India. Geological Society, London, Special Publications vol 206, pp 145–168. https://doi.org/10.1144/GSL.SP.2003.206.01.09

Fountain DM, Salisbury MH (1981) Exposed cross-sections through the continental crust: implications for crustal structure, petrology, and evolution. Earth Planet Sci Lett 56:263–277. https://doi.org/10.1016/0012-821X(81)90133-3

Ghosh G, Bose S, Das K, Dasgupta A, Yamamoto T, Hayasaka Y, Chakrabarti K, Mukhopadhyay J (2016) Transpression and juxtaposition of middle crust over upper crust forming a crustal scale flower structure: insight from structural, fabric, and kinematic studies from the Rengali Province, eastern India. J Struct Geol 83:156–179. https://doi.org/10.1016/j.jsg.2015.12.006

Ghosh G, Bose S, Guha S, Mukhopadhyay J, Aich S (2010) Remobilization of the southern margin of the Singhbhum craton, eastern India during the Eastern Ghats orogeny. Indian J Geol 80:97–114

Gibb RA (1983) Model for suturing of Superior and Churchill plates: an example of double indentation tectonics. Geology 11:413–417. https://doi.org/10.1130/0091-7613(1983)11%3c413:MFSOSA%3e2.0.CO;2

Gibb RA, Thomas MD, Lapointe PL, Mukhopadhyay M (1983) Geophysics of proposed proterozoic sutures in Canada. Precambrian Res 19:349–384. https://doi.org/10.1016/0301-9268(83)90021-9

Gupta S (2012) Strain localization, granulite formation and geodynamic setting of 'hot orogens': a case study from the Eastern Ghats Province, India. Geol J 47:334–351. https://doi.org/10.1002/gj.1328

Gupta S, Bhattacharya A, Raith M, Nanda JK (2000) Contrasting pressure–temperature–deformation history across a vestigial craton–mobile belt boundary: the western margin of the Eastern Ghats Belt at Deobhog, India. J Metamorph Geol 18:683–697. https://doi.org/10.1046/j.1525-1314.2000.00288.x

Harley SL (1987) Precambrian geological relationships in high-grade gneisses of the Rauer Islands, East Antarctica. Aust J Earth Sci 34:175–207. https://doi.org/10.1080/08120098708729404

Harley SL, Fitzsimons ICW, Zhao Y (2013) Antarctica and supercontinent evolution: historical perspectives, recent advances and unresolved issues. Geol Soc Lond Spec Publ 383:1–34. https://doi.org/10.1144/SP383.9

Hirth G, Tullis J (1992) Dislocation creep regimes in quartz aggregates. J Struct Geol 14:145–159. https://doi.org/10.1016/0191-8141(92)90053-Y

Katz MB (1985) The tectonics of Precambrian craton—mobile belts: progressive deformation of polygonal miniplates. Precambrian Res 27:307–319. https://doi.org/10.1016/0301-9268(85)90091-9

Khoza TD, Jones AG, Muller MR, Evans RL, Miensopust MP, Webb SJ (2013) Lithospheric structure of an archean craton and adjacent mobile belt revealed from 2-D and 3-D inversion of magnetotelluric data: example from southern Congo craton in northern Namibia. J Geophys Res Solid Earth 118:4378–4397. https://doi.org/10.1002/jgrb.50258

Li ZX, Bogdanova SV, Collins AS, Davidson A, De Waele B, Ernst RE, Fitzsimons ICW, Fuck RA, Gladkochub DP, Jacobs J, Karlstrom KE, Lu S, Natapov LM, Pease V, Pisarevsky SA, Thrane K, Vernikovsky V (2008) Assembly, configuration, and break-up history of Rodinia: a synthesis. Precambr Res 160:179–210. https://doi.org/10.1016/j.precamres.2007.04.021

Mahalik NK (1996) Lithology and tectonothermal history of the Precambrian rocks of Orissa along the eastern coast of India. J SE Asian Earth Sci 14:209–219. https://doi.org/10.1016/S0743-9547(96)00059-1

Mahapatro SN, Pant NC, Bhowmik SK, Tripathy AK, Nanda JK (2012) Archaean granulite facies metamorphism at the Singhbhum Craton-Eastern Ghats Mobile Belt interface: implication for the Ur supercontinent assembly. Geol J 47:312–333. https://doi.org/10.1002/gj.1311

Mandal A, Gupta S, Mohanty WK, Misra S (2015) Sub-surface structure of a craton-mobile belt interface: evidence from geological and gravity studies across the Rengali Province-Eastern Ghats Belt boundary, eastern India. Tectonophysics 662:140–152. https://doi.org/10.1016/j.tecto.2015.01.016

Mezger K, Cosca MA (1999) The thermal history of the Eastern Ghats Belt (India) as revealed by U-Pb and 40Ar/39Ar dating of metamorphic and magmatic minerals: implications for the SWEAT correlation. Precambr Res 94:251–271. https://doi.org/10.1016/S0301-9268(98)00118-1

Mishra S, Deomurari MP, Wiedenbeck M, Goswami JN, Ray S, Saha AK (1999) 207Pb/206Pb zircon ages and the evolution of the Singhbhum Craton, eastern India: an ion microprobe study. Precambr Res 93:139–151. https://doi.org/10.1016/S0301-9268(98)00085-0

Misra S, Gupta S (2014) Superposed deformation and inherited structures in an ancient dilational step-over zone: post-mortem of the Rengali Province, India. J Struct Geol 59:1–17. https://doi.org/10.1016/j.jsg.2013.11.004

Mukhopadhyay J, Ghosh G, Zimmermann U, Guha S, Mukherjee T (2012) A 3.51 Ga bimodal volcanics-BIF-ultramafic succession from Singhbhum Craton: implications for Palaeoarchaean geodynamic processes from the oldest greenstone succession of the Indian subcontinent. Geol J 47:284–311. https://doi.org/10.1002/gj.1314

Nash CR, Rankin LR, Leeming PM, Harris LB (1996) Delineation of lithostructural domains in northern Orissa (India) from Landsat Thematic Mapper imagery. Tectonophysics 260:245–257. https://doi.org/10.1016/0040-1951(95)00187-5

Olierook HKH, Clark C, Reddy SM, Mazumder R, Jourdan F, Evans NJ (2019) Evolution of the Singhbhum Craton and supracrustal provinces from age, isotopic and chemical constraints. Earth Sci Rev 193:237–259. https://doi.org/10.1016/j.earscirev.2019.04.020

Pant NC, Dasgupta S (eds) (2017) Eastern Ghats Mobile Belt. Crustal evolution of India and Antarctica: the supercontinent connection

Ranjan S, Upadhyay D, Abhinay K, Pruseth KL, Nanda JK (2018) Zircon geochronology of deformed alkaline rocks along the Eastern Ghats Belt margin: India-Antarctica connection and the Enderbia continent. Precambrian Res 310:407–424. https://doi.org/10.1016/j.precamres.2018.04.005

Rivers T, Martignole J, Gower CF, Davidson A (1989) New tectonic divisions of the Grenville Province, Southeast Canadian Shield. Tectonics. https://doi.org/10.1029/TC008i001p00063

Saha AK (1994) Crustal evolution of Singhbhum-North Orissa. Geological Society of India, Eastern India

Sawant AD, Gupta S, Clark C, Misra S (2017) The Rauer-Rengali connection in the Indo-Antarctica amalgam: evidence from structure, metamorphism and geochronology. Geol Soc Lond Spec Publ 457(SP457):9. https://doi.org/10.1144/SP457.9

Stipp M, Stünitz H, Heilbronner R, Schmid SM (2002) Dynamic recrystallization of quartz: correlation between natural and experimental conditions. Geological Society, London, Special Publications vol 200, pp 171–190. https://doi.org/10.1144/GSL.SP.2001.200.01.11

Tesha AL, Nyblade AA, Keller GR, Doser DI (1997) Rift localization in suture-thickened crust: evidence from bouguer gravity anomalies in northeastern Tanzania, East Africa. Tectonophysics 278:315–328. https://doi.org/10.1016/S0040-1951(97)00110-8

Upadhyay D, Raith MM (2006) Intrusion age, geochemistry and metamorphic conditions of a quartz-monzosyenite intrusion at the craton–Eastern Ghats Belt contact near Jojuru, India. Gondwana Res 10:267–276. https://doi.org/10.1016/j.gr.2006.02.011

Vernon RH (2004) A practical guide to rock microstructure. Cambridge University Press. https://doi.org/10.1017/CBO9780511807206

Zhao G, Sun M, Wilde SA, Li S (2004) A Paleo-Mesoproterozoic supercontinent: assembly, growth and breakup. Earth-Science Rev 67:91–123. https://doi.org/10.1016/j.earscirev.2004.02.003

A Preliminary Study on Earthquake Source Properties Based on Geochemistry, Shear Resistance and Melt Pressure of Pseudotachylites, Gangavalli Fault, South India

Bhuban M. Behera, V. Thirukumaran, Neeraj Kumar Sharma and Tapas Kumar Biswal

Abstract Voluminous pseudotachylites occur along NE-SW trending Gangavalli sinistral strike-slip fault in the Southern Granulite Terrane of South India. We made a field, microscope and geochemistry study of the pseudotachylites, and determined the parent rock composition, frictional shear resistance (τ_f) and melt pressure (P_m). Based on the result, we made a preliminary interpretation of source property of the earthquake. The pseudotachylite veins belong to two types, (i) the fault veins were produced by in-situ melting, these were used to compute the τ_f during coseismic slip, and (ii) the injected veins formed due to dilation of the pre-existing weak planes, these have been used to calculate the P_m. The pseudotachylites bear chemical similarity with charnockites in that they show andesite to granite composition in TAS diagram, calc-alkaline trend in AFM plot, possess REE fractionation with LREE enrichment and lack Eu anomaly. Hence, pseudotachylites were derived from melting of the charnockite. Further, the pseudotachylites are dominated by hexagonal β-quartz clasts that suggest the maximum temperature of melting was at 1550 °C. The fault veins exhibit thickness/displacement ratio varying between 0.03 and 0.1. Assuming that there is no loss of melt from fault veins, maximum shear resistance is estimated at 48.95 MPa, characteristic of large magnitude-earthquake. The injected veins are thicker, show varied geometry and dominantly are aligned in NE-SW direction. The melt pressure $P_m > \sigma_2$, stress ratio $\Phi = 0.87$ and driving pressure $R' = 0.9$ were calculated from stereoplot and 3D Mohr circle. Higher stress ratio indicates $\sigma_2 \approx \sigma_1$ that leads to flip-flop of σ_1 from horizontal to vertical. This was probably the result of stress drop during stick-slip mechanism.

Keywords Pseudotachylite · Gangavalli strike-slip fault · Fault vein · Injected vein · Shear stress resistance · Melt pressure

---

B. M. Behera (✉) · N. K. Sharma · T. K. Biswal
Department of Earth Sciences, Indian Institute of Technology Bombay, Mumbai 400076, India

T. K. Biswal
e-mail: tkbiswal@iitb.ac.in

V. Thirukumaran
Government Arts College, Salem 636007, India

T. K. Biswal et al. (eds.), *Structural Geometry of Mobile Belts of the Indian Subcontinent*, Society of Earth Scientists Series,
https://doi.org/10.1007/978-3-030-40593-9_8

1 Introduction

The pseudotachylites have two modes of formation (i) impact melting (Spray and Thompson 1995) and (ii) frictional melting during coseismic slip at high strain rate as high as 1–10 m/s (Sibson 1975; Magloughlin 1992). We are here studying coseismic slip-pseudotachylites. The pseudotachylite melt quenches in contact with the cold host rock that's why the pseudotachylite veins appear as a dark ink mark on the white rock (Kirkpatrick et al. 2012). Devitrification of pseudotachylites produces microcrystals like microlites and spherulites that provide an evidence of melt origin of pseudotachylite (Philpotts 1964; Spray 1992). Hence, pseudotachylite veins are unequivocally considered as direct field evidence to recognize a paleo-earthquake in the geological records. Researches over last four/five decades dealt extensively on the physico-dynamic process involving the production of pseudotachylite melts and correlate them with source properties of earthquake (Di Toro et al. 2005; Beeler et al. 2016). In this paper, we are studying the pseudotachylites of the Gangavalli fault in the Southern Granulite Terrane (SGT), South India (Fig. 1a and inset). We are focusing on the physical and chemical properties of the pseudotachylite veins, determine the parent rock composition, temperature of melting, compute frictional shear resistance (τ_f) and melt pressure (P_m). We used these parameters to have a preliminary knowledge on the source of the earthquake. We classified the pseudotachylite veins into fault veins which are used for computing the τ_f and injected veins to estimate the P_m.

2 Geological Setting

The Gangavalli fault is a NE-SW striking sinistral strike-slip fault in the eastern part of the SGT (Fig. 1a, Behera et al. 2017). The SGT is an Archean terrane that was completely reworked at 2.6 Ga by granulite facies metamorphism, charnockite intrusion and charnockitisation, granite-alkali-anorthosite-basic-ultrabasic rock intrusions. Several greenschist facies shear zones containing mylonite/ultramylonite rocks divide the terrane into distinct blocks. The shear zones are namely the Salem-Attur shear zone, the Palghat-Cauvery shear zone, the Karur-Kambam-Painavu-Trichur shear zone and the Achankovil shear zone (Fig. 1b inset) and the intervening blocks are namely the Salem-Namakkal block, the Madurai block and the Trivandrum block. The Gangavalli fault occurs in the Salem-Namakkal block which had multiple phases of deformation from D_1 to D_4 (Behera et al. 2019). The first deformation D_1 was coeval with granulite facies metamorphism and occurred at 2.5–2.3 Ga. It produced a recumbent fold and S_1 horizontal granulitic fabric which were overprinted by NE-SW trending upright to isoclinal fold during the D_2 deformation. The D_3 deformation is mostly characterized by E-W running ductile shear zone which occurred at 2.0 Ga. The Salem-Attur shear zone is one of such D_3 shear zone that shows N-verging thrust kinematics. The granulites were deformed by brittle strike-slip and

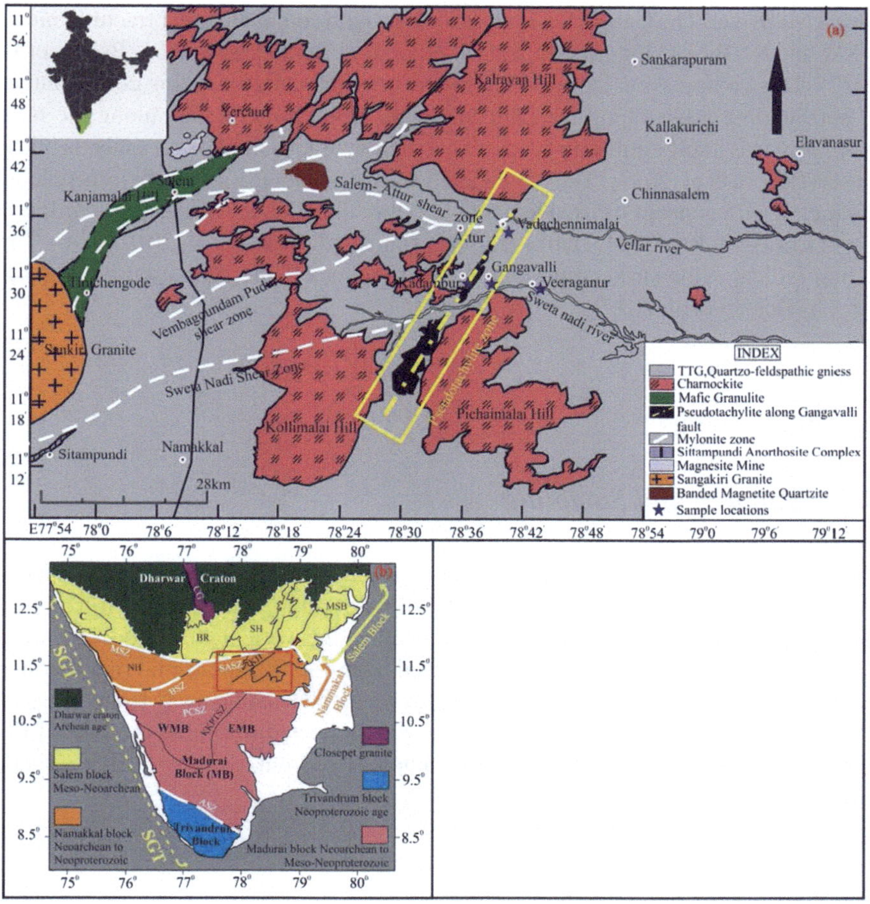

Fig. 1 **a** Geological map of the Gangavalli fault (Yellow box) in the Salem-Namakkal block in the Southern Granulite Terrane (SGT), India (Location map in the inset) (after Sundaralingam et al. 2017). Sample locations are indicated by place names as Vadachennimalai, Gangavalli, Veeraganur and Kadambur, **b** Different blocks in the SGT. ASZ—Achankovil shear zone, BR—Biligirirangan, BSZ—Bhabani shear zone, C—Coorg, CG—Closepet granite, KH—Koli hill, KKPTSZ—Karur-Kambam-Painavu-Trichur shear zone, MSB—Madras block, MSZ—Moyar shear zone, PCSZ—Palghat Cauvery shear zone, SASZ—Salem-Attur shear zone, SH—Shevroy hill, WMB-Western Madurai block, EMB-Eastern Madurai block

normal faults during the D_4 deformation. The Gangavalli fault is a D_4 strike-slip fault (Behera et al. 2019).

The Gangavalli fault stands out as an elongated ridge marked by a 1–5 km wide and 50 km long cataclasites and pseudotachylites (Fig. 1a). It has sheared the 2.6 Ga old charnockites, quartzofeldspathic gneisses of Kollimalai and Pichaimalai hills, 2.0 Ga old mylonites of the Salem-Attur shear zone and 1.9 Ga old granite intrusions. Its northern end is demarked by Vellar River and the southern end is bounded by

Sweta Nadi river. The Gangavalli fault associates with conjugate shear fractures along NNW and NNE directions that indicate N-S horizontal compression (σ_1, Behera et al. 2017). This compression led to a high rate of friction and devastating comminution of surrounding rocks along the fault interface. This size reduction along the fault proximity was a self-similarity process specified by fractal dimension analysis of the clasts (Behera et al. 2017). The fractal analysis and roundness study of clasts reflects the melt origin concept of pseudotachylite rather than impact origin (e.g., Philpotts 1964; Lin 1999; Ray 2004; Deb et al. 2015; Han 2017; Sarkar et al. 2019). ^{40}Ar-^{39}Ar dating indicates the strike-slip faulting and pseudotachylite formation were at ca. 1.8 Ga (Behera et al. 2017).

3 Methodology

3.1 Mapping and Sampling

We have carried out field mapping of the Gangavalli fault using enlarged Survey of India Toposheet No. 58 I/10 and 58 I/11 in 1:10,000 scale. Structural data pertaining to fractures, type of pseudotachylite veins, their orientation and width were collected in the field. Samples were collected from four locations and thin sections were prepared (Fig. 1a). Charnockite samples were collected far away from the pseudotachylite zone. Pseudotachylite samples were collected from massive part of the veins. We separated pure pseudotachylite samples free from clasts by verifying under microscope.

3.2 Geochemical Analysis

We analysed the charnockite and the pseudotachylite samples geochemically. Major, trace and REE were analyzed (result, Table 1) through an Inductively Coupled Plasma Atomic Emission Spectrometer (ICP-AES, Instrument: Jobin Yvon Ultim-2) (for procedure, Sheth et al. 2011). Matrix, microlite and individual clasts of the fault vein were analysed (result, Table 2) under Electron Probe Microanalyser (EPMA) point analysis method (for procedure, Behera et al. 2019). Further, we analysed individual clasts under XRD by separating the clasts manually. All these analyses were carried out at the Department of Earth Sciences, IIT Bombay.

Table 1 ICP-AES chemical data and CIPW norm of charnockite (Ch) and pseudotachylite (Ps) veins, Gangavalli fault, South India

Samples No >	10 (Ch)	16 (Ch)	22 (Ch)	23 (Ch)	3L30 (Ch)	3L33 (Ch)	2 (Ps)	16 (Ps)	22 (Ps)	23 (Ps)	3L30 (Ps)	3L33 (Ps)
Oxides												
SiO_2	61.77	67.12	70.39	71.17	65.06	66.79	68.53	61.39	67.99	71.91	70.19	65.46
TiO_2	0.7	0.19	0.04	0.26	0.8	0.49	0.34	1.94	0.28	#REF!	0.36	0.51
Al_2O_3	17.27	16.99	16.46	15.78	14.03	15.07	15.67	13.81	15.81	14.66	14.91	15.22
Fe_2O_3*	6.35	3.06	1.82	1.99	5.36	4.25	3.21	6.17	3.55	1.94	3.8	4.64
MnO	0.09	0.07	0.04	0.03	0.13	0.05	0.04	0.25	0.06	0.02	0.04	0.05
MgO	2.63	1.21	0.78	0.68	2.69	2.42	1.54	4.98	1.33	0.68	1.34	2.45
CaO	5.02	5.75	4.44	3.43	5.13	4.47	4.33	8.11	4.59	3.44	3.51	4.89
Na_2O	4.45	4.92	4.58	4.69	3.57	3.96	4.53	2.43	4.13	4.49	4.58	3.97
K_2O	0.87	0.68	1.02	1.67	1.35	1.91	1.38	0.65	1.1	1.23	1.15	2.52
P_2O_3	0.13	0	0.08	0.04	0.15	0.37	0.11	0.08	0.14	0.01	0.1	0.29
Total	99.29	99.99	99.67	99.74	98.28	99.78	99.69	99.81	98.98	98.62	99.99	99.99
CIPW												
Q	18.246	23.137	29.52	29.358	26.365	24.806	26.289	23.466	28.755	33.362	29.929	20.728
Or	5.141	4.018	6.027	9.869	8.096	11.287	8.155	3.841	6.559	7.387	6.796	14.892
Ab	37.655	41.631	38.924	39.77	30.716	33.593	38.416	20.554	35.285	38.501	38.755	33.593
An	24.055	22.265	21.415	16.804	18.624	17.74	18.438	24.936	21.579	16.458	16.728	16.266
Di	0	4.571	0.146	0	3.099	0.434	1.048	7.082	0	0.025	0	3.517
Hy	6.55	0.894	1.874	1.693	5.387	5.851	3.349	9.145	3.337	1.706	3.337	4.471
Ilm	0.132	0.149	0.076	0.064	0.278	0.107	0.085	0.534	0.128	0.042	0.085	0.107
Ap	0.3012	0	0.185	0.092	0.347	0.857	0.254	0.185	0.324	0.023	0.231	0.671

(continued)

Table 1 (continued)

Samples No >	10 (Ch)	16 (Ch)	22 (Ch)	23 (Ch)	3L30 (Ch)	3L33 (Ch)	2 (Ps)	16 (Ps)	22 (Ps)	23 (Ps)	3L30 (Ps)	3L33 (Ps)
Crn	0.191	0	0	0.121	0	0	0	0	0.061	0	0.305	0
Rt	0	0	0	0.226	0	0	3.22	6.18	3.59	1.97	3.8	4.64
Mag	0.091	0	0.014	0	0	0	0.723	4.069	0.37	0.533	0.022	1.113
Hem	6.287	3.06	1.819	2	5.45	4.26						
Spn	0	0.278	0	0	1.628	1.064						
Trace elements												
Ba	312.493	162.411	265.304	616.724	428.161	736.596	401.461	261.698	349.732	419.969	429.219	720.279
Co	16.277	5.903	1.659	4.643	14.232	10.132	8.507	43.361	7.038	4.912	185.693	12.031
Cr	21.57	10.115	16.628	14.07	32.855	19.802	14.423	17.019	10.819	9.532	13.507	22.23
Ni	53.771	52.213	39.45	52.836	85.118	48.092	60.82	84.859	56.553	73.495	151.283	90.083
Sc	12.818	5.989	2.976	2.179	11.844	7.177	6.508	38.776	5.829	2.729	5.036	7.222
Y	10.784	8.212	6.862	0.631	10.548	9.522	3.947	31.858	8.422	1.675	5.559	9.177
Zn	1383.24	2094.33	1588.93	2260.48	1136.51	1044.71	2280.64	2583.72	2013.66	3660.14	2749.57	3096.42
V	85.175	32.234	19.942	18.483	83.719	80.735	43.502	409.431	38.493	20.014	33.083	55.863
Zr 339.198	104.813	123.932	215.421	79.669	102.062	171.377	131.063	133.842	107.626	89.876	127.201	134.651
Sr	427.832	280.954	304.529	455.317	235.252	659.67	393.003	172.345	298.29	407.118	275.963	583.256
Pb	3.342	2.571	3.13	2.273	3.421	3.288	3.472	2.584	3.53	2.368	3.08	2.439
REE data												
La	0.34	0.1	0.35	0.2	0.35	0.86	0.55	0.22	0.4	0.22	0.35	0.61
Ce	0.56	0.2	0.74	0.31	0.86	2	0.34	0.66	0.88	0.37	0.72	1.51

(continued)

Table 1 (continued)

Samples No >	10 (Ch)	16 (Ch)	22 (Ch)	23 (Ch)	3L30 (Ch)	3L33 (Ch)	2 (Ps)	16 (Ps)	22 (Ps)	23 (Ps)	3L30 (Ps)	3L33 (Ps)
Pr	0.16	0.05	0.19	0.08	0.23	0.56	0.09	0.2	0.23	0.09	0.19	0.41
Nd	0.32	0.1	0.25	0.09	0.37	0.78	0.14	0.44	0.32	0.1	0.25	0.63
Sm	0.07	0.01	0.05	0.01	0.08	0.15	0.03	0.13	0.06	0.02	0.05	0.13
Eu	0.02	0.01	0.02	0.01	0.02	0.03	0.01	0.04	0.02	0.01	0.01	0.03
Gd	0.07	0.03	0.03	0.01	0.06	0.08	0.03	0.14	0.04	0.01	0.03	0.08
Tb	0.01	0.01	0.01	0	0.01	0.01	0.01	0.03	0.02	0.01	0.01	0.01
Dy	0.06	0.04	0.03	0.01	0.05	0.05	0.02	0.15	0.04	0.01	0.02	0.05
Ho	0.01	0.01	0.01	0	0.01	0.01	0	0.03	0.01	0	0.01	0.01
Er	0.13	0.11	0.06	0.02	0.06	0.04	0.06	0.18	0.04	0.06	0.03	0.06
Tm	0.01	0	0	0	0.01	0	0	0.01	0	0	0	0
Yb	0.03	0.03	0.02	0	0.03	0.02	0.01	0.09	0.02	0.01	0.01	0.02
Lu	0	0	0	0	0	0	0	0.01	0	0	0	0

Table 2 EPMA Chemical analysis of the pseudotachylite clast (1–5), microlite (6–9) and matrix (20–24) of fault vein, Gangavalli fault, South India (Fig. 3)

Points >	1	2	3	4	5	6	7	8	9	20	21	22	23	24
Oxides														
SiO_2	100.05	97.85	100.25	99.15	99.36	68.83	61.1	62.86	64.01	70.1	69.79	62.9	67.89	67.02
TiO_2	0.07	0.05	0	0.01	0	0.25	0.46	0.5	0.55	1.06	0.94	1.23	0.44	0.46
Al_2O_3	0.12	0.93	0.02	0.03	0.05	19.08	19.77	19.14	19.16	8.93	8.44	10.47	18.94	19.2
FeO	0	0.35	0	0.01	0.05	1.14	4.42	3.73	4.53	8.26	8.57	10.63	1.75	2.03
MnO	0	0	0	0	0	0	0.07	0.09	0.07	0.03	0.07	0.03	0	0.03
MgO	0	0.2	0	0	0	0.52	2.27	1.91	2.35	6.05	6.1	7.52	0.94	1.05
CaO	0.02	0.7	0.01	0	0.05	4.58	5.25	4.67	4.64	0.34	0.2	0.24	5.54	5.64
Na_2O	0.04	0.37	0	0	0	7.31	7.6	7.97	7.52	0.14	0.03	0.03	4.77	4.76
K_2O	0	0.03	0	0	0	0.28	0.28	0.26	0.42	4.83	4.77	6.03	1.15	1.16
P_2O_3	0	0	0.02	0.02	0.01	0.05	0.11	0.14	0.13	0.1	0.12	0.16	0.1	0.12
Total	100.35	100.48	100.33	99.25	99.52	102.07	101.37	101.3	103.46	99.97	99.14	99.43	101.59	101.51
CIPW														
Q	99.774	94.11	99.95	99.92	99.729	14.615	0	1.161	3.119	35.245	35.997	20.524	21.288	20.204
Or	0	0.1773	0	0	0	1.595	1.654	1.536	2.422	1.184	28.484	35.931	6.677	6.855
Ab	0.3385	0.7881	0	0	0	60.586	64.309	66.593	61.516	28.586	0.253	0.253	39.77	40.278
An	0.0992	3.1308	0	0	0.136	18.087	19.002	15.476	16.716	1.033	0.208	0.145	26.434	27.196
Di	0	2.1279	0	0	0.041	3.222	5.282	5.409	3.93	0	0	0	0	0
Hy	0	0	0	0.002	0.069	1.303	8.39	8.119	10.988	28.586	29.789	36.555	4.764	5.639
Ilm	0	0.095	0	0.019	0	0.474	0.873	0.931	1.006	2.013	1.804	2.355	0.816	0.873
Ol	0	0	0	0	0	0	1.563	0	0	0	0	0	0	0

(continued)

Table 2 (continued)

Points >	1	2	3	4	5	6	7	8	9	20	21	22	23	24
Ap	0	0	0.046	0.046	0.023	0.115	0.254	0.324	0.301	0.231	0.278	0.37	0.231	0.278
Crn	0.0178	0	0.02	0.03	0	0	0	0	0	3.092	3.177	3.866	0.017	0.146
Rt	0.007	0	0	0	0	0	0	0	0	0	0	0	0	0

3.3 Frictional Stress Measurement

Frictional stress measurement is based on the fact that the volume of melt produced during a frictional event is guided by the work-energy principle. The energy released during frictional work was interpreted to have utilized in melting the country rock (Han 2017; Scholz 2019). But loss of melt through injected vein may produce undesired ambiguity in the determination of frictional shear resistance (Di Toro et al. 2005). Frictional stress measurement was done on fault veins. These veins occur along fracture planes that show displacement. Pseudotachylite vein-dimension such as thickness and fault displacements, are important field parameters to investigate the work energy principle of rock melting. The two parameters are related by following empirical equation (Sibson 1975).

$$d = 436 \times a^2. \tag{1}$$

a = thickness, d = displacement.

According to the above relationship, a 5 m long displacement may produce nearly one cm uniform width of pseudotachylite vein along the fault plane.

The work done by displacement is converted to heat energy that is used for melting of the rock. Scholz (2019) documented work-energy balance equation for an earthquake associated faulting as follows,

$$W_f = Q + E_s + U_s + W_g \tag{2}$$

where, W_f is work done during faulting, Q is heat generated along friction surface, E_s seismic energy released in form of elastic wave, U_s work required for fault gauge formation and W_g work done against gravity. However, E_s is negligible when compared with the total energy released during the earthquake (McGarr 1999), U_s is also negligible with respect to the work done by friction (Scholz 2019) and W_g has no significant value for a strike-slip fault (Gangavalli strike-slip fault). Hence, Eq. 2 is reduced as follows,

$$W_f \approx Q. \tag{3}$$

The work-done, W_f, that produces the pseudotachylite melt is resulted by the frictional heating, Q (Jeffreys 1942; McKenzie and Brune 1972). So, a minimum frictional shear resistance has to overcome before the onset of melting. Therefore, it is important to calculate the shear resistance (τ_f) of the rock that onset melting. Here, we have used a formula suggested by Di Toro et al. (2005) to calculate the τ_f as follows,

$$\tau_f = \rho\left[(1 - \phi)H + c_p(T_i - T_{hr})\right]t/d, \tag{4}$$

where the thickness and displacement (*t/d*) can be measured in the field in a centimeter scale. The thickness was measured at the center of the vein. The ϕ—volume/area fraction or ratio of clasts and melt, was measured on a 2-D plane of photomicrograph using the 'image J' software. The ρ is the density of the rock. The thermal parameters H—latent heat of fusion and c_p—specific heat capacity, T_i—initial melt temperature and T_{hr}—host rock temperature were known parameters. The ϕ was calculated for sixteen samples (Locations are after the name of local places, Fig. 1a).

3.3.1 Limitations

The estimation of an earthquake parameter such as friction shear resistance (τ_f) by using the thickness-displacement model is challenging since some of the melt may be lost as injected vein from the sliding surface. To minimize such error, we have selected those veins which occur independently, confined to a straight fracture without any branches into the country rock.

3.4 Melt Pressure Measurement

Dykes and veins intrude when the magma/melt pressure, P_m exceeds the tectonic stress σ_n across the weak plane such as the fracture plane. This can be denoted as $P_m \geq \sigma_{\mathrm{n}}$, where P_m stands for magma/melt pressure and σ_n stands for normal stress acting on the weak plane (Baer et al. 1994; Jolly and Sanderson 1997). The fractures perpendicular to σ_3 were most dilated since σ_3 is the least tectonic stress. Depending upon the value of P_m relative to three principal stresses ($\sigma_1 > \sigma_2 > \sigma_3$), veins of different orientations develop. When plotted in a stereonet, the veins will produce different distribution patterns. If $P_m \geq \sigma_2$, the girdle distribution pattern is produced. σ_1, σ_2 and σ_3 orientations are determined from the stereoplot (Jolly and Sanderson 1997; Mondal and Mamtani 2013; Martínez-Poza et al. 2014; Tiwari and Biswal 2019). The minimum angle between the dilated fracture and the σ_1, σ_2 and σ_3 directions (θ_1, θ_2, θ_3, respectively) can be measured from the stereoplot. θ_1, θ_2 and θ_3 values are used to construct the 3D Mohr plot to compute the P_m values relative to σ_1, σ_2 and σ_3.

Furthermore, the stress ratio 'Φ' for different stress combinations is computed from the following equations (Jolly and Sanderson 1997; Mondal and Mamtani 2013; Martínez-Poza et al. 2014).

For $P_m > \sigma_2$,

$$\Phi = (\sigma_2 - \sigma_3)/(\sigma_1 - \sigma_3) = 1 - \{(1 - \cos 2\theta_2)/(1 - \cos 2\theta_3)\}. \quad (5)$$

The driving pressure ratio R' that describes the magnitude of the magma/melt pressure relative to the tectonic stress is given by the formula

$$R' = (P_m - \sigma_3)/(\sigma_1 - \sigma_3) = (1 + \cos 2\theta_2)/2. \text{ (Jolly and Sanderson, 1997)} \quad (6)$$

This has been used by several researchers (André et al. 2001; McKeagney et al. 2004; Mazzarini and Isola 2007; Mazzarini et al. 2011; Mondal and Mamtani 2013; Martínez-Poza et al. 2014; Lahiri and Mamtani 2016).

4 Result

4.1 Megascopic Study of Pseudotachylite Veins

The pseudotcahylite veins occur within extremely fractured and brecciated charnockite host rock. Most of the fractures lie in two dominant directions, NE-SW and NW-SE. The pseudotachylite veins are straight and have a dominant NE-SW strike. The veins vary in thickness from millimeters to few centimeters. Generally, two kinds of veins are recognized in the field such as fault veins and injection veins. A common observation is that the fault veins are straight, thin and sometimes difficult to measure their thickness. They occur independently inside a fracture and have no branching into country rock. These veins are undoubtedly identified as fault veins because they preserve an offset marker along the slip surface (Fig. 2a). These markers are gneissic layer or different generation of pseudotachylite vein and/or it may be an offset of older fracture. The pinch-and-swell structures are also noticed in fault veins which result due to uneven distribution of viscosity in the melt (Fig. 2a) (e.g., Gardner et al. 2016). However, the injected veins produce anastomose structures with irregular thickness and some are associated with dendritic structures. Different veins geometries are observed such as bleb, lens, sigmoidal, network, and octopus shape (Fig. 2b). A characteristic feature of injected veins is marked by its tapering end. It indicates the direction of melt migration where the flow is ceased at the tapering end with no driving stress of fluid remains.

There are many fractures observed within the pseudotachylite vein itself. These fractures are orthogonal to the vein wall suggesting a chilled effect as soon as it comes in contact with the surrounding low-temperature country rock. The evidence of the comminution process can be found from the country rock fragments known as lithic clasts which are incorporated during brittle deformation. Extensive amount of monomineralic clasts are present in the pseudotachylite vein. The rounded edges of the fragments/clasts are achieved by high-temperature melting at the corner and the clasts appear like "quasi-conglomerate".

Fig. 2 Field photos and photomicrographs of pseudotachylite veins. **a** Displacement by a fault vein is marked by offset of an older vein. The coin diameter is 2.5 cm, **b** Highly irregular injection vein, **c** Chilled margin of pseudotachylite vein wall, **d** Displacement of clasts within injection vein, **e** Acicular microlites wrap around clasts and oriented in the direction of flow of melt, **f** Sheaf microlites in the pseudotachylite matrix, inset: plumose structure at both ends of the microlite

4.2 Microscopic Study of Pseudotachylite Veins

4.2.1 Matrix

Pseudotachylite veins contain an ultrafine to glassy matrix (Fig. 2c) with a chilled margin at the contact with host rock. Microscopic analysis of matrix particularly of the fault veins, in plane polarizer, indicates different colour shades from dark brown to grey brown within a single vein. The matrix shows a flow folding structure which is produced during melt flow.

4.2.2 Clast

The clasts stand out brighter than the surrounding matrix under crossed Nicol. Some clasts are sub-round with decrepitating feature along the boundary (e.g., Ray 2004). Clasts are dominantly quartz and show wavy extinction. Some of them show micro fault (Fig. 2d). Elongated clasts show preferred orientation along melt flow direction. A wide range of clast size distribution is observed in the pseudotachylite veins.

4.2.3 Microlite

Microlites are profusely developed in the pseudotachylite veins. Their size increases from margin to center of the vein. The microlites are acicular and sheaf microlites. Acicular microlites are needle-shaped and simple group of microlites (Fig. 2e). They are wrapped around the clasts and aligned to a particular orientation giving rise a directive texture and flow direction. The sheaf microlites are spherulite group of microlites (Fig. 2f). They consist of a central bar and plumose fibers at both ends (Fig. 2f inset). These sheaf microlites are concentrated at the center of the vein where complexity in the structure of microlite is more.

4.3 Geochemical Analysis

4.3.1 Charnockite and Pseudotachylite

The SiO_2 varies from 61.77 to 71.17% in charnockite, 61.39 to 71.91% in injected pseudotachylite and 61.1 to 70.1% in fault pseudotachylite veins. The Al_2O_3 % in charnockite is comparable with that of pseudotachylite. Charnockite is quartz, K-feldspar, plagioclase (albite), hypersthene and diopside normative like the injected pseudotachylites (Table 1). Geochemical plots such as SiO_2/total alkali (Fig. 3b) indicates andesite to rhyolite (granite) composition of charnockite and pseudotachylite;

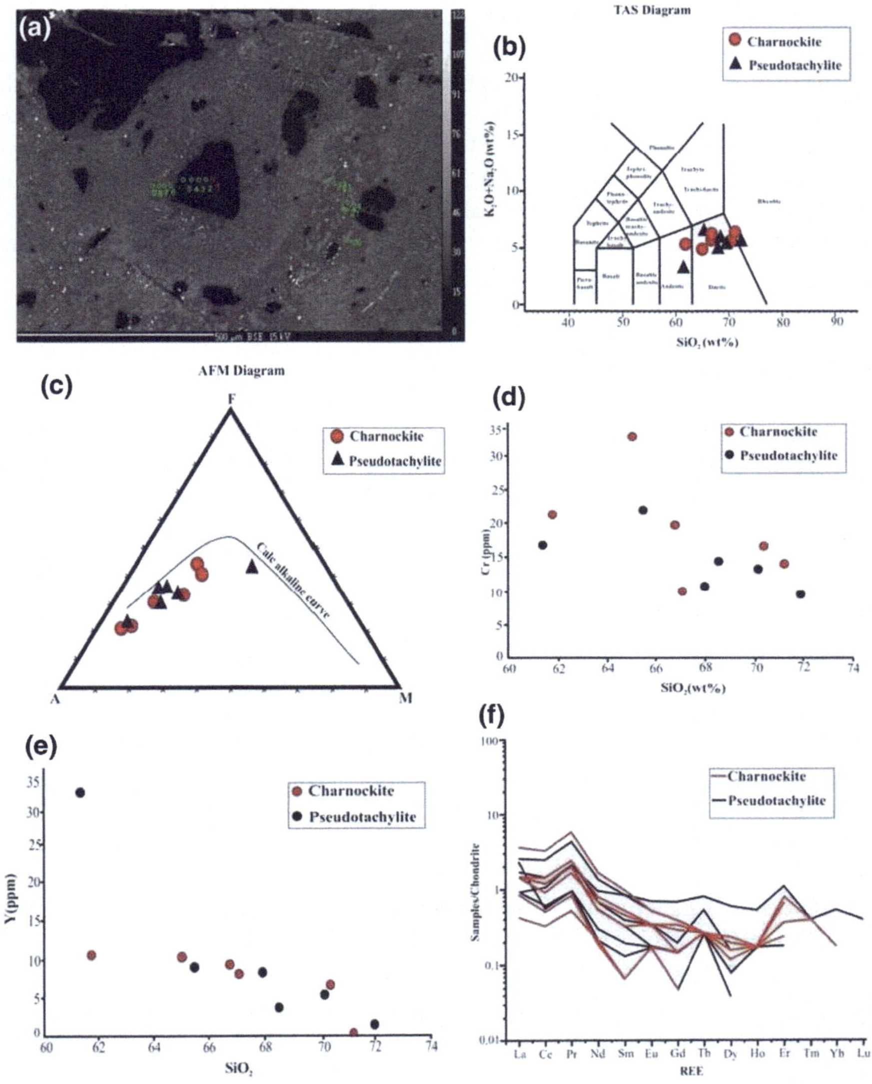

Fig. 3 **a** Back scatter electron image of fault vein with clasts, microlites and two melts of different chemical composition, points are analysed under EPMA. The dark clasts represent quartz minerals (analysis are made at point 1–5), sheaf microlites are albitic composition (analysis are made at point 6–9), melt having high concentration of FeO(T)-MgO-K_2O is analyzed at point 20–22 and high concentration of Na_2O-CaO at point 23–24 (Table 2), **b** TAS diagram for charnockite and pseudotachylite, they indicate andesite-dacite-rhyolite composition, **c** AFM diagram for charnockite and pseudotachylite, they show calc alkaline differentiation trend, **d** and **e** Cr and Ni versus SiO_2 plot showing negative correlation for both charnockite and pseudotachylite, **f** Chondrite normalized REE plot show differentiation trend, LREE enrichment and absence of Eu anomaly

and total alkali-Fe–Mg plot shows calc-alkaline trend (Fig. 3c). Y and Cr show negative correlation with SiO_2 in both (Fig. 3d, e). Chondrite normalized REE plot shows differentiation trend with LREE enrichment, and absence of Eu anomaly for both charnockite and pseudotachylite (Fig. 3f). The EPMA data of fault pseudotachylite veins show (Fig. 3a) that the points 23 and 24 are low in Feo and MgO values and high in CaO and MgO values. Point 20, 21 and 22 have high Feo and Mgo content and low Na_2O and CaO content (Fig. 3a, Table 2). Points 23 and 24 have more normative albite and anorthite while points 20, 21 and 22 have more normative hypersthene (Table 2).

4.3.2 Clast and Microlite

The points 1–5 in EPMA image show quartz clasts (Fig. 3a; Table 2). XRD analysis of clasts shows peak values at three major d-spacing of 4.25, 3.34 and 1.81 Å (Fig. 4a). The clasts are identified as hexagonal β-quartz. The points 6–9 show albite microlite.

4.4 Shear Stress Estimation

The ϕ values vary between 0.1 and 0.25 (Table 3). The thickness (t) of the veins is in the range of 0.2–0.65 cm. Displacement (d) along the fracture hosting the fault vein varies between 0.6 and 10 cm. The t/d ratio is between 0.026 and 0107. The thickness increases with displacement (Fig. 4b). For calculation of "τ_f" we have taken density (ρ) of the charnockite as 2850 kg m^{-3}. The latent heat of fusion (H) is same as granite and may be taken a value of 80 cal/gram $\approx 3.34 \times 10^5$ J/Kg at about 1050 °C (Heuze 1983). Specific heat c_p is taken as ~1000 J/Kg °C for most rock type (Sibson 2003). The minimum initial temperature of melting (T_i) can be considered as 1550 °C which is inferred from the melting of hexagonal β-quartz. The ambient temperature of host charnockite rock (T_{hr}) is taken as a maximum of 250 °C. Using the above conditions and data, we have calculated the frictional shear resistance (τ_f) that varies between 11.97 and 48.95 MPa with an average of 30.70 MPa (Table 3). The τ_f versus displacement shows a negative correlation (Fig. 4c).

4.5 Melt Pressure Estimation

The Fig. 4d shows the stereoplot (240 poles) for pseudotachylite injection veins. Maximum number of plots is in the NW and SE quadrants and very few data lie in the NE and SW quadrants. There is an elliptical void area at the centre. This sort of distribution pattern is known as girdle pattern. The centre of the void area indicates σ_1 position, because, planes normal to σ_1 are least dilated. The median great circle is drawn at the middle of the elliptical void area. The pole of the median great circle

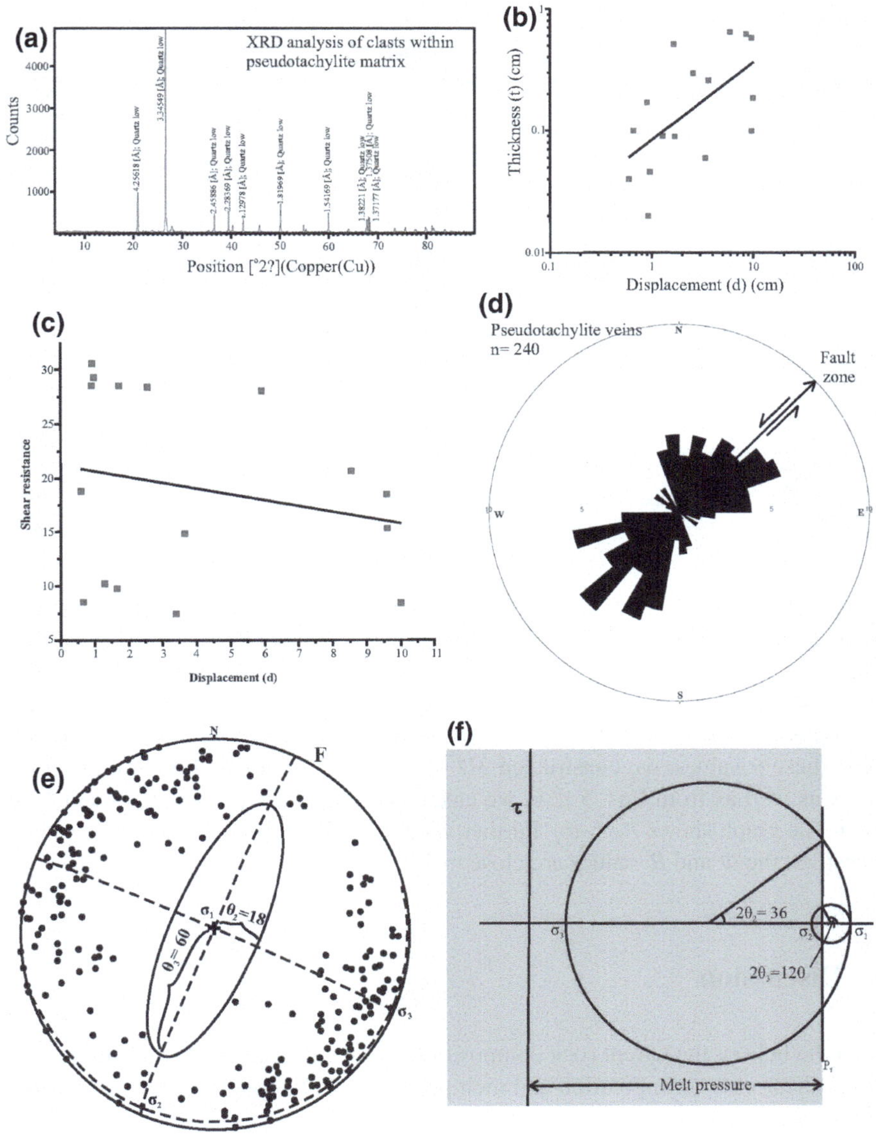

Fig. 4 **a** XRD analysis of quartz clasts which are physically separated from the pseudotachylite vein. All clasts indicate hexagonal β-quartz in albitic matrix, **b** Thickness versus displacement plot for fault veins shows positive correlation, **c** Shear resistance versus Displacement plot for fault vein shows negative correlation, **d** Stereoplot for injected veins, shows girdle distribution, for detail see text, **e** 3D Mohr plot for $P_m > \sigma_2$

Table 3 Estimation of frictional shear resistance (τ_f) from fault veins pseudotachylite of Gangavalli fault, South India. Highest and lowest values are marked in shades of grey

Locations	Locality	Lat./Lon.	ϕ = (clast/melt)	Thickness (*t*) in cm	Displacement (*d*) in cm	*t*/*d*	Shear stress τ_f = ρ*Em*(*t*/*d*) in MPa
L1	Kadambur	N11°29'29.46" E78°37'9.96"	0.13	0.04	0.6	0.067	30.27
			0.14	0.1	0.96	0.104	47.12
L2	Vadachennimalai	N11°34'57.36" E78°39'57.60"	0.1	0.17	1.7	0.1	45.62
			0.14	0.02	0.66	0.030	13.71
			0.1	0.05	1.29	0.036	16.27
L3	Gangavalli	N11°29'23.50" E78°38'29.80"	0.1	0.09	0.9	0.1	45.62
			0.13	0.52	9.6	0.054	24.55
			0.14	0.09	3.4	0.026	11.97
			0.14	0.3	10	0.03	13.57
L4	Veeraganur	N11°28'49.36" E78°43'10.16"	0.25	0.06	1.66	0.036	15.97
			0.16	0.26	2.56	0.102	45.75
			0.24	0.65	8.55	0.076	33.67
			0.11	0.63	9.6	0.065	29.68
			0.1	0.58	5.9	0.098	44.84
			0.11	0.1	0.93	0.107	48.95
			0.1	0.19	3.64	0.052	23.69
Average			0.14	0.24	3.87	0.07	30.70

defines the σ_3 position. Maximum number of data points lie near σ_3. The σ_1 and σ_2 lie on the median great circle. The σ_1 lies near vertical and the σ_2 at the periphery. This is characteristic of normal fault, but we are dealing with Gangavalli fault which is a strike-slip fault. In a strike-slip fault the σ_1 is horizontal. This discrepancy will be discussed later in Sect. 5.4. The θ_2 and θ_3 values were measured as 18° and 60°. With these θ values, we constructed 3D Mohr circle for stress (Fig. 4e). From the circle as well as from Eqs. 5 to 6, we calculated the values of Φ as 0.88 and R' as 0.90. The graph shows $P_m > \sigma_2$. Further, σ_1 and σ_2 are approaching similar values, therefore, the Φ and R' values are close to 1.0.

5 Discussion

With the help of the parent rock composition, clast and microlite composition, frictional shear stress computation and melt pressure calculation, the source properties of the earthquake are to be interpreted.

5.1 *Parent Rock and Coseismic Melting Temperature of Pseudotachylite*

The Gangavalli fault is a Proterozoic exhumed strike-slip fault. The pseudotachylites generated by coseismic slip occur as thin to very thick veins within charnockite host rock. The charnockites show similar chemical composition as the pseudotachylites with regards to major element, trace element and REE pattern. Based on this similarity, we interpreted that charnockite was the parent rock of pseudotachylite formation. The veins have a chilled margin and contain clasts that follow a power-law size distribution indicating frictional melt origin of the pseudotachylites (Behera et al. 2017). The clasts are dominantly β-quartz clasts that indicate the melting temperature must have reached above 1550 °C. The veins are devitrified into albite microlites which indicate 1100 °C of cooling.

The charnockites of the Salem-Namakkal block show a temperature of metamorphism at 800–900 °C (Mukhopadhyay and Bose 1994) that corresponds to a depth of 25–30 km. The age of metamorphism was at ca. 2.5 Ga (Clark et al. 2009; Sato et al. 2011; Behera et al. 2019). The charnockite was exhumed to the upper crust (above 10 km) by thrusting along the Salem-Attur shear zone at 2.0 Ga (Ghosh et al. 2004; Behera et al. 2019). The rock was deformed by strike-slip faulting within the brittle crust and generated coseismic pseudotachylite at 1.8 Ga (Behera et al. 2019). Therefore, the prevailing temperature of surrounding rock during the formation of pseudotachylite melt at the upper crust was about 250 °C. So the frictional stress made huge temperature rise over the ambient temperature resulting in the melting of the rocks.

Charnockite underwent an incongruent melting as inferred from heterogeneity of melt composition in the fault veins. Some spots are Fe–Mg rich and the others are Ca–Na rich. This variation indicates that the minerals in charnockite were melted in batch manner with increase in temperature. Different minerals melt at different temperatures, e.g., biotite at ~800 °C, plagioclase at ~1400 °C and finally quartz at ~1700 °C (Jiang et al. 2015). Biotite melted first to produce Fe–Mg rich melt while K-feldspar-plagioclase melted later to produce Ca–Na rich melt. With large amount of melting the melt attains the composition of the parent rock.

5.2 *Frictional Shear Stress Estimation and Implication About Magnitude of Earthquake*

Pseudotachylites have been used to measure the source properties of the paleoseismicity, like shear stress, stress-drop etc. (Beeler et al. 2016). The thickness, displacement and composition of the melt, clasts and microlites are the parameters for such measurement. We have applied such criteria for the Gangavalli fault. The thickness of fault veins increases with displacement (Fig. 4b) similar to observation made by many researchers (Sibson 1975; Di Toro et al. 2005; Barker 2005). An average displacement of 3.87 cm obtained from our study suggests that even a small slip with very high rate of friction may be able to produce melt with equivalent vein thickness as thin as 0.3 mm. This is in agreement with the study of McKenzie and Brune (1972) that a millimeter scale thin pseudotachylite vein is resulted from a moderate to large earthquake. However, a maximum displacement of 10 cm in the Gangavalli area were recorded which was possibly caused by a large earthquake that produced 3 mm thick vein (Table 3, location L3). Further, clasts to melt ratio has a low value that varies between 0.1 and 0.25 (like that of 0.1–0.7; O'Hara 2001). This lower value indicates that the pseudotachylite veins are rich in melt proportion than clasts. Here, we infer that the higher frictional resistance along the fault surface leads to produce more melts. The process that involves very little melt or no melt as in case of cataclasite is resulted from less frictional resistance as compared to the pseudotachylite melt formation. Therefore, the area fraction of clasts (ϕ) can be used as one of a proxy to distinguish between two processes of slow rate and fast rate of slip. In our case, it was a very fast rate of slip that generates large volume of pseudotachylite in the area.

In general, the shear stress for a dry and hard acidic rock like granite or charnockite varies within 50 MPa (Di Toro et al. 2005). Our estimation of highest value of shear resistance was 48.95 MPa close to that. However, Sibson (1975) obtained a higher value that exceeds up to 100 MPa. The frictional shear resistance diminishes gradually with the displacement (Fig. 4c). A millimeter scale thick pseudotachylite vein can be produced from a high friction resistance >100 MPa whereas a meter scale vein may produce from a low value of 10 MPa (Sibson and Toy 2006). Our result indicates a high magnitude earthquake.

5.3 Melt Pressure and Flip-Flop Between σ_1 and σ_2, Implication for Stick-Slip and Stress Drop

The stereoplot and Mohr plot for 3D stress indicate $P_m > \sigma_2$ and σ_1 is near vertical, the stress tensors correspond to a normal fault. Analysis of conjugate fractures of the Gangavalli fault indicated that the σ_1 was horizontal in NNE-SSW direction and σ_2 was vertical which correspond to strike-slip faulting (Behera et al. 2017). We are interpreting this discrepancy to flip-flop of principal stress axis during coseismic slip. Though the seismicity is broadly a single stage event, it undergoes several stick-slip phases (e.g., Brace and Byerlee 1966). With each jerk, stress drop occurs. As a result the σ_1, which was horizontally directed, was reduced in magnitude. This leads σ_2, which was vertically oriented, to act as σ_1. This happens more often, as for the Gangavalli fault, when Φ value is close to 1.0 ($\sigma_2 \approx \sigma 1$). We didn't calculate the magnitude of stress drop; a stress drop within a range of 3–20 MPa has been reported for a large earthquake (McKenzie and Brune 1972).

In addition to this, synchrony in strike-slip and normal faulting has been described from plate margin as well as within plate. For a strike-slip fault to happen, a normal transfer fault is necessary, and vice versa (Ramsay and Huber 1987). The plate margin is always marked by combination of all three types of faults. The Gangavalli fault (equivalent of Jhavadi Fault) probably defines close to the Precambrian plate margin fault as earlier described (Sharma et al. 2015; Behera et al. 2017). Further, a strike-slip fault associate with extension, transtension and releasing bends where normal faults develop together with strike-slip faults (Sylvester 1988). Therefore, we argue that the strike-slip and normal fault kinematics associate in the Gangavalli fault.

6 Conclusions

Our study on the pseudotachylites of the Gangavalli fault brings out the following conclusions regarding the source properties of the earthquake.

(i) The earthquake occurred within 10 km of the crust.
(ii) The frictional resistance was as high as 48.95 MPa that raised the temperature up to 1550 °C causing melting of the rock. This was a large magnitude earthquake.
(iii) The earthquake followed a stick-slip mechanism resulting in stress drop. The strike-slip kinematics incidentally flip-flopped to normal kinematics with each jerk.

Acknowledgements The authors are thankful to the Department of Earth Sciences, IIT Bombay for providing support for the field work. We convey our gratitude to prof. S. C. Patel and Mr. Anil Kanta Champati for providing us EPMA instrument for chemical analysis and Mrs. Trupti V. Chandrasekhar for providing XRD facility. We thank D. Selvaganapathi for extensive help during fieldwork in Salem, Tamil Nadu.

References

André AS, Sausse J, Lespinasse M (2001) New approach for the quantification of paleostress magnitudes: application to the Soultz vein system Rhine graben, France. Tectonophysics 336:215–231

Baer G, Beyth M, Reches Z (1994) Dikes emplaced into fractured basement, Timna igneous complex. Israel J Geophys Res 99:24039–24051

Barker SL (2005) Pseudotachylyte-generating faults in central Otago, New Zealand. Tectonophysics 397:211–223

Beeler NM, Di Toro G, Nielsen S (2016) Earthquake source properties from pseudotachylite. Bull Seismol Soc Am 106:6. https://doi.org/10.1785/0120150344

Behera BM, Thirukumaran V, Soni A, Mishra PK, Biswal TK (2017) Size distribution and roundness of clasts within pseudotachylytes of the Gangavalli Shear Zone, Salem, Tamil Nadu: an insight into its origin and tectonic significance. J Earth Syst Sci 126:46

Behera BM, Waele BD, Thirukumaran V, Sundaralingam K, Narayanan S, Sivalingam B, Biswal TK (2019) Kinematics, strain pattern and geochronology of the Salem-Attur shear zone: tectonic implications for the multiple sheared Salem-Namakkal blocks of the Southern Granulite Terrane, India. Precambr Res 324:32–61

Brace WF, Byerlee JD (1966) Stick-slip as a mechanism for earthquake. Science 153:990–992

Clark C, Collins AS, Timms NE, Kinny PD, Chetty TRK, Santosh M (2009) SHRIMP U-Pb age constraints on magmatism and high-grade metamorphism in the Salem block, southern India. Gondwana Res 16:27–36

Deb T, Bhattacharyya T, Matin A, Sensarma S (2015) Origin of pseudotachylite based on clast size frequency distribution in Bundelkhand craton central India. J Geol Soc India 85:5

Di Toro G, Pennacchioni G, Teza G (2005) Can pseudotachylytes be used to infer earthquake source parameters? An example of limitations in the study of exhumed faults. Tectonophysics 402:3–20

Gardner RL, Piazolo S, Daczko NR (2016) Shape of pinch and swell structures as a viscosity indicator: application to lower crustal polyphase rocks. J Struct Geol 88:32–45

Ghosh JG, De Wit MJ, Zartman RE (2004) Age and tectonic evolution of Neoproterozoic ductile shear zones in the southern granulite terrain of India, with implications for gondwana studies. Tectonics 23 TC3006

Han R (2017) Pseudotachylytes and seismic fault slip. J Geol Soc Korea 53:159–171

Heuze FE (1983) High-temperature mechanical, physical and thermal properties of granitic rocks—a review. Int J Rock Mech Min Sci Geomech Abstr 20(1):3–10

Jeffreys H (1942) On the mechanics of faulting. Geol Mag 79:291–295

Jiang H, Lee CTA, Morgan JK, Ross CH (2015) Geochemistry and thermodynamics of an earthquake: a case study of pseudotachylites within mylonitic granitoid. Earth Planet Sci Lett 430:235–248

Jolly RJH, Sanderson DJ (1997) A Mohr circle reconstruction for the opening of a pre-existing fracture. J Struct Geol 19:887–892

Kirkpatrick JD, Dobson KJ, Mark DF, Shipton ZK, Brodsky EE, Stuart FM (2012) The depth of pseudotachylyte formation from detailed thermochronology and constraints on coseismic stress drop variability. J Geophys Res Solid Earth 117:B6. https://doi.org/10.1029/2011JB008846

Lahiri S, Mamtani MA (2016) Scaling the 3-D Mohr circle and quantification of paleostress during fluid pressure fluctuation–application to understand gold mineralization in quartz veins of Gadag (southern India). J Struct Geol 88:63–72

Lin A (1999) Roundness of clasts in pseudotachylytes and cataclastic rocks as an indicator of frictional melting. J Struct Geol 21:473–478

Magloughlin JF (1992) Microstructural and chemical changes associated with cataclasis and friction melting at shallow crustal levels: the cataclasite and pseudotachylyte connection. Tectonophysics 204:243–260

Martínez-Poza AI, Druguet E, Castano LM, Carreras J (2014) Dyke intrusion into a pre-existing joint network: the Aiguablava lamprophyre dyke swarm (Catalan Coastal Ranges). Tectonophysics 630:75–90

Mazzarini F, Isola I (2007) Hydraulic connection and fluid overpressure in upper crustal rocks: evidence from the geometry and spatial distribution of veins at Botrona quarry, southern Tuscany, Italy. J Struct Geol 29:1386–1399

Mazzarini F, Musumeci G, Cruden AR (2011) Vein development during folding in the upper brittle crust: the case of tourmaline-rich veins of eastern Elba Island, northern Tyrrhenian Sea, Italy. J Struct Geol 33:1509–1522

McGarr A (1999) On relating apparent stress to the stress causing earthquake slip. J Geophys Res 104:3003–3011

McKeagney CJ, Boulter CA, Jolly RJH, Foster RP (2004) 3-D Mohr Circle analysis of vein opening, Indarama lode-gold deposit, Zinbabwe: implications for exploration. J Struct Geol 26:1275–1291

McKenzie D, Brune JN (1972) Melting on fault planes during large earthquakes. Geophys J Int 29:65–78

Mondal TK, Mamtani MA (2013) 3-D Mohr circle construction using vein orientation data from Gadag (southern India)—implications to recognize fluid pressure fluctuation. J Struct Geol 56:45–56

Mukhopadhyay B, Bose MK (1994) Transitional granulite-eclogite facies metamorphism of basic supracrustal rocks in a shear zone complex in the Precambrian shield of south India. Mineral Mag 58:97–118

O'Hara K (2001) A pseudotachylyte geothermometer. J Struct Geol 23:1345–1357

Philpotts AR (1964) Origin of pseudotachyletes. Am J Sci 262:1008–1035

Ramsay JG, Huber MI (1987) The techniques of modern structural geology: folds and fractures. Academic Press, London

Ray SK (2004) Melt-clast interaction and power-law size distribution of clasts in pseudotachylytes. J Struct Geol 26:1831–1843

Sarkar A, Chattopadhyay A, Singh T (2019) Roundness of survivor clasts as a discriminator for melting and crushing origin of fault rocks: a reappraisal. J Earth Syst Sci 128:51

Sato K, Santosh M, Tsunogae T, Chetty TRK, Hirata T (2011) Laser ablation ICP mass spectrometry for zircon U-Pb geochronology of metamorphosed granite from the Salem Block: implication for Neoarchean crustal evolution in southern India. J Mineral Petrol Sci 106:1–12

Scholz CH (2019) The mechanics of earthquakes and faulting, 3rd edn. Cambridge University Press, New York. https://doi.org/10.1017/9781316681473

Sharma SD, Prathigadapa R, Kattamanchi S, Ramesh DS (2015) Seismological mapping of a geosuture in the southern granulite province of India. Lithosphere 7(2):144–154

Sheth HC, Choudhary AK, Bhattacharyya S, Cucciniello C, Laishram R, Gurav T (2011) The Chogat-Chamardi subvolcanic complex, Saurashtra, northwestern Deccan traps: geology, petrochemistry, and petrogenetic evolution. J Asian Earth Sci 41:307–324

Sibson RH (1975) Generation of pseudotachylyte by ancient seismic faulting. Geophys J Int 43:775–794

Sibson RH (2003) Thickness of the seismic slip zone. Bull Seismol Soc Am 93:1169–1178

Sibson RH, Toy VG (2006) The habitat of fault-generated pseudotachylite: presence versus absence of friction-melts. Geophys Monogr Ser Am Geol Union 170:153–166

Spray JG (1992) A physical basis for the frictional melting of some rock forming minerals. Tectonophysics 204(3–4):205–221

Spray JG, Thompson LM (1995) Friction melt distribution in a multi-ring impact basin. Nature 373:130–132

Sundaralingam K, Biswal TK, Thirukumaran V (2017) Strain analysis of the Salem Attur shear zone of Southern Granulite Terrane around Salem, Tamil Nadu. Geol Soc India 89(1):5–11

Sylvester AG (1988) Strike-slip faults. Geol Soc Am Bull 100:1666–1703

Tiwari SK, Biswal TK (2019) Palaeostress and magma pressure measurement of granite veins in the Neoproterozoic Ambaji granulite, South Delhi terrane, Aravalli-Delhi mobile belt, NW India: implication towards the extension-driven exhumation of the middle–lower crustal rocks. J Earth Syst Sci 128(6):150

Anisotropy of Magnetic Susceptibility (AMS) Studies on Quartzites of Champaner Group, Upper Aravallis: An Implication to Decode Regional Tectonics of Southern Aravalli Mountain Belt (SAMB), Gujarat, Western India

Aditya U. Joshi and Manoj A. Limaye

Abstract Here we present, Anisotropy of Magnetic Susceptibility (AMS) studies on quartzites of Champaner Group, Gujarat, Western India. As quartzites are dominantly present within each formation of the Champaner Group, they have been selected for the study. Our study on AMS signifies two prominent striking planes of magnetic foliation within the rocks. The rocks have (i) ENE-WSW to E-W and (ii) N-S to NE-SW trends. The former trend matches with the regional magnetic foliation of Godhra Granites (GG) and neighboring Banded Gneisses (BG), while the later one does not match with any of the trends resulted due to last phase of deformation within Southern Aravalli Mountain Belt (SAMB). Such heterogeneity among the later trends signifies further continuation of emplacement of GG after syn-tectonic pulse and regional deformation. This latter phase of granite led to the development of broad open N-S trending folds within the supracrustals along with its basement and doming up of sequences towards the eastern periphery of the Champaner Group.

Keywords AMS · Champaner Group · Godhra Granites · SAMB

1 Introduction

Rock samples when subjected to external magnetic field, certain minerals within the sample show same strength of induced magnetization in all directions and are considered as "magnetically isotropic". On the contrary, there are few minerals in the rocks, which have different strength of induced magnetization in preferred directions are called as "magnetically anisotropic". Such minerals have an Anisotropy of Magnetic Susceptibility (AMS). Variation in susceptibility in preferred directions creates susceptibility ellipsoid (Tarling and Hrouda 1993), having three principal axes—long, intermediate and short. These axes are the principal susceptibility axes represented

A. U. Joshi · M. A. Limaye (✉)
Department of Geology, Faculty of Science, The Maharaja Sayajirao University of Baroda, Vadodara 390002, India

T. K. Biswal et al. (eds.), *Structural Geometry of Mobile Belts of the Indian Subcontinent*, Society of Earth Scientists Series,
https://doi.org/10.1007/978-3-030-40593-9_9

in SI units as K1, K2 and K3, where K1 $\geq$ K2 $\geq$ K3. The axes of susceptibility ellipsoid generally show a preferred orientation of principal axes of the strain ellipsoid in tectonically deformed rocks (Hrouda and Janak 1976; Rathore 1979, Borradaile and Tarling 1981, 1984; Hrouda 1982; Borradaile and Mothersill 1984; Borradaile 1987, 1991; Borradaile and Alford 1987; Borradaile and Henry 1997).

In the present paper, we provide AMS data generated on quartzites of Champaner Group belonging to the Aravalli Supergroup with an aim of demonstrating the last deformation imprint recorded within this rock. With the help of the AMS data, strong correlation has been established between the deformation signatures pertaining to the Champaner Group along with its basement and the neighboring Pre-Cambrian stratigraphic units. The present attempt is to explore the AMS fabrics from Champaner Group of rocks, particularly quartzites, as emphases regarding the emplacement record of granite, to understand the SAMB tectonics operated at its southern margin.

2 Geological and Structural Setting

The Champaner Group, a part of upper Aravallis is exposed at the eastern most fringe of Gujarat, India (Gupta et al. 1997) (Fig. 1). It bears an important link between the main Aravalli domains with the Central Indian Tectonic Zone (CITZ). The Champaner Group consists variety of rock types such as quartzites of manganiferous and non-manganiferous variety, polymict and oligomict metaconglomerates, carbonaceous phyllites and schists, spotted slates, hornfelses, manganiferous phyllites, biotite schists, dolomitic limestones, calc-silicate rock and meta-greywackes, which overlies the Paleo-Proterozoic Pre-Champaner Gneissic Complex (PCGC) located at Chhota Udepur region–i.e. “considered basement”, comprising granite-gneisses, quartzites and pelitic gneisses (Gupta et al. 1992, 1997; Merh 1995; Srikarni and Das 1996; Karanth and Das 2000; Das 2003) (Fig. 2). The Champaner rocks show greenschist facies metamoprhism and hornfelses are developed at the contact aureole with granite intrusive. (Das et al. 2009; Joshi and Limaye 2018; Joshi 2019a, b). Granites located in and around Champaner Group indicate syn-to-post emplacement record. Timing of granites suggest a wide span of dates from 1.1 to 0.93 Ga (Joshi and Limaye 2018).

Structurally, the rocks including quartzites, show poorly deformed signatures having three phases of deformation, viz. D_1, D_2 and D_3. The three phases of deformation have resulted into F_1, F_2 and F_3 folds, respectively. D_1-D_2 resulted into F_1-F_2 are coaxial having ESE-WNW and E-W trend, respectively, while the D_3 shows folds having NNW-SSE to NNE-SSW trend. Combination of $F_{1\text{-}3}$ folds have resulted into various interference fold patterns on regional as well as on the outcrop scale. The Type-III interference fold pattern has formed on account of regional deformation resulted due to combinations of $F_{1\text{-}2}$ folds, whereas the Type-I interference pattern superimposed orthogonally over Type-III, which led to the development of domes. This last axial trace generated on account of superimposition is produced by later granitic pulse, represents F_3 folds (Joshi 2019a, b). In addition, there has

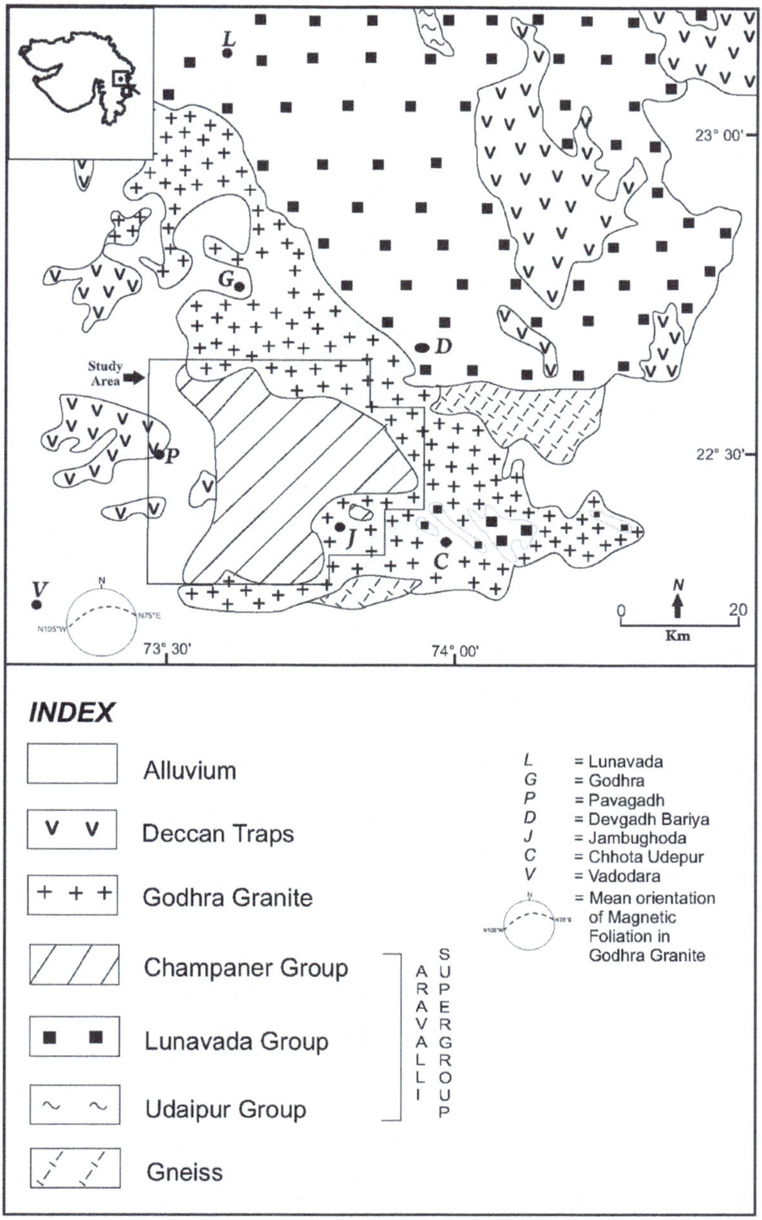

Fig. 1 lithostratigraphic map of SAMB, NW, India, specifying location of the Champaner Group. Modified after Mamtani and Greiling (2005). Square indicates the study area magnetic foliation data presented within Godhra granite is after Mamtani (2014)

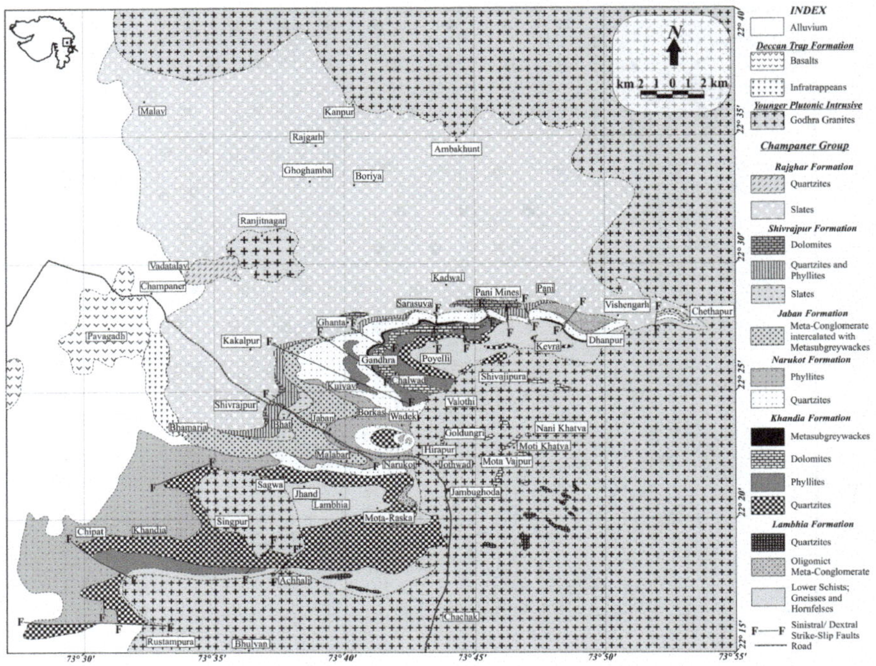

Fig. 2 Geological map of the Champaner Group, after Joshi 2019b

been observed evidences of out-of-sequence deformation (i.e. trends, which not concomitant with F_1, F_2 and F_3) demonstrate varied trends and meso-scale interference patterns inside the detached and isolated patches of calc-silicates, caught up within the granites (Joshi and Limaye 2018).

3 Anisotropy of Magnetic Susceptibility Studies on Quartzites, Principle and Methodology

Quartzites of the Champaner Group are hard, massive, compact; at times flaggy and whitish to dark grey and/or black in colour. They are highly jointed with variable grain size from fine to coarse and display varieties of primary sedimentary structures at different locations of the study area. By and large quartzites of the study area exhibit E-W to NW-SE strikes with variable dip directions. Petrographic study of quartzites reveal granoblastic polygonal, mosaic with two granular boundary varieties (i.e.) in-equigranular interlobate and seriate interlobate. The former type is observed in the quartz grains showing sutured boundaries with large scale variability in grain size, whereas the latter shows less size variation with sutured grain boundaries. At places the grains show inclusions of mica flakes especially Mn-rich-muscovite having colourless to faint green pleochroism (Patel et al. 2016).

The present section deals with the Anisotropy of Magnetic Susceptibility (AMS) studies carried out on quartzites of the Champaner Group. Oriented samples were collected at 12 localities across the Champaner Group and more than 60 cores were analyzed for the AMS study. These measurements were done using KLY-4 Kappabridge at Department of Geology and Geo-physics, Indian Institute of Technology, Kharagpur. The instrument gave the results for the following parameters: (1) the magnitude and orientations of three principal axes of the susceptibility ellipsoid viz. K1, K2 and K3, where K1 $\geq$ K2 $\geq$ K3; (2) mean susceptibility (km); (3) degree of anisotropy (Pj); (4) shape parameter (T); (5) magnitude of foliation (F) and (6) lineation (L) (Table 1). The quartzites of the study area reflect low Km values, having the mean susceptibility as 47.37×10^{-6} SI units. Some quartzites reveal exceptionally high mean susceptibility values more than 250×10^{-6} SI units, due to the presence of biotite content.

4 Result and AMS Data Interpretation

The orientations of all three principal axes K1, K2 and K3 have been plotted using lower hemisphere stereographic projection. The plane passes through K1-K2 and essentially matching with K3 pole defines the magnetic foliation (F). The F plane in the given figure is plotted with the dotted line. The stereographic net is also aided with the planar structure S_0 (bedding plane) acquired through field and has been depicted by means of continuous line (Fig. 3). There are total 12 samples, viz. SN1-4, SN6, CPR1, CPR3-5 and CPR7-9, out of which 61 cores have been analyzed.

As mentioned in the geological and structural setting section, the Champaner Group shows three phases of deformation, viz. D_1, D_2 and D_3, of which first two phases were co-axial (F_1 ~ ESE-WNW; F_2 ~ E-W) and the last phase of deformation has N-S trend. The response of all three phases of deformation has been recorded within the quartzites of the Champaner Group. The samples collected from the southern part of the study area, viz. CPR1, 3, 4 and 5 show dominance of F_1 and F_2 folds. The CPR1 sample has a dominance of F_1 fabric on an outcrop scale due to the presence of mesoscale folds, but shows dominance of magnetic foliation in nearly E-W direction. The superimposition of F_2 fabric over F_1 is clearly seen in the sample collected from NW of Masabar (CPR1). CPR3 sample from the F_2 limb portion show oblique relationship of the magnetic foliation with the bedding. CPR4 and 5 collected from F_2 limb, depicts similar trend of bedding and magnetic foliation. These samples show the effect of F_2 folds present in the southern part of the study area. The signatures of F_3 fold or preferred orientation along N-S direction are not seen in the present samples. This suggests that the last episode of deformation did not show any mineral alignment in the southern part of the study area.

CPR 7 and SN2 sample collected from the eastern part of the study area from Narukot dome clearly signifies similar relationship of S_0 and S_1 in these quartzites.

Table 1 AMS data of the quartzites of the Champaner Group

Sample No.	K_m (10^{-6} SI units)	F	L	P_j	T	K_{max}	K_{int}	K_{min}	K_{max}—D/I	K_{int}—D/I	K_{min}—D/I
SN1	10.80	1.623	1.722	1.101	0.107	1.074	0.967	−0.573	284/45	15/2	107/45
SN2	6.67	6.676	1.619	1.302	0.335	1.158	0.883	−0.644	319/78	134/12	225/01
SN3	2.30	1.912	1.092	1.021	0.726	1.215	1.041	0.744	280/07	13/18	170/71
SN4	95.40	1.030	1.042	1.074	−0.084	1.037	0.989	0.974	275/09	19/55	179/33
SN6	180.0	1.057	1.026	1.087	0.349	1.035	1.008	0.957	300/35	58/36	178/37
CPR1	162.0	1.122	1.015	1.153	0.775	1.045	0.999	0.956	86/44	279/46	182/07
CPR3	20.00	2.331	1.113	1.412	0.419	1.021	0.569	0.723	281/28	13/03	109/62
CPR4	0.0833	1.193	0.957	1.403	−0.052	−0.030	−0.289	−0.450	75/21	320/48	181/34
CPR5	19.20	1.717	1.104	2.014	0.553	1.015	0.652	0.732	292/39	77/45	187/18
CPR7	7.41	1.393	1.817	2.642	0.109	−0.353	−0.472	−0.975	265/80	63/09	154/04
CPR8	145.00	1.050	1.049	1.107	−0.084	1.019	1.066	0.974	154/52	16/31	273/21
CPR9	28.70	1.090	1.042	1.143	0.010	1.023	0.235	0.863	310/35	128/55	219/01

K_m = mean susceptibility; F = magnetic foliation; L = magnetic lineation; P_j = degree of anisotropy; T = shape parameter; K_{max}, K_{int}, K_{min} = represent magnitudes of maximum, intermediate and minimum axis of the magnetic ellipsoid, D/I = the direction and inclination of principle axes of magnetic ellipsoid in degrees

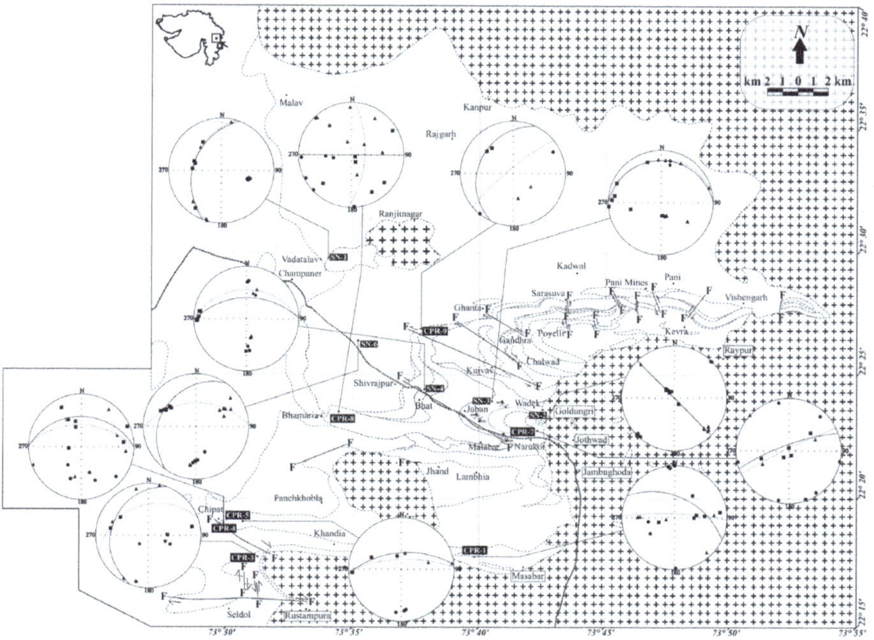

Fig. 3 AMS orientation data plotted as lower hemisphere projections for the Champaner Group. Dotted great circles represent the magnetic foliation, whereas continuous great circles represent the S_0 reading recorded from the field. Filled square, triangle and circle indicate orientations of K1, K2 and K3 respectively

The SN2 sample collected from F_1 fold limb and CPR 7 sample from F_2 limb gives nearer results of F_1 and F_2 fabric. It suggests that although, there is a dominance of N-S deformation in the eastern part of the study area, the preferred magnetic fabric parallel to F_3 is absent. Such condition has occurred due to the presence of F_2 fold dominance in the Narukot area. Further to the west of the Narukot dome, the samples collected along the axial trace of the F_2 fold (SN3, SN4 and SN6), show prominence of E-W magnetic foliation. The D_2 imprint recorded within these rocks clearly suggests that the central part of the Champaner Group has a dominance of N-S shortening. The western and northern part of the study area shows the magnetic fabric parallel to the D_3 deformation of the Champaner Group in the samples viz. CPR 8, CPR 9 and SN1. This magnetic fabric ranges from N-S to NE-SW; however there is no mesoscopic scale imprint of the late deformation in the present localities.

Jelinek plot (Pj vs. T) (Fig. 4), suggest that majority of the cores fall in the oblate field (flattening field) and few cores from the quartzites show their affinity towards prolate field (constrictional field). In order to examine the type of strain, the available AMS data has been plotted as Flinn diagram (F vs. L), which holds good correlation with the Jelinek plot. Most of the AMS data of cores from quartzites of the Champaner Group represented in the Flinn diagram concentrated in the oblate

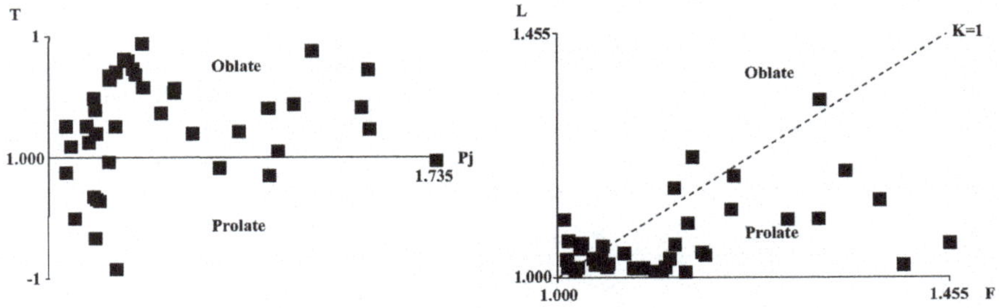

Fig. 4 Jelinek plots (at left) and Flinn diagram (at right) for the quartzites of the Champaner Group

field and few around K = 1 line, which separates the prolate and the oblate field. This is an indication of the plane strain condition (Fig. 4).

5 Discussions

5.1 Significance of AMS Study for the Study Area

The AMS data results from quartzites of Champaner Group indicate that there exist two prominent striking planes of magnetic foliation, viz (i) ENE-WSW to E-W and (ii) N-S to NE-SW. The former range of strikes represent the first two phases (i.e. D1 = F1 ~ ESE-WNW and D2 = F2 ~ E-W), while the later orientation matches with the third deformational phase (i.e. D3 = F3 ~ NNW-SSE to NNE-SSW) recorded within the Champaner Group along with its basement.

Advancement in the present technique indicate that the Anisotropy of Magnetic Susceptibility (AMS) studies have been used to deduce the time relationship between emplacement/fabric development and regional deformation (Mamtani and Greiling 2005); in kinematic analysis and vorticity quantification of syn-tectonic granitoids (Tarling and Hrouda 1993; Benn et al. 1998, 1999, 2001; Majumder and Mamtani 2009; Mamtani et al. 2013); to capture imprints of superposed deformation in granitoids (Mondal 2018; Mamtani et al. 2019). In the light of this extensive work on Godhra and other granitoids, authors posit their AMS data set generated on quartzites by correlating it with the existing literature on structural and magnetic records related to the SAMB.

5.2 AMS of Adjoining Areas

The structural records from Pre-Cambrian stratigraphic units of SAMB, viz. (1) the Banded Gneisses (BG); (2) the Lunavada Group (LG); (3) the Champaner Group

(CG) along with its basement (Pre-Champaner Gneissic Complex-PCGC); (4) the Godhra Granites (GG); signifies that each of the unit that belongs to the SAMB, except GG have underwent at least three phases of deformation (Table 2). Moreover, the magnetic records derived through BG, LG and GG, by the earlier workers state that these magnetic trends are coeval with the last phase of deformation (having strikes ~ WNW to WSW and plunge ~ NW to W), recorded within the SAMB (Mamtani and Greiling, 2005). Unlike in the case of CG, where the latest magnetic foliation (having strikes ~ N-S to NE-SW), demonstrating the last phase of deformation within CG, show a distinct heterogeneity in terms of trends recorded during the waning phase of SAMB. In addition, the strain estimation derived with the help of degree of magnetic anisotropy value (P′ or P_j) from Godhra granite samples, suggest that southern part of GG, in proximity to CITZ tends to show inflated P′ or P_j values, implying the rocks have accommodated higher strain condition than their northern

Table 2 Summary of deformation events recorded in neighbouring Precambrian Supracrustals and its relationship with the Champaner Group along with its basement

The Banded Gneisses (BG)	The Lunawada Group (LG)	The Pre-Champaner b Complex (PCGC)	The Champaner Group (CG)	Godhra granite (GG)
D_1 and D_2 coaxial having axial trace trend NE-SW to ENE-WSW	D_1 and D_2 coaxial having axial trace trend NE-SW	D_1 having axial trace trend N-S of recline folds	–	–
D_3 having axial trace trend ESE-WNW to ENE-WSW	D_3 having axial trace trend E-W to NW-SE	D_2 and D_3 having axial trace trend E-W with different fold morphology (a. recline & b. upright)	D_1 and D_2 coaxial having axial trace trend F1 ~ ENE-WSW F2 ~ E-W	Preferred orientation of feldspar laths trend ESEWNW; magnetic foliation trend ESE-WNW to ENE-WSW
–	–	*D_4 having axial trace trend N-S upright warps and kinks*	*D_3 having axial trace trend NNW-SSE to NNE-SSW + out-of-sequence deformation trend NE-SW*	–

Structural records from the Banded Gneisses and the Lunawada Group (Mamtani 1998; Mamtani and Greiling 2005); the data from the Pre-Champaner Gneisses (Karanth and Das 2000); Godhra granite magnetic data and preferred orientation of feldspar laths (Mamtani and Greiling 2005; Mamtani et al. 2002; Sen and Mamtani 2006). The columns in italics indicate correlatable axial trace trends within Champaner Group and its basement, which are unmatched with the last phase of deformation associated to the Southern Aravalli Mountain Belt (SAMB). Modified after Joshi 2019b counter parts (Mamtani et al. 2019). However, the quartzite samples collected from the eastern part of the Champaner Group, having predominantly N-S to NE-SW magnetic foliation reflects higher P' or P_j value, suggests that the eastern part of the CG have accommodated high strain condition than any of the other direction (Table 1).

In order to understand the heterogeneity in terms of magnetic records, the study of granite located in and around the CG becomes vital. On the basis of structural data proposed by Joshi and Limaye (2018), suggests that the granites located in and around the Champaner region shows evidences of prolong emplacement records. One such evidence has been proposed by a field photograph published in Joshi and Limaye (2014), where, the coarse grained later pulse having intrusive relationship with the comparative fine grained granite variety. Also, the evidences suggested by Joshi (2019a) on genesis of interference fold patterns within CG, explained forceful emplacement of granite deforming the country rocks along N-S trend, on the basis of the model suggested by He et al. (2009) for Fangshan pluton, SW Beijing, China. In addition to that, the existing geo-chemical data of granites located in and around CG is in favor of two pulses revealing prior 'S' type granite (Goyal et al. 1997) and later the 'A' type variety representing transitional or post-orogenic uplift (suggested by Maithani et al. 1998 and Goyal et al. 2001). Finally, the Geophysical studies carried out within this terrain corroborate to the fact that there lies a sub-surface pluton beneath the CG, which has eventually up rooted the country rocks along with the pre-existing granitic pulse during the later stages (Joshi et al. 2018).

Hence based on the above facts it can be stated that the possibility of prolong emplacement record, which governed the tectonics of SAMB from its southern margin cannot be ruled out. At the end of regional deformation the granite pulse was still active at the southern margin deforming the Champaner metasediments along with its basement, to give rise different trends. These trends being restricted to the Champaner Group and its basement, have not been recorded in the neighboring supracrustals and thus how show heterogeneity in terms of last deformation imprint within the SAMB (Table 2).

6 Conclusion

1. The AMS studies on quartzites from Champaner Group indicate two prominent striking planes of magnetic foliation, viz (i) ENE-WSW to E-W and (ii) N-S to NE-SW. These trends matches with the three deformational phases occurred within the present group, along with its basement (i.e. D1 = F1 ~ ESE-WNW and D2 = F2 ~ E-W; D3 = F3 ~ NNW-SSE to NNE-SSW).
2. Correlating these magnetic fabrics of CG with the available structural and magnetic records from SAMB, suggest that there lies a distinct heterogeneity in case of latest trends of CG and SAMB.

3. Degree of magnetic anisotropy value (P′ or P_j) from quartzite samples indicate that the eastern part of the Champaner Group has accommodated high strain condition than any of the other directions.
4. The above facts are in favor of indicating the prolong emplacement record of GG, which governed the tectonics of SAMB from its southern margin.
5. During the waning phase of regional deformation of CG, the emplacement was still continued to deformed the Champaner metasediments along N-S trend, along with its basement.
6. As this latest trend was confined to the CG and its basement, have not been recorded with the neighboring supracrustals of SAMB.

Acknowledgements Authors are thankful to Prof. Manish A. Mamtani for providing unconditional support to carry out AMS analysis at Department of Geology and Geophysics, Indian Institute of Technology (IIT), Kharagpur, West-Bengal. Authors are also grateful to Prof. L. S. Chamyal, (Former Head, Department of Geology, The M. S. University of Baroda, Vadodara) for providing necessary facilities. Authors are grateful to Prof. T. K. Biswal, IITB, for inviting to contribute in IGC 2020 special volume and handling the manuscript. We are also thankful to anonymous reviewer for providing valuable suggestions.

References

Benn K, Ham MN, Pignotta GS, Bleeker W (1998) Emplacement and deformation of granites during transpression: magnetic fabrics of the Archean Sparrow pluton Slave Province, Canada. J Struct Geol 20:1247–1259

Benn K, Paterson SR, Lund SP, Pignotta GS, Kruse S (2001) Magmatic fabrics in batholiths as markers of regional strains and plate kinematics: example of the cretaceous Mt. Stuart Batholith. Phys Chem Earth Part—Solid Earth Geodesy 26(4–5):343–354

Benn K, Roest WR, Rochette P, Evans NG, Pignotta GS (1999) Geophysical and structural signatures of syntectonic batholiths construction: the south mountain batholith meguma terrane, Nova Scotia. Geophys J Int 136:144–158

Borradaile GJ (1987) Anisotropy of magnetic susceptibility: rock composition versus strain. Tectonophysics 138:327–329

Borradaile GJ (1991) Correlation of strain with anisotropy of magnetic susceptibility (AMS). Pure appl Geophys 135:15–29

Borradaile GJ, Alford C (1987) Relationship between magnetic susceptibility and strain in laboratory experiments. Tectonophysics 133:121–135

Borradaile GJ, Henry B (1997) Tectonic applications of magnetic susceptibility and its anisotropy. Earth-Sci Rev 42:49–93

Borradaile GJ, Mothersill JS (1984) Coaxial deformed and magnetic fabrics without simply correlated magnitudes of principal values. Phys Earth Pl Sci Int 35:294–300

Borradaile GJ, Tarling DH (1981) The influence of deformation mechanisms on magnetic fabric in weakly deformed rocks. Tectonophysics 77:151–168

Borradaile GJ, Tarling DH (1984) Strain partitioning and magnetic fabrics in particulate flow. Can J Earth Sci 21:694–697

Das S (2003) Deformation and metamorphic history of the precambrian rocks in north eastern part of Vadodara district, Gujarat with a reference to the stratigraphy and tectonics. Unpublished Ph.D. Thesis, The M. S. University of Baroda, Vadodara, pp 1–110

Das S, Singh PK, Sikarni C (2009) A preliminary study of thermal metamorphism in the Champaner Group of rocks in Panchmahals and Vadodara districts of Gujarat. India J Geosci 63:373–382
Goyal N, Pant PC, Hansda PK, Pandey BK (2001) Geochemistry and Rb–Sr age of the late Proterozoic Godhra granite of central Gujarat, India. J Geol Soc India 58:391–398
Goyal N, Varughese SK, Hansda PK, Ramachandran S, Singh R (1997) Geochemistry of granites of Jambughoda, Panchmahals district, Gujarat and uranium mineralization in Champaner Group. J Geol Soc India 50:769–778
Gupta SN, Arora YK, Mathur RK, Iqbaluddin Prasad B, Sahai TN, Sharma SB (1997) The Precambrian geology of the Aravalli region, southern Rajasthan and NE Gujarat. Mem Geol Surv India 123:1–262
Gupta SN, Mathur RK, Arora YK (1992) Lithostratigraphy of Proterozoic rocks of Rajasthan and Gujarat—a review. Rec Geol Surv India 115:63–85
He B, Xu YG, Paterson S (2009) Magmatic diapirism of the Fangshan pluton, southwest of Beijing, China. J Struct Geol 31:615–626
Hrouda F (1982) Magnetic anisotropy of rocks and its application in geology and geophysics. Geophys Surv 5:37–82
Hrouda F, Janak F (1976) The changes in shape of the magnetic susceptibility ellipsoid during progressive metamorphism and deformation. Tectonophysics 34:135–148
Joshi AU (2019a) Fold interference patterns in Meso-Proterozoic Champaner fold belt (CFB) Gujarat, western India. J Earth Syst Sci. https://doi.org/10.1007/s12040-019-1075-z
Joshi AU (2019b) Structural evolution of precambrian rocks of Champaner Group, Gujarat, Western India. Unpublished Ph.D. Thesis, The Maharaja Sayajirao University of Baroda, pp 1–190
Joshi AU, Limaye MA (2014) Evidence of syndeformational granitoid emplacement within Champaner Group, Gujarat. J Maharaja Sayajirao Univ Baroda 49:45–54
Joshi AU, Limaye MA (2018) Rootless calc-silicate folds in granite: an implication towards syn to post plutonic emplacement. J Earth Syst Sci 127(5):ID67
Joshi AU, Sant DA, Parvez IA, Rangarajan G, Limaye MA, Mukherjee S, Charola MJ, Bhatt MN, Mistry SP (2018) Sub-surface profiling of granite pluton using microtremor method: Southern Aravalli, Gujarat, India. Int J Earth Sci 107:191–201
Karanth RV, Das S (2000) Deformational history of the pre-Champaner gneissic complex in Chhota Udepur area, Vadodara district, Gujarat. Indian J Geol 72:43–54
Maithani PB, Rathaiah YV, Varughese SK, Singh R (1998) Granites of Zoz area, Baroda District, Gujarat and its economic significance. J Geol Soc India 51:201–206
Majumder S, Mamtani MA (2009) Magnetic fabric in the Malanjkhand Granite (central India)—implications for regional tectonics and Proterozoic suturing of the Indian shield. Phys Earth Planet Inter 172:310–323. https://doi.org/10.1016/j.pepi.2008.10.007
Mamtani MA (1998) Deformational mechanisms of the Lunavada Pre-Cambrian rocks, Panchmahal district, Gujarat; Unpublished Ph.D. thesis, M.S. University of Baroda. pp 1–268
Mamtani MA (2014) Magnetic fabric as a vorticity gauge in syntectonically deformed granitic rocks. Tectonophys 629:189–196
Mamtani MA, Bhatt S, Rana V, Sen K, Mondal TK (2019) Application of AMS in understanding regional deformation fabric development and granite emplacement: examples from Indian cratons. Geol Soc, London, Spec Publ 489. https://doi.org/10.1144/SP489-2019-292
Mamtani MA, Greiling RO (2005) Granite emplacement and its relation with regional deformation in the Aravalli Mountain Belt (India)—Inferences from magnetic fabric. J Struct Geol 27:2008–2029
Mamtani MA, Karmakar B, Merh SS (2002) Evidence of polyphase deformation in gneissic rocks around Devgadh Bariya: implications for evolution of Godhra Granite. Gondwana Res 5:401–408
Mamtani MA, Pal T, Greiling RO (2013) Kinematic analysis using AMS data from a deformed granitoid. J Struct Geol 50:119–132. https://doi.org/10.1016/j.jsg.2012.03.002
Merh SS (1995) Geology of Gujarat. Geol Soc India Publ, Bangalore, p 244
Mondal TK (2018) Evolution of fabric in Chitradurga granite (south India) – A study based on microstructure, anisotropy of magnetic susceptibility (AMS) and vorticity analysis. Tectonophysics 723:149–161. https://doi.org/10.1016/j.tecto.2017.12.013

Patel D, Joshi AU, Limaye MA (2016) Sequential development of microstructures in quartzites of Champaner Group, Gujarat. J Geosci Res 1(2):101–104
Rathore JS (1979) Magnetic susceptibility anisotropy in the Cambrian slate belt of North Wales and correlation with strain. Tectonophysics 53:83–97
Sen K, Mamtani MA (2006) Magnetic fabric, shape preferred orientation and regional strain in granitic rocks. J Struct Geol 1870–1882
Srikarni C, Das S (1996) Stratigraphy and sedimentation history of Champaner Group, Gujarat. J Indian Assoc Sedim 15:93–108
Tarling DH, Hrouda F (1993) The magnetic anisotropy of rocks. Chapman and Hall, London

Tectonic Implications of Small-Scale Structures in the Main Central Thrust Zone of Garhwal Higher Himalaya

Sayandeep Banerjee and Hari B. Srivastava

Abstract In the Garhwal Himalaya, the Main Central Thrust Zone (MCTZ) is bounded by the Main Central Thrust (MCT) in the south and Vaikrita Thrust (VT) in the north. The MCTZ along the Alakhnanda, Bhagirathi and Yamuna valleys is marked by different small-scale structures and are classified as early structures i.e. pre-thrusting, structures due to progressive ductile shearing (syn-thrusting) and late structures (post-thrusting). The detailed study of small-scale structures/kinematic indicators such as sigmoidal foliations, shear band (S-C) structures, asymmetric porphyroclast tails, intergranular faults, folded layering and sheath folds suggest top to SSW sense of movement in the rocks of MCTZ of Garhwal Himalaya. The strain studies and crystallographic preferred orientation of quartz reveal that mesoscopic ductile shear zones grow in a narrow zone in response to very high strain, which deformed the internal crystallographic fabric of rock. The study of compressive stress from several mesoscopic sigmoidal foliations suggest that, in MCTZ the maximum compressive stress direction acted in the NNE-SSW direction horizontally synchronous to the northward movement of Indian Plate.

Keywords Garhwal higher himalaya · MCT · Top-to-SSW thrusting · NNE-SSW compression

1 Introduction

Small-scale structures are commonly considered as a manifestation of various factors like mode of deformation, type of metamorphism, rheology of the rock, etc., and thus, the investigation of the geometric patterns of small-scale structures is a fundamental step towards the understanding of complex geologic processes. Over the years, the study of small-scale structures has become a powerful tool used by the structural geologists to determine the geometry and orientation of major structures (Cosgrove 1980). From long back, such attempts on the identification, elucidation, and analyses of these small-scale geometric patterns are regarded as crucial information in Geosciences, and especially in understanding large-scale geodynamic processes.

S. Banerjee · H. B. Srivastava (✉)
Department of Geology, Banaras Hindu University, Varanasi 221005, India

T. K. Biswal et al. (eds.), *Structural Geometry of Mobile Belts of the Indian Subcontinent*, Society of Earth Scientists Series,
https://doi.org/10.1007/978-3-030-40593-9_10

The relationships between minor structures and major structures have been examined in detail by Leith (1923), Wilson (1961), amongst others. However, the major paradigm shift in the understanding of deformation took place with the introduction of infinitesimal and finite strain theory (Ramsay 1967; Jaeger 1969; Means 1976; Hobbs et al. 1976) that facilitated the correlation of different small-scale structures with progressive accumulation of strain and consequently, progressive deformation. For decades now, small-scale structures, as observed in foliated rocks have played an important role in the understanding of deformation patterns in orogenic belts (Ramsay and Huber 1987; Marshak and Mitra 1988; Price and Cosgrove 1990; Twiss and Moores 2007). Several workers have attempted to portray the process of progressive deformation with the help of effective recognition and analyses of small-scale structures (e.g., Brun and Pons 1981; Platt and Behrmann 1986; Ghosh and Sengupta 1987; Passchier 1990; Gray and Mitra 1993, Talbot and Sokoutis 1995; Carreras et al. 2005; Bell and Bruce 2007; Gomez-Rivas et al. 2007; Druguet et al. 2009; Bons et al. 2012; Matin et al. 2012; Banerjee and Matin 2013; Lisle 2013). While, studying the small-scale structures, it has also been observed that earlier group of minor structures are frequently overprinted by the later structures which may in turn be affected by an even later deformation. Superposition of minor structure can either by sequential development of structures with progressive deformation during single phase of deformation or as a result of two or more separate tectonic events. Despite the importance of small-scale structures, it turns out to be a challenging task to correlate the different stages of small-scale structures since; the progressive development and migration of deformation patterns often obliterate the records of early formed small-scale structures (Marcos and Arboleya 1975; Bell and Bruce 2007). It is also important to note that each phase of deformation is generally associated with a group of minor structures, some of which may be deformed by small-scale structures, which develop at slightly later stage.

The MCTZ of the Garhwal Higher Himalaya is a ductile shear zone of high heterogeneous strain localized in a narrow planar zone in medium to high-grade crystalline rocks. Small-scale structures in the vicinity of the shear zone bear the imprints of deformation and provide valuable information to demarcate the history of progressive deformation in crystalline rocks that have undergone natural deformation at moderate to high temperature (e.g., Choukroune and Gapais 1983; Ramsay and Allision 1979; Srivastava et al. 1995). Shear zones can be best studied in sections perpendicular to the plane of shear and parallel to the shear direction (XZ section of the strain ellipsoid). The increase of the foliation intensity can be related to an increase in strain intensity and well developed crystallographic fabric (Ramsay and Graham 1970; Huddleston 1977; Simpson 1986; Srivastava et al. 2000 and Srivastava and Tripathy 2007). Thus, shear zone provides useful information such as shear strain (Ramsay 1967; Srivastava et al. 2000), amount and direction of tectonic transport (with the help of different kinematic indicators), percentage of volume loss (Srivastava et al. 1995) and compressive stress direction from conjugate shear zones (Srivastava and Tripathy 2007).

In the present work, an attempt has been made to study the small-scale structures occurring in the MCTZ of the Garhwal Higher Himalaya, and a detailed picture of major deformational phases is presented. Each major phase has also been determined using criteria such as extent and intensity of deformation.

2 Geological Setting

The continent–continent collision in the Himalaya is well documented by Late Cenozoic intra-continental crustal shortening and southward thrusting along the MCT and associate thrusts over the Lesser Himalayan sediments (Le Fort 1975). The Garhwal Himalaya exposes all the four major tectonic sub-divisions of Himalaya from south to north (Fig. 1). The Outer most part of the Garhwal Himalaya (1. molassic Siwalik group of Mio-Pliocene age), is demarcated by NW-SE striking Main Boundary Thrust (MBT). The Siwaliks are separated in the north from the (2) Lesser Himalaya—a thick pile of highly folded Upper Precambrian(?)—Paleozoic sedimentary units with a few outcrops of crystalline rocks. Lesser Himalaya units are separated in the north by the Main Central Thrust (MCT) from the (3) Higher Himalaya (massive north dipping pile of metamorphic rocks that constitutes the Central Crystalline Zone). The Central Crystallines has a faulted contact (South Tibet Detachment, STD) with the Tethyan Himalaya (a thick sedimentary pile of Cambrian to Lower Eocene rocks).

The MCT is a north dipping shear zone formed as a result of northward emplacement of a part of deep rooted crust that now constitutes the Central Crystalline Zone

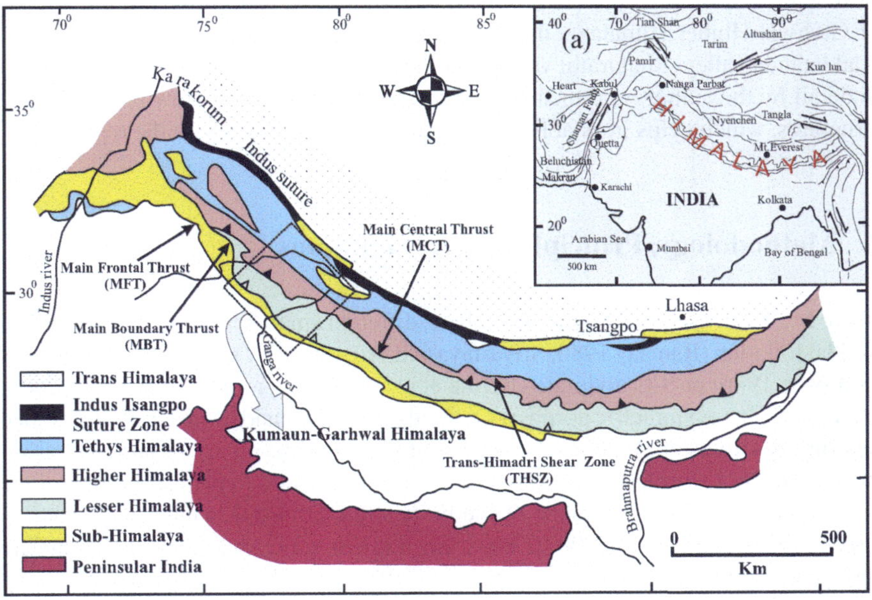

Fig. 1 Simplified map showing the major tectonic sub-divisions of the Himalaya (modified after Gansser 1964). The dashed rectangle marks the region of Kumaun-Garhwal Himalaya

of the Higher Himalaya over a less metamorphosed sedimentary rocks of the Lesser Himalaya (Bhattacharya and Weber 2004). The south directed ductile shearing has affected rocks exposed over about 7–10 km wide region in the area of study. The rocks of both hanging wall and footwall have been affected by ductile deformation due to thrusting along the MCT. Earlier studies of Bouchez and Peacher (1981), Singh and Thakur (2001), Metcalf (1993), Searle et al. (1993), Bhattacharya and Weber (2004), and Srivastava and Tripathy (2007), have interpreted that MCT is not a plane but a several kilometer-thick crustal scale high strain ductile shear zone referred to as the MCTZ. The lower limit of the MCTZ is known as the MCT (Heim and Gansser 1939) or Munsiari Thrust (Valdiya 1980). In the Kumaun Garhwal Himalaya, the north dipping MCT separates the crystalline rocks of Munsiari Group (Valdiya 1980) from the quartzites and metavolcanics of Garhwal Group (Jain 1971). Further north the rocks of Munsiari Group, are separated by the rocks of Vaikrita Group (Fig. 2) of Higher Himalaya along the Vaikrita Thrust (VT). The MCTZ is mainly constituted of the lithological units like schist, gneisses amphibolites and migmatites. The rocks of the MCTZ of Garhwal Higher Himalaya reveal a dip of about 40–55° in NE to ENE direction and exhibit mylonitic foliation along both hanging wall and footwall. Bedding plane (S_1) is though rarely seen exhibit isoclinal folding in the quarzo-feldspathic layers in the gneisses of the area. In the MCTZ, the mylonitic foliation exhibit variation in strike from NNW to WNW mainly dipping towards northern direction (Fig. 2a–c). The linear fabrics are characterized by stretching lineations, slickenlines and the fold hinges. Stretching lineations are trending mainly in N to NE direction with low to moderate amount of plunge (Fig. 2d–f). The schists and gneisses exhibit several mesoscopic ductile shear zones and S-C fabrics. In the MCT Zone of Garhwal Higher Himalaya, the deformational features in three river valleys i.e. Alakhnanda valley, Bhagirathi valley and Yamuna Valley (Fig. 2) have been studied in detail by the authors and their co-workers, from the point of view of small-scale structures, which helps in establishing the deformational history of the region.

3 Methodology, Principle and Assumptions

In the MCTZ of Garhwal Higher Himalaya, different small-scale structures like Sigmoidal foliation (Ramsay 1980; Srivastava and Tripathy 2007); Shear band structures (Simpson 1986) or S-C mylonite (Berthe et al. 1979; Lister and Snoke 1984); intergranular faults; asymmetric porphyroclast tails (Passchier and Simpson 1986); folded layering (Carreras et al. 2005); have been utilized to determine the sense of movement in the rock. Stretching lineations are extremely useful in the recognition of the X-axis that lies on the XZ plane of the finite strain ellipsoid in case of high strain zones (Twiss and Moores 2007). Thus, for a thrust zone, stretching lineations are often parallel to regional transport (Fossen 2016). Though, fold axes and stretching lineations are genetically different, a complicated relationship between these persists in case of a thrust sheet (Rykkelid and Fossen 1992). Sheath folds generally occur

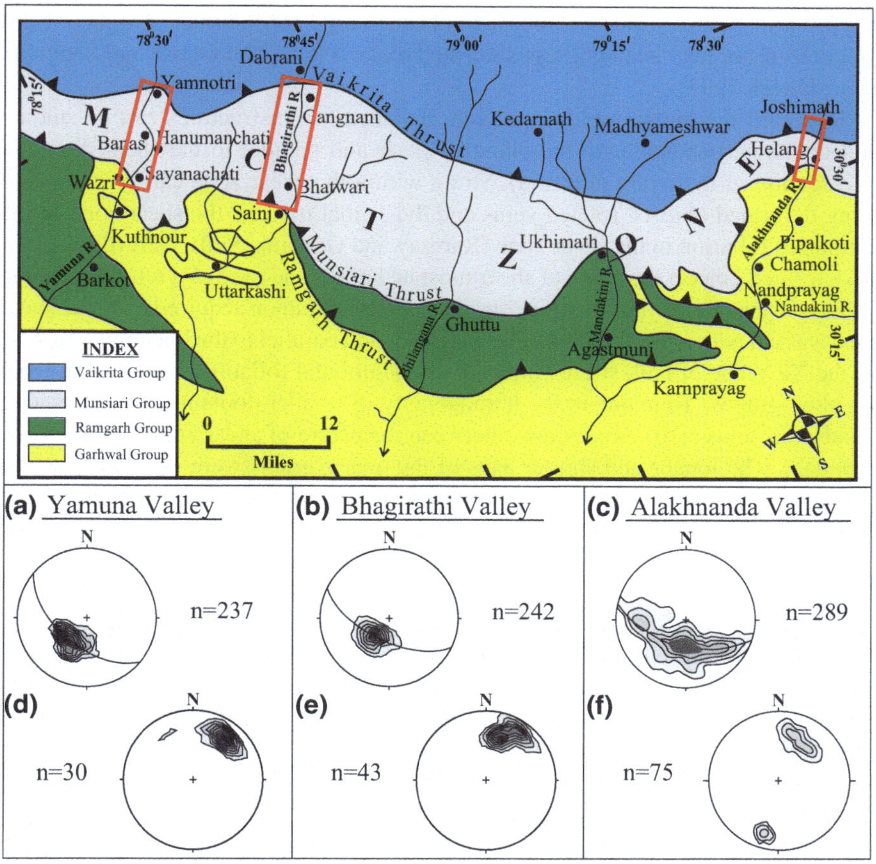

Fig. 2 (Inset) Simplified tectonic map of Garhwal Higher Himalaya (modified after Metcalf 1993). Stereoplots of poles to mylonitic foliations (**a**, **b**, **c**) and stretching lineations (**d**, **e**, **f**) in the MCTZ of Yamuna valley, Bhagirathi valley and Alakhnanda valley, respectively

in zones of high shearing (Cobbold and Quinnquis 1980; Malavieille 1987). Sheath fold has its name because, the hinge of the fold is itself folded into the axial plane of the fold, so that, the folded surface has the form of a sheath, such as it would fit around a sword. It has long been recognized that within zones of intense non-coaxial deformation, fold axes may rotate progressively towards the transport direction. In most ductile shear zones, the axes of the dominant set of folds are sub-parallel to the mylonitic lineation. The fold hinges are nearly parallel to the direction of maximum stretching and the axial surfaces are at low angle to the shear zone boundary. Thus, when the ductile shear develops in connection with a thrusting movement the most dominant folds are reclined (Ghosh 1993). The developments of sheath folds have been explained by accentuation of initial irregularities of the bedding or the foliation surface by a strongly non-coaxial deformation (Ramsay 1980; Cobbold and Quinquis 1980). The cross sectional morphology of shear zone folds, including sheath folds, is typical of buckling folds and they are not necessarily concentrated at sight of local irregularities or near island of less deformed rocks (Platt 1983; Ghosh and Sengupta 1987; Ghosh 1993).

To record the maximum deformation, the shear zone structures can be studied in sections perpendicular to the plane of shear and parallel to the shear direction (XZ section of the strain ellipsoid). Strain within the shear zone can be computed using deformed objects, rotated veins and dykes that transect the shear zone or the attitude of foliation in the shear zone (Ramsay and Graham 1970). Nevertheless, for all deformed objects the plane of shear must be known and used as a reference frame for measuring angles. To determine strain, the rock specimen acquired from the shear zone was cut perpendicular to the plane of shear and parallel to the lineation direction on the XZ plane of bulk strain ellipsoid. The sigmoidal foliations were traced from the photographs. To maintain the homogeneity in smaller domains (sub-areas a, b, c and d) isogons at 10° were drawn between the centre of shear zone to shear zone boundary. The longer and shorter axes of the quartz grains were measured in four sub-area (a, b, c and d) under the microscope and orientation of the longer axes of the grain with respect to the schistosity (at centre of shear zone) were recorded and the Rs. values (following Lisle 1985) were determined for different sub-areas.

In the last three decades, quartz has received much attention for the fabric and strain studies. This is probably due to its importance in controlling the rheology of large portions of the earth's crust, but also partly due to the large variety of fabric type found under different deformational and metamorphic conditions (Law 1990). Quartz is the most abundant mineral available in nature, which records the deformation on grain scale at different deformational and metamorphic conditions. The quartz c-axis patterns that were measured from non-coaxially deformed mylonites are commonly asymmetric with respect to the main foliation and lineation of the rock but are often symmetric with respect to the flow plane. The pattern is said to be single girdle if the c-axes form a single diffuse band of points across the stereo net and crossed girdle if the main girdle has minor girdles branching from it. Maxima may occur near the centre of the diagram (Type II crosses girdle; Lister 1977), near the pole to the flow plane (Type I crosses girdle; Lister 1977) or midway between, depending upon the predominant system of crystallographic plane and directions that have accommodated the shape changes of the crystal (Simpson 1986).

The study of crystallographic fabric in one of the specimen of gneissic rock collected from the MCTZ of Bhagirathi valley near village 'Sainj', utilized for the quartz c-axis analysis under universal stage (Srivastava et al. 2000). The crystallographic preferred orientation of quartz was measured manually, from three thin sections at three locations cut parallel to lineation and normal to foliation (XZ plane), on universal stage using the method suggested by Turner and Weiss (1963) and are plotted and contoured on the lower hemisphere of equal area net.

In order to determine the compressive stress direction responsible for shearing, the strike and dip direction of the several small-scale sigmoidal foliations were measured in the field for the Bhagirathi, Alakhananda and Yamuna Valleys and were plotted on the stereographic net and on rose diagram. At several places, the strike trend of sigmoidal foliations cross cut each other. Hence, it is assumed that cross cutting shear zones are part of conjugate pair which developed during the same episode of deformation. Ramsay (1980) and Ramsay and Huber (1987) are also of the opinion that this is the general relationship between simultaneously developed shear zones.

The pole to the sigmoidal foliations were plotted on the lower hemisphere of equal area net and contoured for Bhagirathi, Alakhananda and Yamuna Valleys. Similarly, the strikes of the sigmoidal foliations were plotted on the rose diagram. On the equal area net the intersection of planes perpendicular to the maxima of shear planes gives σ_2 direction. In contrast to brittle deformation (Anderson 1951) the obtuse bisector between the shear planes in ductile deformation parallels the maximum compressive stress direction (σ_1). The minimum compression direction (σ_3) is given by the acute bisector and the zone of intersection parallels to the intermediate compression direction (σ_2).

4 Results

4.1 Deformation Structures

The rocks of the MCTZ are marked by the well developed planar and linear fabrics. Bedding planes are characterized by the lithological layering, which are preserved at a few places. The schists, gneisses and migmatites of the area display a number of small-scale structures ranging in size from microscopic scale to outcrop scale. The structures observed can be broadly grouped into, Pre-, Syn- and Post- thrusting structures.

4.1.1 Pre-Thrusting Structures

In the MCTZ of Garhwal Himalaya two generations of coaxial folding (F_1 and F_2) can be observed. The F_1 folds are tight to isoclinal in nature (Fig. 3a), and gently plunge towards NNE direction with low to moderate amount of plunge. The axial planes of F_1 folds exhibit well developed axial plane cleavage (S_2). The F_1 folds are further folded to F_2 folds (Fig. 3b), which is the most dominant planar fabric in the area and plunge towards NNE direction, are coaxial to F_1 folds.

4.1.2 Syn-Thrusting Structures

All the three types i.e. ductile, brittle ductile and brittle shear zones are commonly observed in the crystalline rocks of Higher Garhwal Himalaya. Variety of mesoscopic shear zones ranging in size from millimeter to meter scale is generally parallel/or at acute angle to the foliation is developed in the crystalline rocks of the MCTZ. In the Garhwal Himalaya, the strike length of mesoscopic shear zones range from a few cm

Fig. 3 Field photographs **a** Isoclinal folding in the migmatite gneiss in Bhagirathi valley, **b** refolded structure developed in the gneisses in Alakhnanda Valley (width of photograph 5mt.)

to ~800 m where the width varies from ~5 to ~20 cm. The MCTZ is characterized by different kinematic indicators like sigmoidal foliation, shear bands, asymmetric porphyroclasts tails, intergranular fault folded layering and sheath folds. The area is also marked by mylonitc foliations and stretching lineations.

Mylonitic Foliation and Stretching Lineation

In the MCTZ the mylonitic foliation exhibit a variation in the strike from NNE to WNW direction mainly dipping towards northern direction (Fig. 2a, b, c). Stretching lineation mainly trend in N to NE direction with low to moderate amount of plunge (Fig. 2d, e, f). The mylonitic gneisses of the MCTZ exhibit ubiquitous development of stretching lineation (Fig. 6a). In simple shear dominated deformation zones like that of MCTZ, the folds usually develop parallel to X direction and also stretching lineation, but rotate toward the horizontal transport direction with increasing strain. Hence, careful determination of stretching lineation not only provides an insight to the regional transport direction, also it helps in the identification of folds related to the syn-thrusting structures.

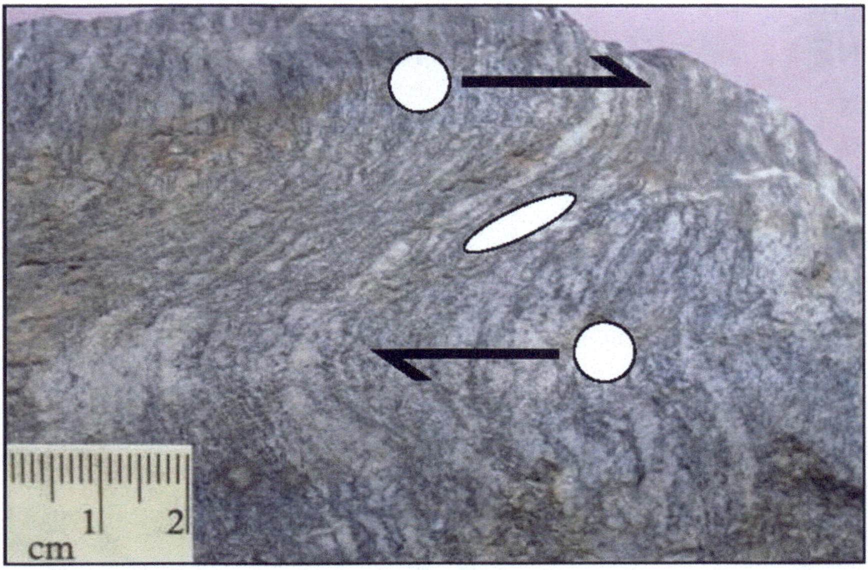

Fig. 4 Photograph of a hands specimen exhibiting sigmoidal foliation on vertical section along the XZ plane

Sigmoidal Foliation

Pre-shear foliations are deflected and are sigmoidally curved in ductile shear zones. Sigmoidal foliations (Fig. 4) are one of most significant shear sense indicators in the crystalline rocks of the Garhwal Himalaya. In the MCTZ, the small-scale sigmoidal foliations range in size from cm to few hundreds of meters. In ductile shear zones the deflection and differential displacement of the wall are accompanied entirely by the ductile flow and on outcrop it exhibits no sharp discontinuity across the zone and the shear magnitude vary smoothly across the zone. In the MCTZ the small-scale sigmoidal foliations exhibit two prominent strike trends, mainly varies from NNE to ENE and from NNW to WNW directions. Hereafter for description purpose the NNE to ENE striking sigmoidal shear zones are described as NE striking shear zone and NNW to WNW striking sigmoidal shear zones are described as NW striking shear zone. At several places, the NE striking and SW striking shear zones form a conjugate pair (Fig. 5a). In some places, NE striking shear zone offset the SW striking shear zone and at places SW striking shear zone displace NE striking shear zone. Hence, the NE striking and NW striking shear zones are inferred as a conjugate pair that developed during same deformational episode. Thus, sigmoidal foliation provides an opportunity to study the crystallographic preferred orientation of quartz, estimation bulk strain and the compressive stress direction, which has been dealt in Sect. 4.2, 4.3 and 4.4 respectively.

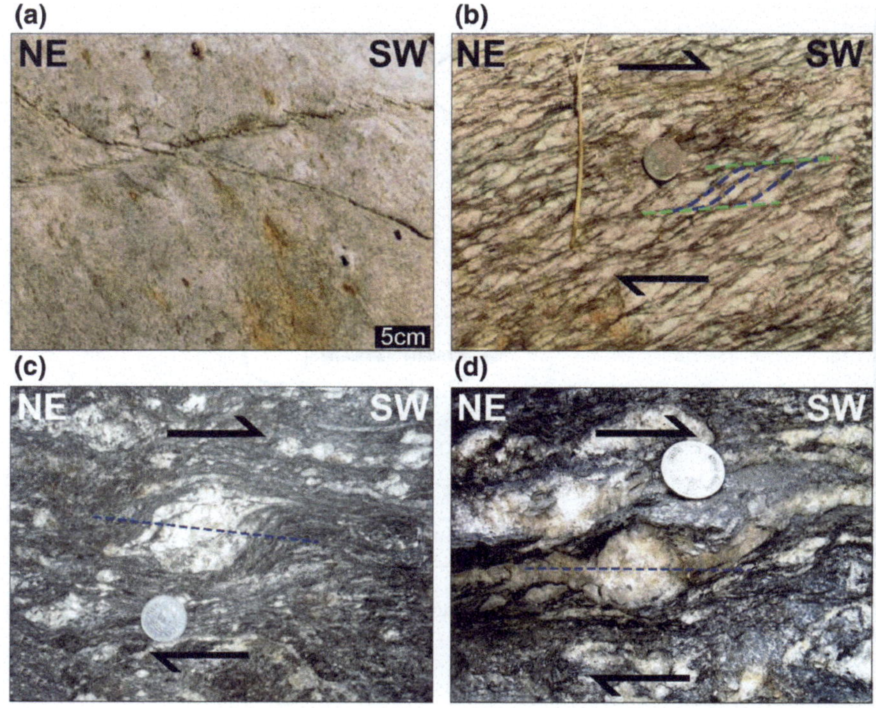

Fig. 5 Field photograph of small-scale structures in the MCTZ **a** Conjugate ductile shear zone showing sigmoidal foliation, **b** shear band structure, **c** σ type porphyroclast, and **d** δ type porphyroclast, depicting top to SSW sense of movement on vertical sections

Shear Bands

In the MCTZ several small-scale ductile shear zones that form within a much wider non-coaxial flow regime such as major shear zone or ductile fault, which are generally sub-parallel to one another, and are evenly spaced across the outcrop that deflect the pre-existing planar fabric with consistent sense of shear. These bands are also known as S-C mylonite. The shear bands (Fig. 5b) are very common in the crystalline rocks of Garhwal Himalaya and exhibit top to the SSW sense of shear.

Asymmetric Porphyroclast Tails

There are two types of porphyroclast trails sigma (σ) and delta (δ) types are commonly observed in the crystalline rocks of MCTZ. Wedge shaped tails of recrystallized materials (Fig. 5c) characterizes the σ type. The median line of the tail lies on opposite side of the reference plane at all points. Sometimes, the tails themselves have a characteristic morphology with a planar shape on the side of the tail farthest from the reference plane and a concave sector on the side nearest to the reference plane. The σ type of porphyroclast system is thought to be the result of high rate of dynamic recrystallization of the mineral with respect to shear strain rate in the rock. In the δ type of porphyroclasts (Fig. 5d) the median line of the tail crosses the reference plane next to the central grain. The δ type porphyroclasts normally develops in very fine grained and highly strained ultramylonites, where the rate of crystallization of the porphyroclast is very low with respect to the shear strain rate. In the crystalline rocks of MCT Zone of Garhwal Higher Himalaya both σ and δ type porphyroclasts have been observed which provide SSW sense of movement (Fig. 5c, d).

Intragranular Faults

The gneisses of the area exhibit a large number of fractures and displaced grain exhibiting synthetic and antithetic sense of movement. In the MCT Zone of Garhwal Higher Himalaya, these fractured and displaced grains show a top to SSW sense of movement (Fig. 6b).

Folded Layering

In shear zones, non-coaxial deformation causes local perturbations in the layering (Fig. 6c). With continued deformation, these perturbations evolve into asymmetric folds whose vergence is consistent with the shear sense within the mylonite zone. In the MCTZ, the folding in the mylonitic layering suggest top to SSW sense of movement.

Sheath Folds

In sheath fold the hinge of the fold is itself folded into the axial plane of the fold so that the folded surface looks in the form of a sheath. The fold hinges are nearly parallel to the direction of maximum stretching axial surfaces are at low angle to the shear zone boundary. Sheath folds are non-cylindrical folded structure developed due to very high strain during strong non-coaxial deformation. In the Alakhanda valley the sheath folds (Fig. 6d) are developed in the quartz mica schist, exposed near the village Helang located 15–20 m north of MCT (Fig. 2). The orientation of the sheath fold could not be corroborated with the orientation of F_1 and F_2 folds of Pre-thrusting deformational event. However, the location of sheath fold suggest that this may be due to rotation of the earlier folds due to strong non-coaxial deformation during the syn-thrusting event or developed due to accentuation of initial perturbations of the bedding or the foliation surface near the MCT.

Fig. 6 Photograph showing **a** stretching lineation on plan in the mylonitic gneisses of the MCTZ, **b** intragranular faults depicting top to the SSW sense of shear, **c** folding of the lithological layering showing top to the SW movement, **d** development of sheath fold on plan in the MCTZ

4.1.3 Post Thrusting Structures

These include brittle deformation structures such as by fractures, joints and faults. In the MCTZ, fracture cleavages are well developed in the schists and gneisses, which cuts across the schistosity and at times axial plane cleavages. In the rocks of the MCTZ, the fracture cleavage shows a variation in strike direction from NNE-SSW to EW and dip ranges from 30 to 70° towards southern and northern direction. Both tension and shear joints have been observed in MCTZ. Tension joints are often found to have filled up with secondary quartz veins and occur in the form of tension gases and occasionally show sigmoidal shape. Shear joints commonly visible in the MCTZ occur in a conjugate system. In general, the joint plane exhibits a variation in their strike from NW-SE to ENE-WSW direction with moderate to high dips towards northern and southern direction. The MCTZ of Garhwal Higher Himalaya is traversed by a number of post thrusting release faults on a variety of scales ranging from minor to small-scale to megascopic faults. Majority of these faults show similar pattern and rough parallelism to the thrust plane (Longitudinal fault), while a few faults are nearly perpendicular to the thrust plane (Transverse fault). In the present work, these later structures have not been dealt in detail.

4.2 *Crystallographic Preferred Orientation of Quartz*

To study the progressive asymmetry of the c-axes from centre to boundary of shear zone three thin sections from three different positions a, b, and c has been studied (Fig. 7). The plots of c-axis reveal a top to SSW sense of movement which was found to be consistent with the sense of shear. Near the shear zone boundary (Fig. 7a), the plot of c-axis fabric exhibits a tendency to lie on a weak girdle where the c-axis maxima make an angle of 40° from the foliation normal and is perpendicular to the shear zone boundary. At location Fig. 7b the quartz c-axis maxima display an angle of 20° from foliation normal and are perpendicular to the shear zone boundary. The sense of obliquity of the girdle distribution of the quartz c-axis is the same as the sense of shear. In the centre of shear zone (Fig. 7c), the c-axis single girdle is well defined and the maxima make an angle of 9° with the foliation normal. Thus, in

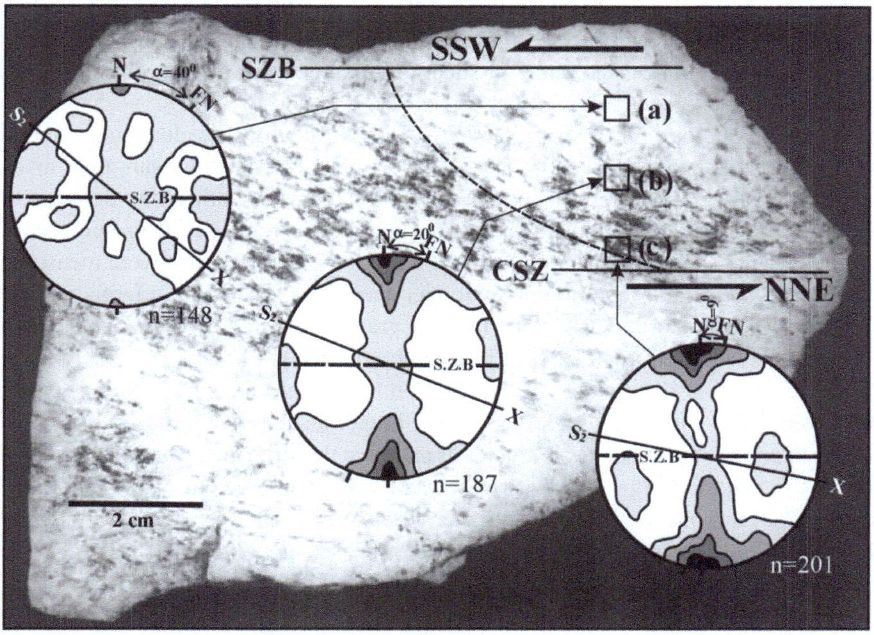

Fig. 7 Photograph of a hand specimen showing sigmoidal foliation in the gneisses of the Bhagirathi valley on vertical section. SZB—shear zone boundary and CSZ—centre of shear zone. Quartz c-axes fabrics from three positions in the ductile shear zone **a** Near the CSZ, **c** near the CZB and **b** in between CSZ and CZB have been carried out. Contour intervals are 2, 4, 6, < 6% per unit area. S_2-foliation; X-lineation FN-Foliation normal

Table 1 Strain data from the shear zone in Fig. 8. N is the number of grains measured, Rs. and φ, the strain ratio and orientation of long axis of strain ellipse derived by Lisle (1985) method. H is the harmonic mean. Difference = Rs/H X 100 (after Srivastava et al. 2000)

Sub-area	N	Rs.	φ	H	Difference
a	31	4.40	4	4.20	4.76
b	47	3.00	14	3.07	2.28
c	39	2.10	25	2.05	2.43
d	83	1.45	35	–	–

a shear zone, with progressive increase in strain the angle between c-axes maxima and the foliation normal reduces, and finally, in the centre of shear zone the c-axes maxima becomes nearly perpendicular to the foliation (Srivastava et al. 2000).

4.3 Estimation of Strain

Strain within ductile shear zone has been determined by using deformed quartz grains under the microscopic scale. Shear zone with sigmoidal trace of foliation reveal an increase in strain intensity from edge towards centre and can be best studied in a section perpendicular to the plane of shear and parallel to the shear direction (XZ plane).

From the photograph of specimen (Fig. 7), the foliations of elongated mineral were traced and the isogons of 10°, 20°, 30° were prepared. The four sub-areas made by the isogons of 10° interval namely a, b, c and d have been utilized to define sub-areas of approximately homogeneous strain, and data were plotted from sub-area a, b c and d (Table 1). The longer and shorter axes of the quartz grains were measured in each sub-area and orientation of the longer axes of the grain with reference to the schistosity were recorded and plotted on the graph suggested by Lisle (1985). The Rs. values were determined from different sub-areas (Fig. 8).

In the sub-area d near the shear zone boundary the quartz grain were weakly deformed and it was difficult to measure the longer and shorter axes of ellipse, and hence, in order to estimate the strain in this sub-area the strain measurement has been done by the method suggested by Srivastava (1995). It has been observed that in the centre of shear zone strain value is higher (Rs. 4.40 in sub-area a) and it gradually decreases (Rs. 1.45 in sub-area d) towards the shear zone boundary (Fig. 8).

4.4 Estimation of Compressive Stress Direction

In the MCTZ, the small-scale sigmoidal foliations range in size from cm to a few hundreds of meters. In ductile shear zones the deflection and differential displacement

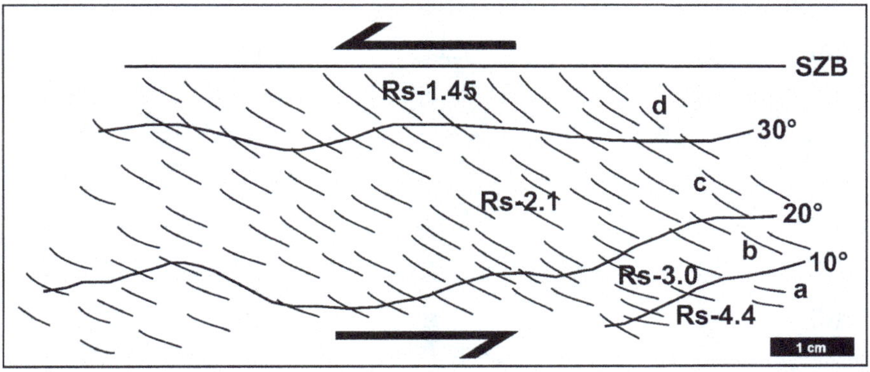

Fig. 8 Isogons of foliation trace from Fig. 7 at 10° interval. Light discontinuous lines show foliation trace. **a, b, c, d** are the sub-areas of shear zone, for the strain estimation. The strain values Rs. are shown in each sub-area. modified after Srivastava et al. 2000

of the wall are accompanied entirely by the ductile flow and on outcrop it exhibits no sharp discontinuity across the zone where the shear magnitude varies smoothly across the zone (Srivastava and Tripathy 2007). In most of the places, where veins are present, the foliations are sigmoidally curved within the shear zone and help in determining the sense of shear. The sigmoidal foliations exhibit moderate to steep dip. To determine the compressive stress direction responsible for shearing, the strike and dip direction of the small-scale sigmoidal shear zones were measured in the field, for the Bhagirathi, Alakhananda and Yamuna Valleys. The pole to the NE striking and SW striking shear planes were plotted on the lower hemisphere of equal area net and contoured for three valleys (Fig. 9a–c). The streoplots suggest that σ_1 in Bhagirathi, Alakhnanda and Yamuna valleys trend in N10°, N02° and N20° respectively (Fig. 9a–c). The strike of sigmoidal foliations was also plotted on the rose diagram (Fig. 9d–f). The rose diagrams depict that the σ_1 direction varies from 005° in the Bhagirathi valley (Fig. 9d), to 003° in Alakhnanda valley (Fig. 9e) to 024° in the Yamuna valley (Fig. 9f). The results obtained from steroplot and the rose diagram from all the three river valleys suggest that maximum compressive stress direction acted in a NNE-SSW direction horizontally and is responsible for the thrusting of the crystalline rocks over the rocks of Garhwal Group.

5 Discussion

The study of small to micro-scale structures observed in the MCT zone of the Garhwal Himalaya reveal that area has suffered four phases of tectonic deformations. The first phase of deformation in the area developed due to development of F_1 isoclinal folding in the lithological layering (S_1) (Fig. 3a). These F_1 folds exhibit a variation in their plunge direction from N20 to N40° E direction. The axial plane of these isoclinals

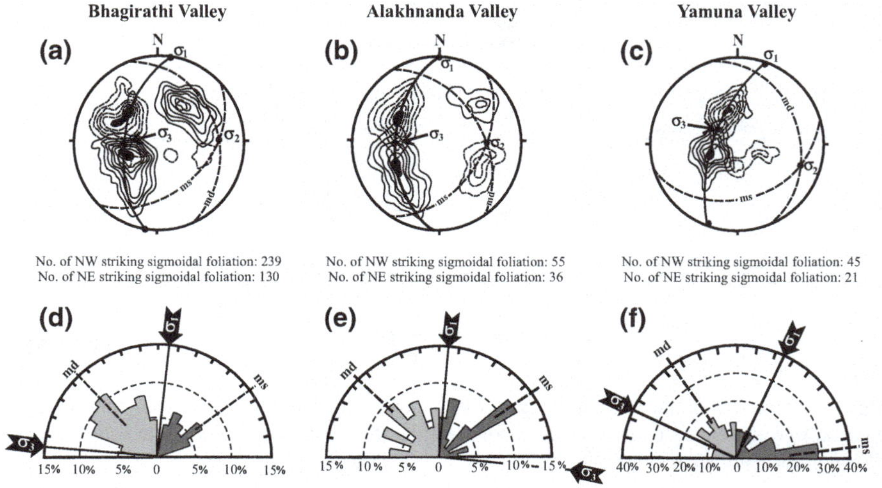

Fig. 9 Plot of pole to sigmoidal foliation on equal area net (**a**, **b**, **c**) and trend of the strike of sigmoidal foliation on rose diagram (**d**, **e**, **f**) for Bhagirathi valley, Alakhnanda valley, and Yamuna valley respectively. Contour intervals. 2, 4, 6, 8, 10 < 10% per unit area. (after Srivastava and Tripathy 2007). Continuous contour lines represent NW striking sigmoidal foliation and dashed contour lines represents NE striking sigmoidal foliation; ms-mean NE striking sigmoidal foliation; md-mean NW striking sigmoidal foliation; σ_1, σ_2, and σ_3 are maximum, intermediate, and minimum compressive directions, respectively

folds exhibit development of axial planar cleavage (S_2) and is the most dominant planar fabric in the area. Second phase of deformation is attributed to the second phase of folding (F_2) that mainly plunges towards NNE direction, coaxial to the F_1 folds and developed on the limbs of the F_1 isoclinal folds (Fig. 3b). The F_1 and F_2 are considered as Pre-Himalayan structures (Jain et al. 2002).

The third phase of deformation is marked by the structures developed due to progressive ductile shearing took place during thrusting of crystalline rocks over the rocks of Garhwal Group. The foliation planes containing NNE plunging stretching lineation expressed by the highly stretched feldspar porphyroclasts in the MCT Zone. Various asymmetric structures such as sigmoidal foliation (Ramsay and Graham 1970), asymmetric recrystallized tails around feldspar (Simpson and Schmidt 1983), S-C structure, sheath folds, folded layering, and intragranular faults in the MCTZ suggest top to SSW directed shear sense (Berthe et al. 1979; Passchier and Trouw 2005). Due to shearing earlier developed structures are reoriented in NE to NNE direction. At places the thrust plane is also folded, this may be explained by the folding during last stage of third phase of deformation. The fourth phase of deformation is characterized by brittle deformation like faults and joints.

The detailed analysis of sigmoidal foliation gave an opportunity to study strain, crystallographic preferred orientation and the compressive stress direction in the MCTZ of Garhwal Higher Himalaya. Strain studies in a shear zone reveal of strain value is higher at the centre of shear zone and it gradually decreases towards the shear zone boundary (Fig. 8). The study of crystallographic preferred orientation of quartz reveal that in a shear zone, with progressive increase in deformation the angle between c-axes maxima and the foliation normal reduces, and finally, in the centre of shear zone the c-axes maxima becomes perpendicular to the foliation. The strain analysis (Fig. 8) and study of crystallographic preferred orientation of quartz (Fig. 7) reveal that small-scale ductile shear zones grow in a narrow zone in response to very high strain, which might deform even the internal fabric of rock. The results obtained from stereoplots and the rose diagrams (Fig. 9) from all the three river valleys suggest that maximum compressive stress direction acted in a NNE-SSW direction horizontally. The movement within the MCTZ reflected by the NNE-SSW directed maximum compression direction could be correlated with the northward crustal movement of the Indian Plate between Tibetan Plate (Bouchez and Pecher 1981).

6 Conclusions

The present study reveals that the rocks of MCT Zone have suffered four phases of deformations where the first two phase of deformation are attributed to folding known as early structures (i.e., pre-thrusting) whereas, the third phase of deformation (syn-thrusting) is attributed to ductile shearing during which the translations of the crystalline rocks over the sedimentary rocks took place. This phase of deformation displays different kinds of shear zone indicators exhibiting sense of movement of the rocks. The NE and NW trending sigmoidal shear zones exhibit a variation in its strike from NNE to ENE and to NNW to WNW direction respectively, and have been utilized for the estimation of compressive stress direction in the region. The composite plot of the strike and dip of the shear zones on stereographic net and on rose diagram from different valleys of Garhwal Higher Himalaya have suggested that the maximum compressive stress (σ_1) acted roughly in NNE-SSW direction horizontally, which has resulted the thrusting of crystalline rocks over the sedimentary rocks along MCT. Further, the different kinematic indicators of the MCT zone developed during different types of contractional and extensional structures in one overall compression regime exhibit top to SSW sense of movement. The direction of shear obtained from stretching lineations and slickenlines consistently indicate a top to south-directed sense of movement and suggest that area has suffered a major ductile shearing. The movement within the MCTZ reflected by NNE-SSW directed maximum compressional direction could be correlated to the north ward crustal movement of the Indian plate.

Acknowledgements The present article is a review of the earlier work carried by the authors and research students and co-workers. The diagrams and text published earlier have been utilized after modification and have been referred in the text. We also acknowledge Prof. T. K. Biswal for improving the quality of the manuscript and efficient editorial handling.

References

Anderson EM (1951) The dynamics of faulting. Oliver and Boyd, Edinburgh
Banerjee S, Matin A (2013) Evolution of microstructures in precambrian shear zones: An example from eastern India. J Struct Geol 50:199–208
Bell TH, Bruce MD (2007) Progressive deformation partitioning and deformation history: evidence from millipede structures. J Struct Geol 29(1):18–35
Berthe D, Choukroune P, Jegouzo (1979) Orthogneiss, mylonite and non-coaxial deformation of granites: the example of the South Armorican shear zone. J Struct Geol 1:31–42
Bhattacharya AR, Weber K (2004) Fabric development during shear deformation in the main central thrust zone, NW Himalaya, India. Tectonophysics 387:23–46
Bons PD, Elburg MA, Gomez-Rivas E (2012) A review of the formation of tectonic veins and their microstructures. J Struct Geol 43:33–62
Bouchez JL, Pecher A (1981) The Himalayan main thrust pile and its quartz rich tectonites in central Nepal. Tectonophysics 78:23–50
Brun JP, Pons J (1981) Strain patterns of pluton emplacement in a crust undergoing non-coaxial deformation, Sierra Morena, Southern Spain. J Struct Geol 3(3):219–229
Carreras J, Druguet E, Griera A (2005) Shear zone-related folds. J Struct Geol 27(7):1229–1251
Choukroune P, Gapais D (1983) Strain pattern in the Aar granite (Central Alps): orthogneiss developed by bulk inhomogeneous flattening. J Struct Geol 5:411–418
Cobbold PR, Quinquis H (1980) Development of sheath folds in shear regime. J Struct Geol 2:119–126
Cosgrove JW (1980) The tectonic implications of some small scale structures in the mona complex of Holy Isle, North Wales. J Struct Geol 2(4):383–396
Druguet E, Alsop GI, Carreras J (2009) Coeval brittle and ductile structures associated with extreme deformation partitioning in a multilayer sequence. J Struct Geol 31(5):498–511
Fossen H (2016) Structural geology. Cambridge University Press, 463p
Gansser A (1964) Geology of the Himalayas
Ghosh SK (1993) Structural geology: fundamentals and modern developments. Pergamon Press, Oxford, 598p
Ghosh SK, Sengupta S (1987) Progressive evolution of structures in a ductile shear zones. J Struct Geol 9:277–288
Gomez-Rivas E, Bons PD, Griera A, Carreras J, Druguet E, Evans L (2007) Strain and vorticity analysis using small-scale faults and associated drag folds. J Struct Geol 29(12):1882–1899
Gray MB, Mitra G (1993) Migration of deformation fronts during progressive deformation: evidence from detailed structural studies in the pennsylvania anthracite region, USA. J Struct Geol 15(3–5):435–449
Heim A, Gansser A (1939) Central Himalaya: geological observation of the swiss expedition in 1936. MemS Soc Helv Sci Nat 73:1–245
Hobbs BE, Means WD, Williams PF (1976) An outline of structural geology. Wiley, New York, 571p
Hudleston PJ (1977) Progressive deformation and development of fabric across the zones of shear in glacial ice. In: Saxena S, Bhattacharjee S (eds) Energetics of geological processes. Springer, New York, pp 121–187
Jaeger JC (1969) Elasticity, fracture, and flow. Methuen, London
Jain AK (1971) Stratigraphy and tectonics of Lesser Himalayan region of Uttarkashi. Garhwal Himalaya Himal Geol 1:25–57
Jain AK, Singh S, Manickavavasagam RM (2002) Himalayan collision tectonics. Gondwana Res Group Mem 7:1–114
Law RD (1990) Deformation mechanisms, rheology and tectonics (Knipe RJ, Rutter EH, eds), vol 74. Geological Society Special Publication, pp 485–509
Le Fort P (1975) Himalayas: the collided range, present knowledge of continental arc. Am J Sci 275A:1–44

Leith CK (1923) Rock cleavage. Bull US Geol Surv 239:1–216
Lisle R (1985) Geological strain analysis: a manual for Rf/φ technique. Pergamon Press, Oxford, 99p
Lisle RJ (2013) Geological strain analysis: a manual for the Rf/ø method. Elsevier
Lister GS (1977) Crossed girdle c-axis fabrics in quartzites plastically deformed by plane strain and progressive simple shear. Tectonophysics 39:51–54
Leister GS, Snoke AW (1984) S-C mylonites. J Struct Geol 6:617–638
Malavieille J (1987) Extensional shearing deformation and kilometer scale "a"-type folds in a cordilleran metamorphic core complex (Raft River Mountain, Northwestern Utah). Tectonics 6:423–448
Marcos A, Arboleya ML (1975) Evidence of progressive deformation in minor structures in Western Asturias (NW Spain). Geol Rundsch 64(1):278–287
Marshak S, Mitra G (1988) Description of mesoscopic structures. In: Marshak S, Mitra G (eds) Book basic methods of structural geology (Part-II Special Topics), pp 213–247
Matin A, Banerjee S, Gupta CD, Banerjee N (2012) Progressive deformation across a ductile shear zone: an example from the Singhbhum shear zone, eastern India. Int Geol Rev 54:290–301
Means WD (1976) Stress and strain. Springer, Berlin
Metcalf RP (1993) Pressure temperature and time constraints on metamorphism across the main central thrust zone of higher himalayan slab in the Garhwal Himalaya. In: Trelor PG, Searle MP (eds) Himalayan tectonics, vol 74. Geological Society of London Special Publication, London, pp 485–509
Passchier CW (1990) Reconstruction of deformation and flow parameters from deformed vein sets. Tectonophysics 180(2–4):185–199
Passchier CW, Simpson C (1986) Porphyroclast system as kinematic indicators. J Struct Geol 8:831–843
Passchier CW, Trouw RAJ (2005) Microtectonics. Springer, Berlin
Platt JP (1983) Progressive refolding in ductile shear zones. J Struct Geol 5:619–622
Platt JP, Behrmann JH (1986) Structures and fabrics in a crustal-scale shear zone, Betic cordillera. SE Spain J Struct Geol 8(1):15–33
Price NJ, Cosgrove JW (1990) Analysis of geological structures. Cambridge University Press, New York, pp 1–502
Ramsay JG, Graham RH (1970) Strain variations in shear belts: Canadian. J Earth Sci 7:786–813
Ramsay JG (1967) Folding and fracturing of rocks. McGraw-Hill, New York, 700p
Ramsay JG (1980) Shear zone geometry: a review. J Struct Geol 2:83–99
Ramsay JG, Allision I (1979) Structural analysis of shear zones in a alpinised hercynian granite, Maggia Nappen Pennine zone, Central Alps. Schweizeizerische Mineral Und Petrogr Mittlungen 59:251–279
Ramsay JG, Huber MI (1987) Techniques of modern structural geology, 2: fold and fractures. Academic Press, New York
Rykkelid E, Fossen H (1992) Composite fabrics in mid-crustal gneisses: observations from the Øygarden complex, West Norway Caledonides. J Struct Geol 14:1–9
Searle MP, Metcalf RP, Rex AJ, Norry MJ (1993) Field relations petrogenesis and emplacement of Bhagirathi leucogranite, Garhwal Himalaya. In: Treloar PJ, Searle MP (eds). Himalayan Tectonics, vol 74. Geological Society of London, pp 429–444
Simpson C, Schmidt SM (1983) An evaluation of criteria to deduce the sense of movement in sheared rocks. Bull Geol Soc Am 94:1281–1288
Simpson C (1986) Determination of movement sense in Mylonites. J Geol
Singh K, Thakur VC (2001) Microstructure and strain variation across the footwall of main central thrust zone, Garhwal Himalaya. J Asian Earth Sci 19:17–29
Srivastava HB (1995) Two dimensional strain estimation from weakly deformed rocks. Ann Tecton, IX, pp 3–6
Srivastava HB, Hudleston PJ, Early D (1995) Strain and possible volume loss in a high grade ductile shear zone. J Struct Geol 17:1217–1231

Srivastava HB, Sahai A, Lal SN (2000) Strain and crystallographic fabric in mesoscopic ductile shear zone of Garhwal Himalaya. Gondwana Res 3:395–404
Srivastava HB, Tripathy NR (2007) Geometrical analysis of mesoscopic shear zones in the crystalline rocks of MCT zone of Garhwal higher Himalaya. J Asian Earth Sci 30:599–612
Talbot CJ, Sokoutis D (1995) Strain ellipsoids from incompetent dykes: application to volume loss during mylonitization in the Singö gneiss zone, central Sweden. J Struct Geol 17(7):927–948
Turner FJ, Weiss LE (1963) Structural analysis of metamorphic tectonites. Mc Gray-Hill, New York
Twiss RJ, Moores EM (2007) Structural geology. WH Freeman and Company, New York, p 532
Valdiya KS (1980) Geology of Kumaun Lesser Himalaya. Wadia Institute of Himalayan Geology, Dehradun, 291p
Wilson G (1961) The tectonic significance of small scale structures and their importance to the geologist in the field. Ann Geol Belg 84:424–548

Thrust Sheets, Tectonic Windows, and Intermontane Basins in the Nepal Himalaya

Megh Raj Dhital and Basanta Raj Adhikari

Abstract The Himalayan Range is generally classified into a number of broad longitudinal tectonic belts. Despite a long history of investigation, some fundamental issues of their stratigraphy and structure are still unresolved. Especially, there has been considerable controversy over delineating the Greater Himalayan and Lesser Himalayan belts of Nepal. The Greater Himalayan thrust sheet represents the hanging wall of the Main Central Thrust. In Nepal, the thrust sheet forms two large open folds: the Great Midland Antiform in the inner zone and the Great Mahabharat Synform in the outer part. The Main Himalayan Thrust and Main Central Thrust constitute respectively the floor and roof of a mega duplex where some detached Lesser Himalayan horses are exposed in various tectonic windows. The Main Himalayan Thrust plays a role of sole thrust in the imbricate stack developed within the foreland fold-and-thrust belt. The key structural and stratigraphic aspects of thrust sheets, tectonic windows, klippen, and intermontane basins are discussed together with the neotectonic activity in the Nepal Himalaya.

Keywords Nepal himalaya · Tectonic window · Thrust sheet · Mega duplex · Active fault · Thakkhola graben

1 Introduction

Although it was almost a century ago that Argand (1924) had proposed a collision of India with Eurasia and the creation of Himalaya, its tectonic features are still not clear. Continental tectonics is complex and mountain ranges differ significantly from one another in their internal structure (Molnar 2015). Continent-continent collision

---

M. R. Dhital
Department of Geology, Tri-Chandra Campus, Tribhuvan University, Kirtipur, Nepal

B. R. Adhikari (✉)
Institute for Disaster Management and Reconstruction, Sichuan University-Hong Kong Polytechnic University, Chengdu, China
e-mail: bradhikari@ioe.edu.np

Department of Civil Engineering, Institute of Engineering, Tribhuvan University, Lalitpur, Nepal

T. K. Biswal et al. (eds.), *Structural Geometry of Mobile Belts of the Indian Subcontinent*, Society of Earth Scientists Series,
https://doi.org/10.1007/978-3-030-40593-9_11

in the Alps developed a number of north-vergent nappes (Heim 1922; Schmid et al. 1996; Ballévre et al. 2018; Manzotti and Zucali 2013; Dal Piaz et al. 2003; Malusá et al. 2011). Detailed field mapping supported by abundant fossils and radiometric age determinations in the Alps made it possible to delineate accurately the geological structures and prepare a comprehensive geological map. Unfortunately, it was not the case with the Himalaya, except for the Tethyan sedimentary sequence, where the fossils abound. Hence, there have been utterly diverse views on Himalayan tectonics.

Since Nepal is centrally located within the Himalayan arc, geological investigations in this segment play a pivotal role in understanding this quintessential collided orogen and consequently deciphering continental plate tectonics. Hagen (1969) published the first tectonic map of Nepal where he proposed about 20 nappes. But Hashimoto et al. (1973) and Talalov (1972) denied almost all of Hagen's nappes, and advocated vertical block faulting. Heim and Gansser (1939), Lombard (1958), Bordet (1961), Gansser (1964), and Stöcklin (1980) drew a single and continuous thrust sheet (the Tibetan Slab or the Greater Himalayan Crystallines) and they consolidated the concept of Main Central Thrust (MCT).

Most of the contemporary geological maps of Nepal result essentially from the amalgamation of previous maps and some new observations with an appealing tectonic model. Such hybrid geological maps differ from author to author. For example, Schelling and Arita (1991) made a geological map of East Nepal and drew a 'balanced' cross-section by incorporating a huge north-dipping fictitious ramp and reckoned 185–245 km of crustal shortening. This model was utilised subsequently by Pandey et al. (1995) in Central Nepal to depict the distribution of earthquake hypocentres and prepare a seismotectonic model for the Nepal Himalaya. Several other researchers (e.g., Upreti and Le Fort 1999; DeCelles et al. 2001; Robinson et al. 2006; Goscombe et al. 2006; Searle et al. 2008) have published different hybrid geological maps to estimate either crustal shortening or deformation mechanism in the Himalaya.

During post-collisional underthrusting of the Indian lithosphere, the overlying crystallines were subjected to partial subduction as well as a complex cycle of prograde and retrograde metamorphism (Bhargava 2000; Webb et al. 2007; Carosi et al. 2018a, b). Owing to continued convergence, the crystallines were ultimately detached from the subducting Indian Plate and propagated to the south in the form of a thrust sheet that dragged through the incompetent sediments composing the Lesser Himalaya. This process resulted into a south-verging mega duplex where the Lesser Himalayan sequence is sandwiched between the Main Himalayan Thrust (MHT), supposed to represent the floor, and the MCT, corresponding to the roof (Figs. 1 and 2). The very thick (more than 15 km) Greater Himalayan Crystallines constitute the hanging wall of the MCT. The thrust sheet is stiff and competent enough not to yield to significant deformation. Since various researchers have defined the MCT (e.g., Heim and Gansser 1939; Gansser 1964; Valdiya 1980; Arita 1983; Amatya and Jnawali, 1994; DeCelles et al. 1998, Searle et al. 2008; Carosi et al. 2018b), here the thrust is taken as the dislocation separating the rocks of Nawakot Complex from the overlying Kathmandu Complex or their counterparts (Stöcklin 1980). The

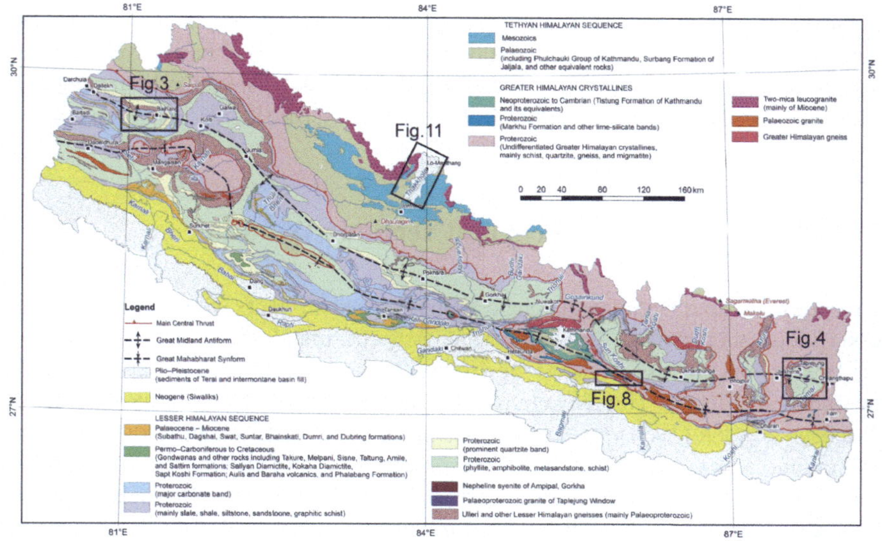

Fig. 1 Generalised geological map of Nepal depicting main litho-tectonic units. *Source* Modified from Dhital (2015)

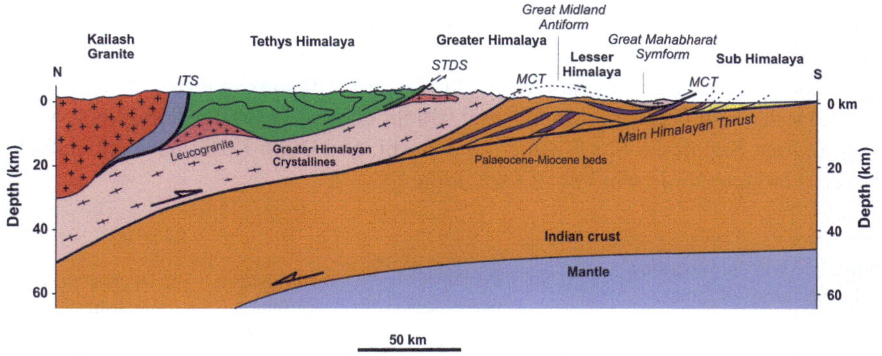

Fig. 2 Schematic geological cross-section across Himalaya showing main structural features. ITS: Indus-Tsangpo Suture; STDS: South Tibet Detachment System; MCT: Main Central Thrust

Greater Himalayan Crystallines are equivalent essentially to the Crystalline Nappes of Fuchs and Frank (1970), Fuchs (1981), or the Tibetan Slab of Lombard (1958), Bordet (1961), and Le Fort (1971, 1975).

In the Nepal Himalaya, the Greater Himalayan thrust sheet gave rise to two large open folds (Fig. 1): the Great Midland Antiform in the inner zone and the Great Mahabharat Synform in the outer belt (Hagen 1969; Dhital 2015). Subsequent erosion has obliterated most part of the antiform and some part of the synform. Consequently, a number of Lesser Himalayan horses detached from the MHT are exposed in various tectonic widows. The horses are forming antiformal stacks in the inner zone, whereas other contractional faults emerging from the MHT have developed secondary duplexes and imbricate fans towards the foreland (Dhital and Kizaki 1987; Johnson 1994; DeCelles et al. 1998; Robinson and Martin, 2014).

2 Great Midland Antiform and Tectonic Windows

In response to rapid Himalayan upheaval, the antecedent rivers were forced to relentlessly incise their course, causing further isostatic imbalance (England and Molnar 1990). This process removed the Greater Himalayan thrust sheet and exhumed the underlying Lesser Himalayan sequence in tectonic windows. In the Northwest Himalaya, Auden (1934) recognised first the tectonic window at Solon where the Pachmunda and the Krol synclines are separated by an anticline. Similarly, West (1939) mapped the tectonic window of Shali (north of Simla). Berthelsen (1951) drew a generalised geological section through the Northwest Himalaya where he showed many nappes and windows in the Sutlej River valley.

In the Nepal Himalaya, the Great Midland Antiform has successive culminations and depressions marked by tectonic windows and bridges (Fig. 1 and Table 1). Hagen (1969) mapped many tectonic windows, including those of Bajhang, Okhaldhunga, Arun, and Taplejung. A solitary small klippe of the Greater Himalayan Crystallines is located to the northwest of the Kali Gandaki–Trishuli confluence (Fig. 1). The klippe was witness to the extensive erosion of the Greater Himalayan thrust sheet in West Nepal by the Kali Gandaki, Marsyangdi, Budhi Gandaki, and Trishuli rivers. Similarly, in East Nepal, the Sun Koshi, Tama Koshi, and Dudh Koshi have collectively carved a large tectonic window of Okhaldhunga.

Table 1 Main culminations and depressions of Great Midland Antiform

S. No	Principal antecedent river	Culmination (tectonic window)	Depression (tectonic bridge)
1	Mahakali (or Kali)	Mahakali	Dallekh
2	Seti	Bajhang	Kolti
3	Karnali	Galwa	Jumla
4	Bheri	Bheri	Dhorpatan
5	Trishuli	Trishuli	Gosainkund
6	Koshi	Okhaldhunga	Bhojpur
7	Arun	Arun	Jirikhimti
8	Tamar	Taplejung	Chyangthapu

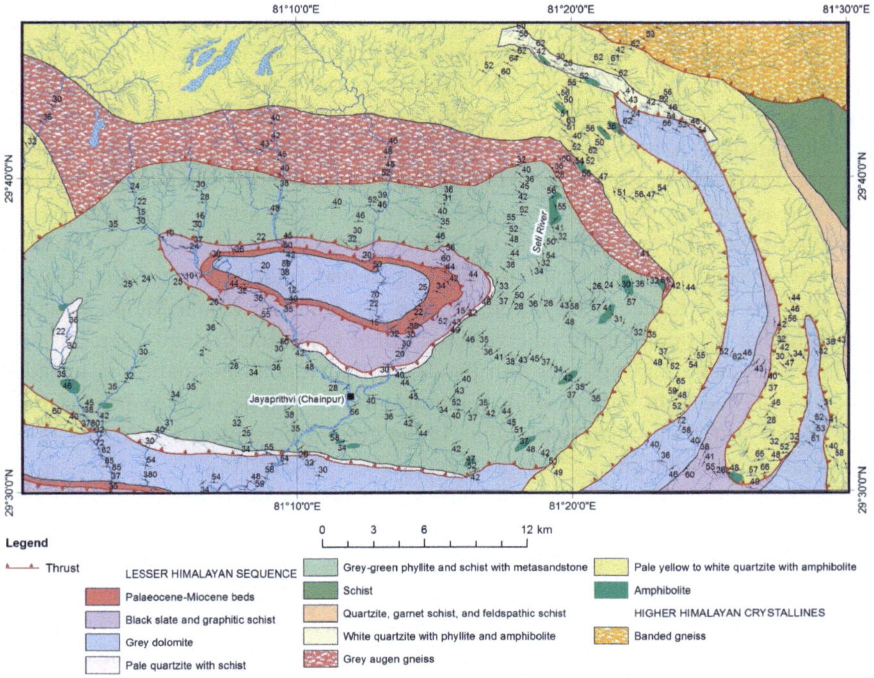

Fig. 3 Geological map of Bajhang tectonic window showing the younger strata in the core. *Source* Modified from Khan (1996, 2000)

2.1 *Tectonic Windows of West Nepal*

Heim and Gansser (1939) and Gansser (1964) investigated the Mahakali tectonic window. Fuchs and Frank (1970) and Fuchs (1977) mapped in detail the Galwa and Bheri tectonic windows in West Nepal. They also discovered the Rukum Nappe in this region and post collisional Palaeocene–Miocene strata in the cores of the two strongly deformed windows. The Bajhang tectonic window (Fig. 3) is a classic example where the Paloaecene–Miocene rocks are exposed in its core. The oldest Lesser Himalayan sequence is represented by phyllites and biotite schists with some orthoquartzite, metasandstone, and amphibolite bands. Augen gneisses are injected in a thick pale quartzite sequence, which is thrust over the dolomites and slates.

2.2 *Tectonic Windows of East Nepal*

The Okhaldhunga, Arun, and Taplejung windows in East Nepal (Fig. 1) contain mainly Palaeoproterozoic schists and quartzites followed by black slates and graphitic schists with a few discontinuous quartzite bands. Sporadic granites and frequently occurring augen gneisses at various stratigraphic levels entail that the gneisses may differ in age. The uppermost part of Lesser Himalayan sequence is represented by amphibolites, feldspathic schists, garnet schists, and crystalline limestones. The Greater Himalayan Crystallines contain banded gneisses, migmatites, and other medium to high-grade rocks with some schist, quartzite, and marble bands.

In the Taplejung window (Figs. 1 and 4), the Palaeoproterozoic Kuncha Formation is composed of grey to dark green-grey *gritty* phyllite, chlorite schist, and biotite schist with thick to very thick quartzite bands and alternations (Fig. 5). The rock is strongly crenulated and contains boudinaged and folded quartz veins. Toward the lower part of Kuncha Formation appears the Subhang Khola Member, which is a lenticular unit containing pale yellow to light grey, fine-grained quartzite with phyllite and schist partings and thin bands. In the Kuncha Formation is intruded about 1.8 Ga old light grey to white medium-grained granite (Auden 1934; Upreti et al. 2003) with biotite and tourmaline. In the granite xenoliths of dark grey schist and quartzite are frequent.

The Kuncha Formation is transitionally succeeded by the Chilingdin Formation (Fig. 5) of grey, light grey, to pale yellow, thick- to very thick-banded quartzite alternating with phyllite, chlorite schist, and biotite schist. Generally the rock is massive. The Chilingdin Formation is followed stratigraphically upwards by the Phyame Formation of grey to dark grey, laminated garnet-biotite schist and graphitic schist with thin quartzite bands (Fig. 4). In this formation towards the southeast end of the map, grey augen gneisses or feldspathic schists are found with thin garnet schist, kyanite schist, and quartzite bands. The overlying Phidim Formation contains from a few metres to tens of metres thick, grey, green-grey to light grey feldspathic schist and quartzite with biotite, garnet, and sporadically kyanite (towards upper part). Extremely rare crystalline limestone bands are also present. In fact, this formation differs from the Phayme Formation mainly in its feldspar content. This formation is also affected by inverted metamorphism and in its upper levels occur light grey to pale quartzites or feldspathic quartzites with amphibolite. The amount and grain size of feldspar (mainly orthoclase) steadily increases from lower to upper horizons.

Above the MCT, the Greater Himalayan Crystallines are represented by the Prangbung Formation (Figs. 4 and 5) of grey to light grey, banded paragneisses with kyanite (in the lower part) and sillimanite. The gneisses contain purple garnets towards their lower horizons. Many pegmatite veins ranging in thickness from 5 cm to 10 m are injected parallel or slightly oblique to foliation. About 10–50 m thick marble and calcareous gneiss bands are sporadically observed in the lower part. The pegmatite veins consist of feldspar, quartz, muscovite, tourmaline, and occasionally gemstones including ruby and sapphire.

In the Taplejung window, inverted metamorphism (Schelling and Arita 1991) is ubiquitous and it gradually increases from the lower stratigraphic levels to the upper horizons. While moving structurally upwards, the grade of metamorphism is marked by the successive appearance of chlorite, biotite, and garnet index minerals in the Lesser Himalayan Sequence and kyanite and sillimanite in the Greater Himalayan Crystallines (Fig. 4). Most of the Kuncha Formation lies in the biotite zone and its upper part passes into the garnet zone. Though kyanite appears mainly in the lower

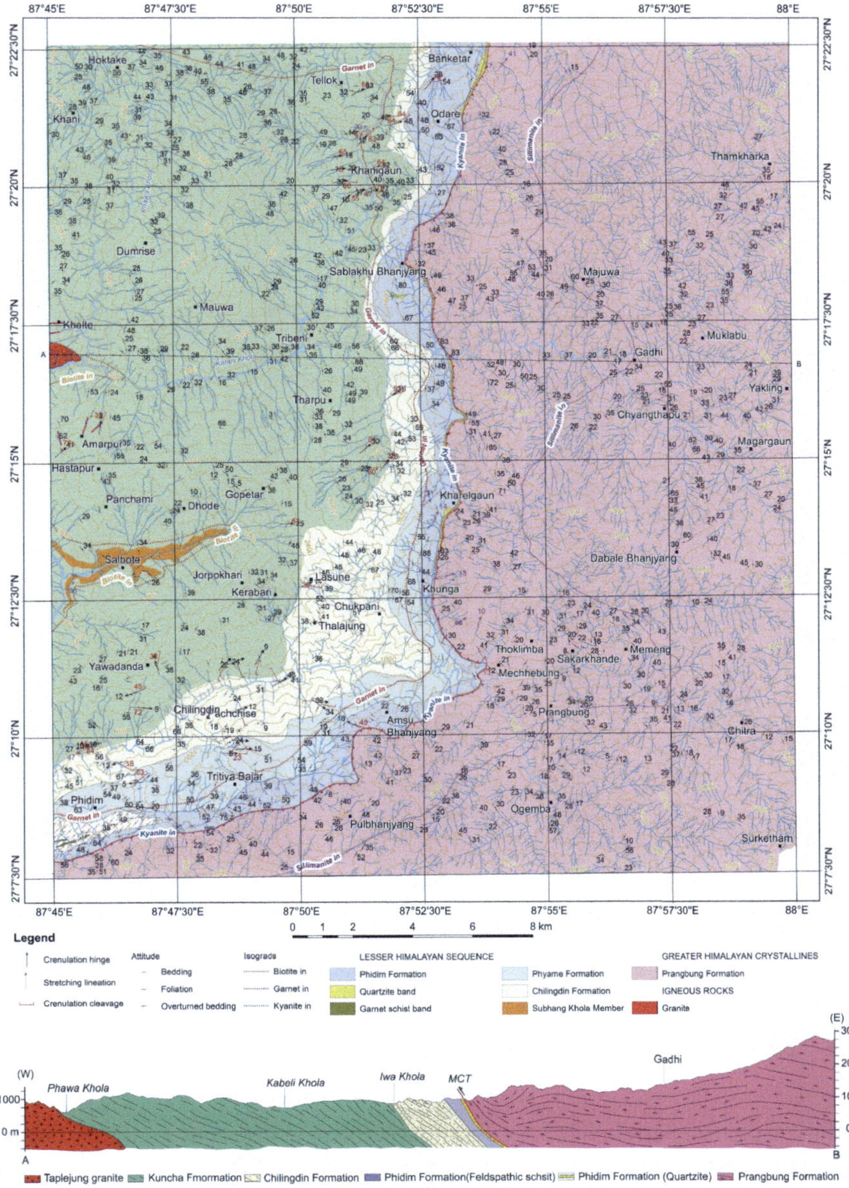

Fig. 4 Geological map and cross-section of part of Taplejung window in East Nepal. *Source* Modified from Dhital (2015)

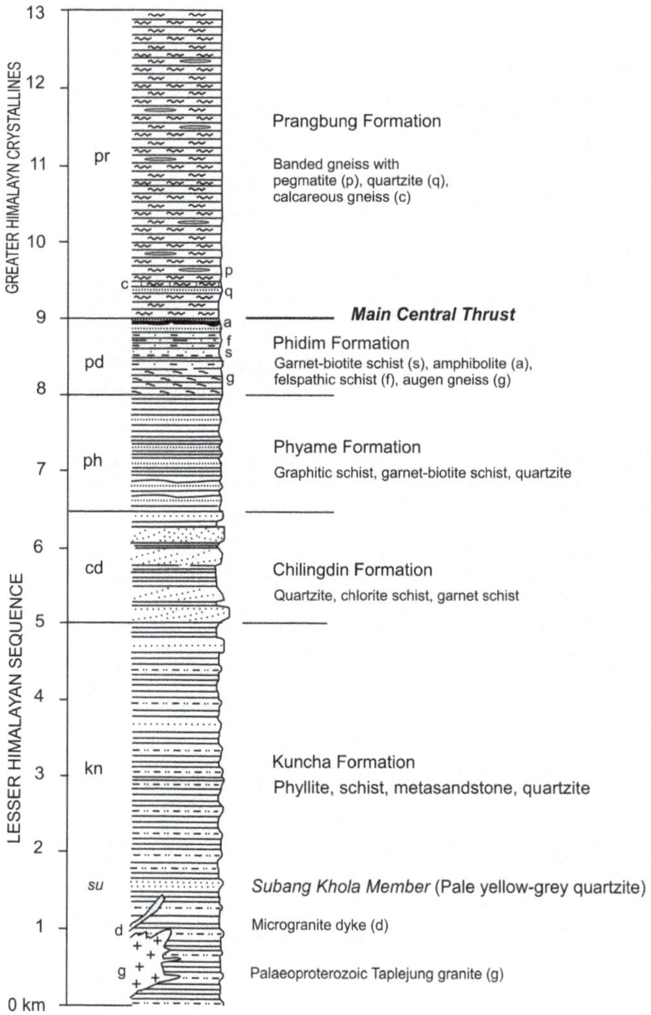

Fig. 5 Tectono-stratigrpahic column of the Taplejung tectonic window and surrounding region

portion of the Greater Himalayan Crystallines, in some instances it already occurs in the upper part of the Phidim Formation. Carosi et al. (2007) have also reported kyanite-bearing quartzites in the Lesser Himalayan sequences. Hence, the isograds obliquely cut the lithological units and they do not define the MCT.

The rocks of Taplejung window exhibit well developed foliation, crenulation cleavage, and stretching lineation (Fig. 4). The stereographic projection of main foliation (S_1) reveals that the window is a non-cylindrical fold (Fig. 6), resembling a large dome.

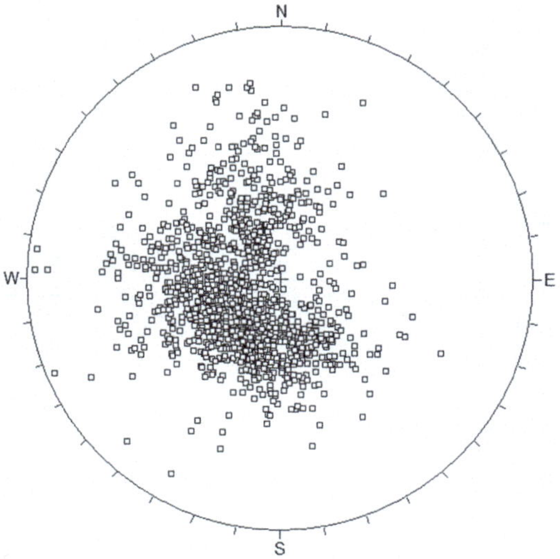

Fig. 6 Stereographic projection of 1339 main foliation (S_1) poles related to a first phase of deformation (D_1). Equal angle, lower hemispherical projection

The crenulation cleavage (S_2) in the Lesser Himalayan Sequence was formed during a second deformation phase (D_2) and its attitude is quite different from the orientation of main foliation (S_1) in the Lesser Himalayan as well as Greater Himalayan rocks. The crenulation cleavage (Fig. 7a, b) is steeply dipping essentially due NNW. Stretching lineations, crenulation hinges (representing F_2 folds), and mineral lineations are also observed in some part of the window (Fig. 4). Most of them are associate with the D_2 phase and their orientation is due ENE or WSW.

Many quartz veins are stretched parallel to S_1 and they have developed a boudinage structure during the D_1 phase. The boudins frequently occur in the metapelites as well as the metasandstones and quartzites of the Kuncha Formation, Chilingdin Formation, and Phidim Formation. These previously boudinaged quartz veins were subsequently folded. As a result, folded boudins are seen in the Kuncha Formation (Fig. 7a, b). The quartz boudins range in size from a few millimetres to a few centimetres. A widespread distribution of folded quartz boudins indicates that the region underwent the D_2 phase intense deformation after their emplacement. Since it is not possible for any progressive deformation (pure shear or simple shear) to switch from extension to compression (Ramsay 1967), the D_1 and D_2 phases must had been quite independent.

Infrequent 'cross-laminae' are observed in the gneisses of Greater Himalayan Crystallines (Fig. 7c) and such structures allude to their sedimentary origin. Due to D_2 phase shearing, sigma- and delta-type structures are well developed in the feldspar porphyroblasts of some banded gneisses (Fig. 7d).

Thus, the investigation of foliation, crenulation cleavage, lineation, and deformed quartz veins reveals at least two distinct phases of deformation in the region. Presumably, the D_1 phase was associated with the movement of MCT and the D_2 phase can be attributed to the formation of dome and its subsequent erosion to expose the Lesser Himalayan Sequence in the Taplejung tectonic window.

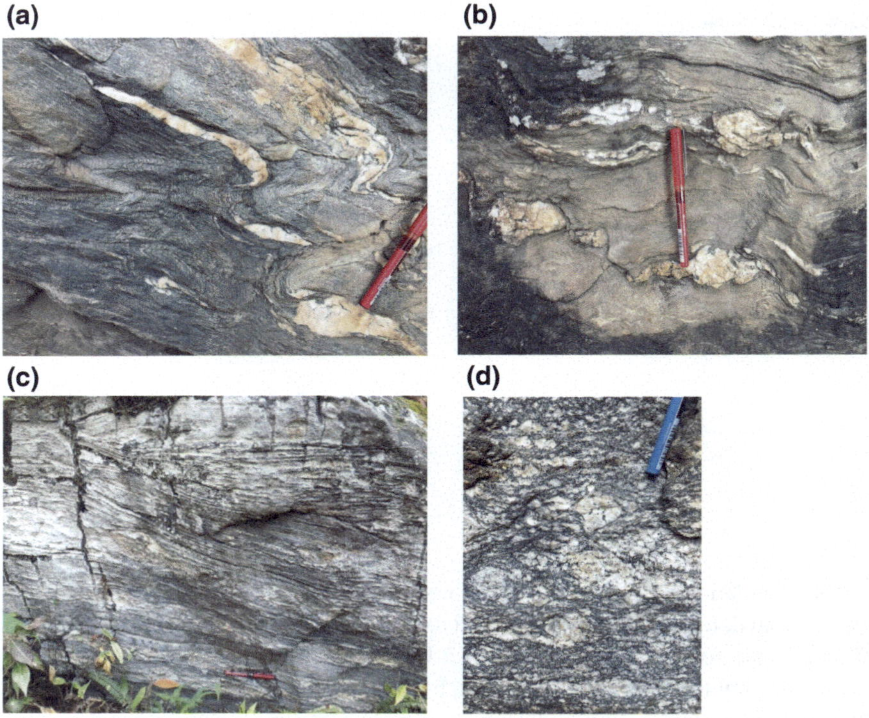

Fig. 7 **a**, **b** Folded boudins of quartz veins in Kuncha Formation. Crenulation cleavage of D_2 deformation phase is also well developed; **c** 'Cross-laminae' in banded paragneiss of Greater Himalayan Crystallines; **d** a delta-type structure in the gneiss of Greater Himalayan Crystallines

3 Great Mahabharat Synform and Klippen

After the seminal work of Heim and Gansser (1939) in the neighbouring tracts of Kumaun, India, many have attempted to decipher the main tectonic and stratigraphic features of Far West Nepal. Remarkably, Hagen (1969) defined from this region his Dadeldhura Autochthonous Zone where augen gneisses and granites preponderate. He also considered it to be connected laterally with the Galwa tectonic window. Fuchs and Frank (1970), Frank and Fuchs (1970) as well as Fuchs (1977, 1981) divided the metamorphic rocks into their Chail Nappes and the overlying Crystalline Nappes. Bashyal (1982, 1986) worked out the structure and stratigraphy of Far West Nepal and mapped the Bajhang Klippe, Parchuni Klippe, and Dadeldhura Nappe, from north to south respectively. He also showed that his Dadeldhura Nappe is the eastern continuation of the Almora Nappe of Gansser (1964) and Valdiya (1980). Since both the Dadeldhura Nappe and Kathmandu Complex (Stöcklin 1980) in Central Nepal are parts of the Great Mahabharat Synform, and contain Palaeozoic granites (Le Fort et al. 1983; Einfalt et al. 1993), they were taken to be equivalent units by a large number of investigators (e.g., Shrestha et al. 1987; Amatya and Jnawali 1994; Beyssac et al. 2004). Upreti and Le Fort (1999) came up with a rather different proposition in this regard. They defined a Lesser Himalayan 'rootless' nappe (comprising the rocks of Dadeldhura and Kathmandu Complex) and another overlying Greater Himalayan Crystalline Nappe rooted to the north. Likewise, DeCelles et al. (2000, 2001) and Robinson et al. (2006) designated their Ramgarh Thrust and overlying Dadeldhura Thrust (or MCT), and extended them throughout Nepal. Thus, the researchers have proposed several stratigraphic schemes and drawn the thrusts of their choice. As a result, the geology of Far West Nepal is tangled in confusion.

Recent field mapping in Far West Nepal revealed that the synformally folded Dadeldhura Nappe and klippen belong to the Lesser Himalayan Sequence of Proterozoic age (Fig. 1). They constitute a single folded roof thrust, consisting of phyllites, schists, quartzites, graphitic schists, amphibolites, and augen gneisses with blue and smoky quartz. In contrast to the Kathmandu Complex, inverted metamorphism profoundly affects the Dadeldhura Nappe and klippen, where it ranges from the chlorite grade at the lower structural level to the garnet or higher grade towards the top. Hence, to attribute these rocks to the Greater Himalayan Crystallines on the basis of the lone argument that they include Palaeozoic granites in the south limb of Great Mahabharat Synform is untenable. The Jutogh thrust sheet in the Simla area (Bhargava et al. 2016) and the Mandi granite (Mehta 1977) in the Himachal Pradesh, India, also occur in a similar setting.

The Mahabharat Synclinorium (Stöcklin 1980) in Central Nepal is part of the Great Mahabharat Synform and is made up of the Kathmandu Complex. It is a folded roof thrust. Erosion has also shredded the Greater Himalayan thrust sheet to yield many klippen resting over the Lesser Himalayan Sequence.

An example of the Great Mahabharat Synform is seen in Central Nepal, east of Kathmandu (Fig. 8), where the Lesser Himalayan Sequence is represented by the Nawakot Complex and the Greater Himalayan Crystallines and Tethyan succession are made up of the Kathmandu Complex (Stöcklin 1980). The Benighats (graphitic schsits with sporadic crystalline limestone bands) of Nawakot Complex are thrust over the Siwaliks (represented by Lower, Middle, and Upper Siwaliks), whereas the MCT brings the Kathmandu Complex over the Nawakot Complex. Here, the Kathmandu Complex commences with the Kalitar Formation (Kyanite schist and

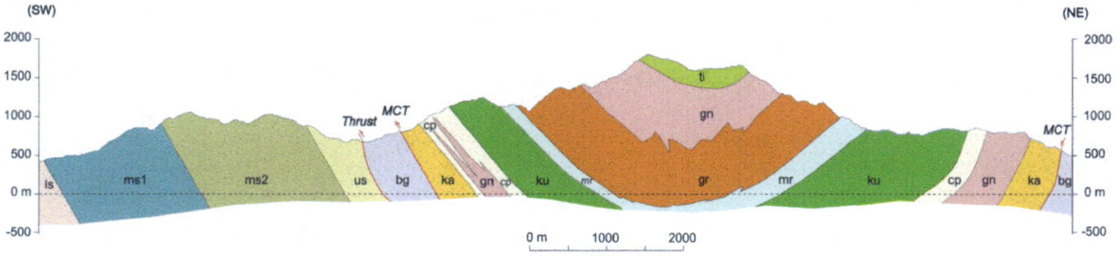

Fig. 8 Geological cross-section across the great mahabharat synform in the sun koshi-marin area of Central Nepal, east of Kathmandu.ls: Lower Siwaliks, ms1: middle siwaliks (lower member), ms^2: middle siwaliks (upper member), us: upper siwaliks, bg: benighats, ka: kalitar formation, gn: augen gneiss, cp: chisapani quartzite, mr: markhu formation, gr: palaeozoic granite, ti: tistung formation, MCT: Main Central Thrust

quartzite) and it is succeeded respectively by the Chisapani Quartzite, Kulikhani Formation (garnet-kyanite schist), Markhu Formation (schist, quartzite, and marble), and Tistung Formation (garnet schist and quartzite).

In Central Nepal, the metasediments of the Kathmandu Complex are affected by a regional metamorphism of Barrovian type (Stöcklin et al. 1982), which increases from the unmetamorphosed sediments on top to the coarsely crystalline garnet schists at the base of the section. In other locations, especially to the north and east of Kathmandu, kyanite- and sillimanite-bearing banded gneisses also appear towards the lower part of the Kathmandu Complex. The banded gneisses have an irregular distribution and they are the product of migmatisation of the metasediments (Stöcklin 1980). Palaeozoic granites intrude the Kathmandu Complex and they are closely associated with augen gneisses. Many of the augen gneisses have a similar granitic appearance and, in the Sun Koshi–Marin Section (Fig. 8) as well as in several other places, the granites and augen gneisses show imperceptible transitions, especially, when they are close to the MCT (Stöcklin et al. 1982).

4 Active Faults and Seismicity

Most of the convergence of India and Eurasia is accommodated within the Himalaya by movement on various active faults (Fig. 9). When the convergence is locked in some sector, the stored elastic energy is ultimately released in the form of tremors

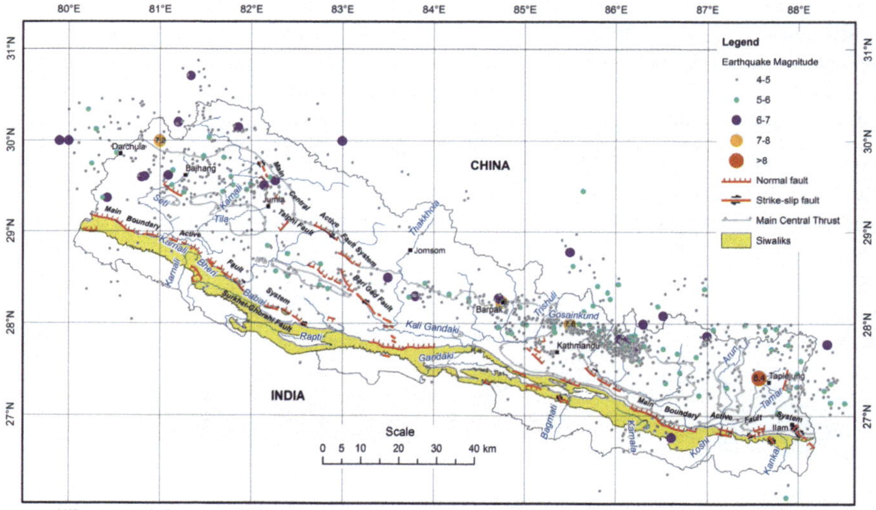

Fig. 9 Major active faults in Nepal (modified from Nakata 1982) and its background seismicity with a magnitude exceeding 4, occurred between 1995 and March 2016. Also are shown some known historic earthquakes exceeding magnitude 6. *Source* The Department of Mines and Geology, Kathmandu

along the mountain range and in its surroundings (Seeber and Armbruster, 1981; Pandey et al. 1995; Bilham et al. 1997). Active faults are found in the inner as well as the outer fold-and-thrust belt of the Himalaya and they are invariably related to the deformation of post-collisional Palaeogene–Neogene sediments. Therefore, post-collisional sediments played a significant role in the tectonic transport and contraction of Lesser Himalaya and Sub Himalaya (or Siwaliks). Shallow seismicity in the Nepal Himalaya is concentrated to the inner active fault zone, but not towards the foreland. Such selective distribution is presumably related to a stiffer roof extending to a depth of 35–40 km and compressed horses or thrust sheets that can store significant elastic energy.

The Himalayan Range and neighbouring tract have experienced many large tremors, namely the Nepal earthquake of 1833 with an estimated magnitude of 7.8 (Bilham 1995), the Shillong earthquake of 12 June 1897 with an estimated magnitude 8.7, the 1905 Kangra earthquake of estimated magnitude 8.0, the 1934 Nepal–Bihar earthquake of estimated magnitude 8.4, and the 1950 Assam earthquake of magnitude 8.5 (Oldham 1899; Middlemiss 1910; Auden 1934; Sharma and Malik 2006). On August 1988 the Udayapur earthquake of magnitude 6.6 struck East Nepal and its focal depth was estimated at 57 km (Dikshit 1991). The Gorkha earthquake of Mw 7.8 occurred on 25 April 2015 in Central Nepal and its focal depth was about 10–15 km (Adhikari et al. 2015; Avouac et al. 2015).

Microseismicity in Nepal is characterised by a narrow belt that follows approximately the front of the Great Himalayan Range. This kind of confined distribution reflects deformation between the upper and lower crusts along the MHT under the Lesser and Greater Himalaya (Pandey et al. 1999; Avouac 2003). Though the entire country is seismically active, there is a significant lateral variation. Microseismic activity is quite intense in East and Far West Nepal, whereas the (Bilham and England 1995) level of seismic activity is low in West Nepal. The earthquakes are generally shallower than 30 km and they are clustered around a depth of about 20 km. The frequency–magnitude relationship of microseismicity follows the Gutenberg–Richter law (Pandey et al. 1999), where the *b*-value varies between 0.75 and 0.95, and does not seem to change significantly over areas.

5 Intermontane Basins

Nepal hosts quite many intermontane basins, including the Thakkhola, Kathmandu, Pokhara, Hetaunda, Chitwan, Dang, Deukhuri, and Surkhet (Figs. 1 and 10). The basin of Thakkhola is a half-graben, the Kathmandu basin displays a classic centripetal drainage pattern and is filled up essentially with Pleistocene lake and fluvial sediments. The Pokhara basin contains the Holocene sediments accumulated by rivers, debris flows, and glacier lake outburst floods. The intermontane basins either within the Siwaliks or between the Lesser Himalaya and Siwaliks are imbricate fault-bound. Such basins are characterised by salients or convex-to-foreland curves and recesses or concave-to-foreland curves (Macedo and Marshak 1999) frequently

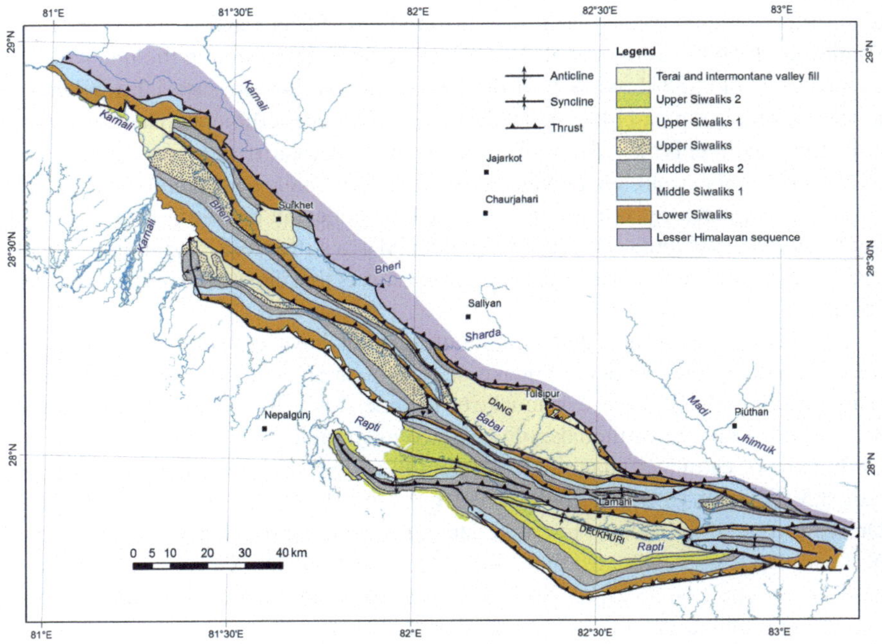

Fig. 10 Geological map of intermontane basins showing salients and recesses in West Nepal. *Source* Modified from Dhital (2015)

observed in the map view pattern of structural trend lines (thrusts, fold hinges, and bedding) in the Himalayan frontal fold-thrust belt of Nepal (Fig. 10). The salients are mostly of basin-controlled type and thickest basin fill is located at the apex of the salient.

5.1 Thakkhola Graben

Graben development in the Himalaya and Tibetan plateau is related to the convective removal of the lower portion of thick Asian lithosphere (Platt and England 1994), which helps east-west extension of the Tibetan plateau during the India-Asia collision (Coleman and Hodges 1995). This Cenozoic extension is responsible for the creation of grabens in Tibet and the northern part of Himalaya (Molnar and Tapponnier 1978). These normal faults have developed since 14 Ma in south Tibet (Coleman and Hodges 1995) and about 8 Ma (Harrison et al. 1997) or 4 Ma in Central Tibet (Yin et al. 1999). Generally, the normal faults either trend perpendicular to the Himalayan arc or strike north–south; but they also exhibit major deviations from these orientations (Kapp and Guynn 2004). The graben spacing decreases from south to the north systematically, which reflects the northward decrease in crustal thickness (Yin 2000).

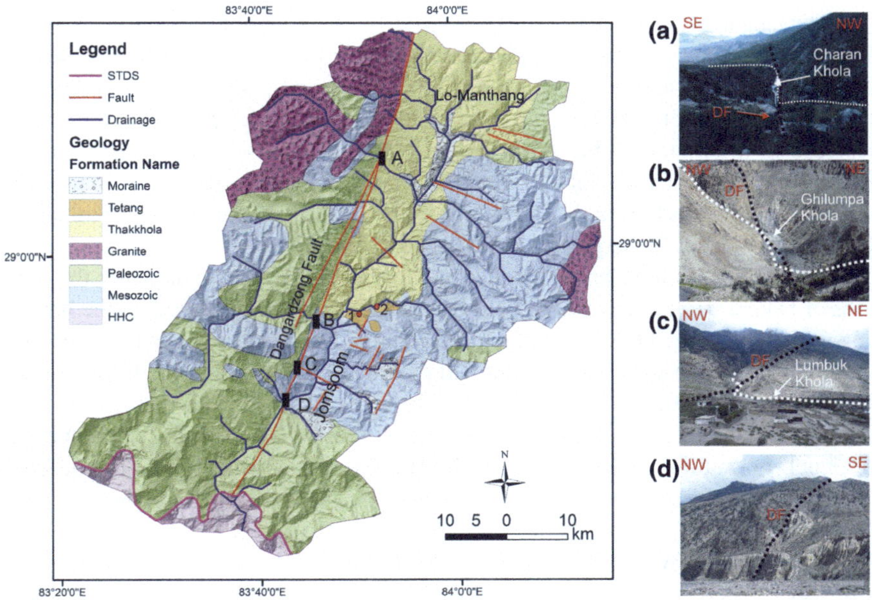

Fig. 11 Geological map of Thakkhola graben and trace of Dangardzong fault at different places along the strike of the fault. **a** Charan Khola changing its direction, **b** sharp bend of Ghilumpa Khola, **c** Lumbuk Khola changing its course, **d** small depression along the Dangardzong fault (DF). *Source* Modified from Adhikari (2009)

The Thakkhola graben (Figs. 1 and 11) is unique in the Himalaya and shows complex kinematic and geometric relationships with the South Tibet Detachment System (STDS) (Brown and Nazarchuk 1993; Patel et al. 1993) and the Dangardzong fault (Hurtado et al. 2001). This asymmetrical graben lies in the Palaeozoic to Cretaceous rocks of the Tethyan sequence between the STDS (Burchfiel et al. 1992) to the south and Indus-Tsngpo suture zone to the north. The graben is bounded by the Mustang-Mugu leucogranites (Le Fort and France-Lanord 1994) of 17.6 ± 0.3 Ma (Harrison et al. 1997) in the northwest of the graben while the basement is composed of a thick and nearly continuous early Palaeozoic to early Tertiary marine sedimentary succession deposited on the northern continental margin of the Indian plate. These rocks are crumpled, stacked, and deformed as a consequence of collision between India and Eurasia in the early Eocene (Garzanti et al. 1987; Searle et al. 1987). The graben is westerly bounded by the Dangardzong fault which was developed synchronously with the motion of the Annapurna Detachment during Miocene time (Hurtado et al. 2001). The minimum age of the east-west extension in this graben is ca. 14 Ma based on ^{49}Ar/^{39}Ar ages of hydrothermal muscovite that crystallized in one of the northeast-striking fractures (Coleman and Hodges 1995). Numerous other normal faults are responsible for the development of the graben along with N20°–40° trending faults (Colchen 1999).

The Dangardzong fault is clearly marked by a topographic depression in the western side of the graben and appears as a cluster of faults. The fault separates the Tethyan sequences and graben-fill sediments in the northern part of the graben and cuts through the Higher Himalayan crystallines on the southern part. The progressive decrease in the metamorphic grade from the biotite zone of greenschist facies to the chlorite zone of the footwall along the fault suggests a decrease of footwall exhumation towards the south (Hurtado et al. 2001). The mapping of this fault along the western margin of the graben is described below.

A sharp offset of the Charan Khola near Ghar Ghumba clearly indicates the presence of Dangardzong fault where fault breccia occurs in the riverbed, at the base of a quartzite cliff (Fig. 11a). Similarly, the Ghilumpa Khola sharply bends southwards and flows along the strike of the fault and then turns eastwards to the Kali Gandaki River making an offset of about 100 m (Fig. 11b). The fault places the grey fine-grained schists of Tilicho Pass Formation on its footwall and Createaceous quartzite of Chuck Formation on its hanging wall. The kinematic indicators in the bedrock show a normal sense of movement with some right-lateral strike-slip displacement (Hurtado et al. 2001).

About 1 km NW of Dongardzong, the Lumbuk Khola makes a sharp bend (Fig. 11c), where the fault strikes N26° E and dips 68° due SE. There are black schists and quartzites on its footwall and quartzites on the hanging wall. The fault has not been active since ca. 5.1 Ma (Hurtado et al. 2001). The footwall is brecciated near the Syang village (Fig. 11d), where its minimum age is estimated at 17.2 ka based on terrace chronology. The Dangardzong Fault terminates around 3.5 km northeast of Lete and 17 km SW of Syang, near Titi. The fault was delineated based on surface morphology, SPOT imagery and interpretation of talus and regolith lithology (Hurtado et al. 2001).

The prominent east-west striking Lupra fault was interpreted as a thrust that was later activated as a normal fault during the graben development (Godin 2003). Different syndepositional growth faults are widespread in the graben-fill sediments (Fig. 12). A northeast-southwest striking small-scale normal fault displaces the sub-

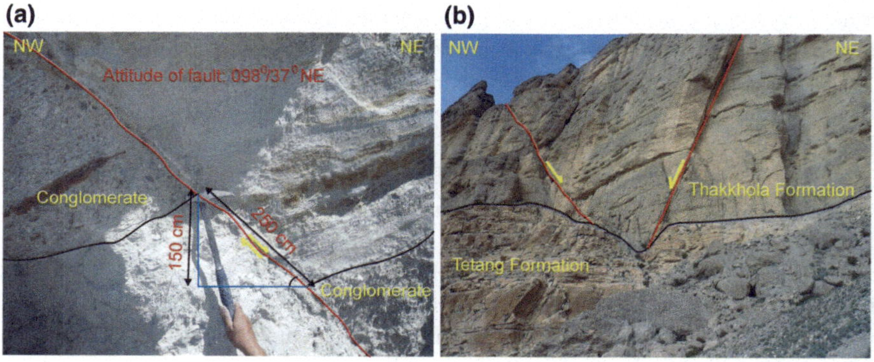

Fig. 12 Normal growth faults in basin-fill sediments. **a** Tetang formation, **b** Thakkhola formation

sequent beds by about 150 cm in the Tetang Formation, indicating the onset of east-west extension (Fig. 12a). Similarly, two east- and west-dipping normal growth faults in the Dhinkyo Khola section have displaced the imbricated conglomerate beds of Thakkhola Formation up to the unconformity between the Tetang and Thakkhola formations (Fig. 12b).

The Palaeozoic to Mesozoic basement rocks of graben are unconformably overlain by the Plio-Pleistocene graben-fill sediments distributed over 90 km in the north-south and 20–30 km east-west directions. They are characterised by more than 840 m of continental debris (Fort et al. 1982, Adhikari and Wagreich 2011b). The oldest sedimentary units of the graben are of the middle Miocene to upper Pliocene Tetang and Thakkhola formations (Fig. 11). Many researches have estimated the age of Tetang Formation at 11–9.6 Ma and the oldest age of Thakkhola Formation at 8 Ma based on magnetostratigraphy (Garzione et al. 2000) and the deposition of the Thakkhola Formation continued until at least 2 Ma (Yoshida et al. 1984). The gently (13–23°) northwest dipping sedimentary strata of Tetang Formation are separated by a low-angle (about 5°) unconformity representing a temporal gap of at least 2.6 Ma, beginning at ca. 9.6 Ma and ending after 7 Ma (Garzione et al. 2000) with the very gently (~13°) northwest dipping Thakkhola Formation (Fort et al. 1982; Adhikari and Wagreich 2011a).

The Tetang Formation is the oldest formation in this graben and it is well exposed around the Tetang village and Dhinkyo Khola (Fig. 11). Its thickness varies from a few metres to more than 200 m and it onlaps against the Cretaceous Chuck Formation. This formation is composed of a fining-upward sequence, consisting of pebbles and gravel of quartzite, shale, sandstone, and carbonates at the bottom and lacustrine sediments containing fine siltstone and limestone on the top. The accommodation space for the sediments may have been created as a response to normal faulting and footwall uplift associated with the STDS (Adhikari and Wagreich 2011a; Garzione et al. 2003). The Thakkhola Formation spreads towards the western and eastern parts of the graben, starting from the Tetang village up to Lomanthang and its distribution was strictly controlled by the Dangardzong fault in the western side. The thickness of the Tetang Formation is more than 620 m in the Chele village. The thickness decreases eastwards where it rests unconformably on the Tetang Formation and the Mesozoic rocks lying farther east (Adhikari 2009). The clast size in both imbricated and massive conglomerates decreases from south to north, but in them the frequency of silt layers increases. The thickness of layers indicates that the velocity and capacity of the river was strong in the central part compared to the northern margin of the basin. These sediments were deposited in alluvial fans, braided river systems, and fluvio-lacustrine to lacustrine environments.

6 Conclusions

The propagation of MCT toward the foreland has brought the Greater Himalayan Crystallines over the Lesser Himalayan Sequence forming a mega duplex, the floor thrust of which is supposed to be the MHT. The competent Greater Himalayan thrust sheet is gently folded to form the Great Midland Antiform towards the hinterland and the Great Mahabharat Synform towards the foreland. The interplay of thrusting, erosion, and folding has created a complex pattern of tectonic windows and klippen in the Nepal Himalaya. Erosion of the thrust sheet by antecedent rivers carved many tectonic windows and klippen. Post collisional Palaeogene-Neogene sediments occur in the outer as well as inner belt and they have played a significant role in duplex formation, neotectonic movement, and seismicity in the Himalaya. There are many intermontane valleys in the Nepal Himalaya that developed after collision of India with Eurasia. The Thakkhola graben is one of them and it was formed during Cenozoic east-west extension of the Himalaya.

Acknowledgements We thank Tribhuvan University and the Department of Mines and Geology, the Government of Nepal, for their support and laboratory facilities.

References

Adhikari BR (2009) Sedimentology and basin analysis of the Thakkhola-Mustang graben, Central Nepal. Unpublished Ph.D. thesis, submitted to Vienna University

Adhikari BR, Wagreich M (2011a) Facies analysis and basin architecture of the Thakkhola-Mustang graben (neogene-quaternary), Central Nepal Himalaya. Austrian J Earth Sci 104:66–80

Adhikari BR, Wagreich M (2011b) Provenance evolution of collapse graben fill in the Himalaya—the miocene to quaternary Thakkhola-Mustang graben (Nepal). Sediment Geol 233:1–14

Adhikari LB, Gautam UP, Koirala BP, Bhattarai M, Kandel T, Gupta RM, Timsina C, Maharjan N, Dahal T, Hoste-Colomer R, Canao Y, Dandine M, Guilherm A, Merrer S, Roudil P, Bollinger L (2015) The aftershock sequence of the 2015 April 25 Gorkha-Nepal earthquake. Geophys J Int 203:2119–2124

Amatya KM, Jnawali BM (1994). Geological map of Nepal, scale: 1:1000,000. Department of Mines and Geology, International Centre for Integrated Mountain Development, Carl Duisberg Gesellschaft e.V., and United Nation as Environment Programme

Argand E (1924) La tectonique de l'Asie. Congrès Géologique International, Belgique, Comptes Rendues de la 13ème Session, en Belgique 1922. H. Vaillant-Carmanne, Liège, pp 171–372

Arita K (1983) Origin of the inverted metamorphism of the lower Himalaya, Central Nepal. Tectonophysics 95:43–60

Auden JB (1934) The geology of the Krol Belt. Rec Geol Surv India LXVII(4):357–454 (with 9 plates, including a geological map in colours, scale: 1 inch = 2 miles)

Avouac J-P (2003) Mountain building, erosion and seismic cycle in the Nepal Himalaya. Adv Geophys 46:1–80

Avouac J-P, Meng L, Wei S, Wang T, Ampuero J-P (2015) Lower edge of locked Main Himalayan thrust unzipped by the 2015 Gorkha earthquake. Nat Geosci 8:708–711

Ballévre M, Manzotti P, Dal Piaz EV (2018) Pre-alpine (variscan) inheritance: a key for the location of the future valaisan basin (Western Alps). Tectonics 37:786–817

Bashyal RP (1982) Geological framework of far Western Nepal. Himal Geol Wadia Inst Himal Geol 12:40–50

Bashyal RP (1986). Geology of lesser Himalaya, far Western Nepal. In: Le Fort P, Colchen M, Montenat C (eds) Évolution des domains orogéniques d'Asie méridionale (de la Turquie à l'Indonésie), Livre jubilaire Pierre Bordet, Sciences de la Terre, Mémoire no. 47, Nancy, pp 31–42

Berthelsen A (1951) A geological section through the Himalaya: a preliminary report. Bull Geol Soc Den 12:102–104

Beyssac O, Bollinger L, Avouac J-P, Goffé B (2004) Thermal metamorphism in the lesser Himalaya of Nepal determined from Raman spectroscopy of carbonaceous material. Earth Planet Sci Lett 225(1–2):233–241

Bhargava ON (2000) The precambrian sequences in the Western Himalaya. Geol Surv India Spec Publ 55:69–84

Bhargava ON, Thöni M, Miller C (2016) Isotopic evidence of early palaeozoic metamorphism in the lesser Himalaya (Jutogh Group), Himachal Pradesh, India: its implication. Himal Geol 37(2):73–84

Bilham R (1995) Location and magnitude of the 1833 Nepal earthquake and its relation to the rupture zones of contiguous great Himalayan earthquakes. Curr Sci 69(2):101–128

Bilham R, Larson K, Freymueller J et al (1997) GPS measurements of present-day convergence across the Nepal Himalaya. Nature 386:61–64

Bilham R, England P (1995) Entertaining a great earthquake in western Nepal: historic inactivity and geodetic tests for the development of strain. J Nepal Geol Soc 11:73–88

Bordet P (1961). Recherches Géologiques dans L'Himalaya du Népal, Région du Makalu, Expéditions Françaises a l'Himalaya 1954–1955. Edition du Centre National de la Recherche Scientifique (C. N. R. S.), p 275 (with two geological maps in colours)

Brown RL, Nazarchuk JH (1993) Annapurna detachment fault in the Greater Himalaya of Central Nepal. Geol Soc Lond Spec Publ 74:461–473

Burchfiel BC, Zhiliang C, Hodges KV, Yuping L, Royden LH, Changrong D (1992) The South Tibetan detachment system, Himalayan Orogen: extension contemporaneous with and parallel to shortening in a collisional mountain belt. Geol Soc Am Spec Pap 269:1–41

Carosi R, Montomoli C, Visoná D (2007) A structural transect in the lower Dolpo: insights on the tectonic evolution of western Nepal. J Asian Earth Sci 29:407–423

Carosi R, Montomoli C, Iaccarino S, Visonà D (2018a) 20 years of geological mapping of the metamorphic core across Central and Eastern Himalayas. Earth Sci Rev 177:124–138

Carosi R, Montomoli C, Iaccarino S, Visonà D (2018b) Structural evolution, metamorphism and melting in the greater Himalayan sequence in central-western Nepal. Geological Society, London, Special Publications no 483. https://doi.org/10.1144/SP483.3

Colchen M (1999) The Thakkhola-Mustang graben in Nepal and the late cenozoic extension in the higher Himalayas. J Asian Earth Sci 17:683–702

Coleman M, Hodges K (1995) Evidence for Tibetan plateau uplift before 14 Myr ago from a new minimum age for east–west extension. Nature 374:49

Dal Piaz GV, Bistachhi A, Massironi M (2003) Geological outline of the Alps. Episodes 26:175–180

DeCelles PG, Gehrels GE, Quade J, Lareau B, Spurlin M (2000) Tectonic implications of U-Pb zircon ages of the Himalayan orogenic belt in Nepal. Science 288:497–499

DeCelles PG, Gehrels GE, Quade J, Kapp PA, Ojha TP, Upreti BN (1998) Neogene foreland basin deposits, erosional unroofing, and the kinematic history of the Himalayan fold-thrust belt, western Nepal. Geol Soc Am Bull 110:2–21

DeCelles PG, Robinson DM, Quade J, Ojha TP, Garzione CN, Copeland P, Upreti BN (2001) Stratigraphy, structure and tectonic evolution of the Himalayan fold–thrust belt in western Nepal. Tectonics 20:487–509

Dhital MR (2015) Geology of the Nepal Himalaya: regional perspective of the classic collided orogen. Springer, Berlin, p 498

Dhital MR, Kizaki K (1987) Structural aspect of the Northern Dang, lesser Himalaya. Bull Coll Sci Univ Ryukyus 45:159–182

Dikshit AM (1991) Geological effects and intensity distribution of the Udayapur (Nepal) earthquake of August 20, 1988. J Nepal Geol Soc 7:1–17

Einfalt HC, Hoehndorf A, Kaphle KP (1993) Rediometric age determination of the Dadeldhura granite, Lesser Himalaya, Far Western Nepal. Schweiz Miner Petrogr Mitt 73:97–106

England P, Molnar P (1990) Surface uplift, uplift of rocks, and exhumation of rocks. Geology 18:1173–1177

Fort M, Freytet P, Colchen M (1982) Structural and sedimentological evolution of the Thakkhola-Mustang Graben (Nepal Himalaya). Z Für Geomorphol 42:75–98

Frank W, Fuchs GR (1970) Geological investigations in West Nepal and their significance for the geology of the Himalayas. Geol Rundsch 59:552–580

Fuchs G (1977) The geology of the Karnali and Dolpo Regions, Western Nepal. Jahrb Geol Bundesanst Wien 120(2):1–103 (with 9 plates)

Fuchs G (1981) Geologic-tectonic map of the Himalaya. Scale: 1:2,000,000. Geol Bundesanst Wien

Fuchs G, Frank W (1970). The geology of West Nepal between the rivers Kali Gandaki and Thulo Bheri. Jahrb Geol Bundesanst 18:1–103 (with a geological map and cross-sections)

Gansser A (1964) Geology of the Himalayas. Interscience, New York, p 289

Garzanti E, Baud A, Mascle G (1987) Sedimentary record of the northward flight of India and its collision with Eurasia (Ladakh Himalaya, India). Geodin Acta 1:297–312

Garzione CN, DeCelles PG, Hodkinson DG, Ojha TP, Upreti BN (2003) East-West extension and miocene environmental change in the southern Tibetan plateau: Thakkhola graben, Central Nepal. Geol Soc Am Bull 115:3–20

Garzione CN, Dettman DL, Quade J, DeCelles PG, Butler RF (2000) High times on the tibetan plateau: paleoelevation of the Thakkhola graben. Nepal Geol 28:339–342

Godin L (2003) Structural evolution of the tethyan sedimentary sequence in the Annapurna area, Central Nepal Himalaya. J Asian Earth Sci 22:307–328

Goscombe B, Gray D, Hand M (2006) Crustal architecture of the Himalayan metamorphic front in eastern Nepal. Gondwana Res 10:232–255

Hagen T (1969) Report on the geological survey of Nepal. Volume 1: preliminary reconnaissance. Denkschiiften der Schweizerischen Naturforschenden Gesellschaft, Band LXXXVI/1, p 185 (with a geological map)

Harrison MT, Lovera OM, Grove M (1997) New insights into the origin of two contrasting Himalayan granite belts. Geology 25:899–902

Hashimoto S, Ohta Y, Akiba C (1973) Geology of the Nepal Himalayas. Saikon Publishing Co. Ltd., Tokyo, p 292 (with 6 plates including a geological map of Nepal (Plates 1 and 2) in colours, scale: 1:500,000)

Heim A (1922) Geologie der Schweiz. Verlag Chr. Herm. Tauchnitz, Leibzig, p 704

Heim A, Gansser A (1939). Central Himalaya: geological observations of the Swiss expedition 1936. Denkschriften der Schweizerischen Naturforschenden Gesellschaft, Band LXXIII, Abh. 1:245 (with geological maps in colors, sections, and plates)

Hurtado JM Jr, Hodges KV, Whipple KX (2001) Neotectonics of the Thakkhola graben and implications for recent activity on the South Tibetan fault system in the Central Nepal Himalaya. Geol Soc Am Bull 113:222–240

Johnson MRW (1994) Volume balance of erosional loss and sediment deposition related to Himalayan uplifts. J Geol Soc Lond 151:217–220

Kapp P, Guynn JH (2004) Indian punch rifts Tibet. Geology 32:993–996

Khan HR (1996) Geology of a part of Bajhang District, Far Western Nepal. Unpublished report (submitted to the Department of Mines and Geology, Kathmandu, Nepal, with a geological map on a scale of 1 inch to 1 mile)

Khan HR (2000) Geology of a part of Bajhang and Bajura Districts, Far Western Nepal. Unpublished report (submitted to the Department of Mines and Geology, Kathmandu, Nepal, with a geological map on a scale of 1 inch to 1 mile)

Le Fort P (1971) Les formations cristallophyliennes de la Thakkhola. In: Bordet P, Colchen M, Krummenacher D, Le Fort P, Mouterde R, Rémi M (eds) Recherches géologiques dans l'Himalaya du Népal, région de la Thakkhola. Centre National de la Reserches Scientifique, Paris, pp 41–48

Le Fort P (1975) Himalaya: the collided range. Am J Sci 275A:1–44

Le Fort P, Debon F, Sonet J (1983) The lower Paleozoic—Lesser Himalayan granitic belt: emphasis on the Simchar pluton of central Nepal. In: Shams FA (ed) Granites of Himalaya, Karakorum, and Hindu-Kush. Institute of Geology, Punjab University, Lahore, pp 235–255

Le Fort P, France-Lanord C (1994) Granites from Mustang and surrounding regions, Central Nepal. J Nepal Geol Soc 10:79–81

Lombard A (1958) Un itinéraire géologique dans l'Est du Népal (Massif du Mont Everest). Mémoires Société Helvétique Sci Natlles 82(1):107

Macedo J, Marshak S (1999) Controls on geometry of fold-thrust belt salients. Geol Soc Am Bull 111:1808–1822

Malusá M, Faccenna C, Garzanti E, Polino R (2011) Divergence in subduction zones and exhumation of high pressure rocks (Eocene Western Alps). Earth Planet Sci Lett 310:21–32

Manzotti P, Zucali M (2013) The pre-alpine tectonic history of the Austroalpine continental basement in the Alpelline unit (Western Italian Alps). Geol Mag 150:153–172

Mehta PK (1977) Rb–Sr geochronology of the kullu-mandi belts: its implications for the Himalaya. Sond Geologic Runds 66:186–195

Middlemiss CS (1910) The Kangra earthquake of 4th April 1905. Memoier Geol Surv India XXXVIII:1–409

Molnar P (2015) Plate tectonics. A Very Short Introduction, Oxford, p 136

Molnar P, Tapponnier P (1978) Active tectonics of Tibet. J Geophys Research Solid Earth 83:5361–5375

Nakata T (1982) A photogrametric study on active faults in the Nepal Himalayas. J Nepal Geol Soc 2:67–80 (Special Issue)

Oldham RD (1899) Report on the great Earthquake of 12th June 1987. Memoier Geol Surv India XXIV:1–379

Pandey MR, Tandukar RP, Avouac J-P, Lavé J, Massot JP (1995) Interseismic strain accumulation on the Himalayan crustal ramp (Nepal). Geophys Res Lett 22:751–754

Pandey MR, Tandukar RP, Avouac J-P, Vergne J, Heritier T (1999) Seismotectonics of Nepal Himalayas from a local seismic network. J Asian Earth Sci 17:703–712

Patel R, Singh S, Asokan A, Manickavasagam R, Jain A (1993) Extensional tectonics in the Himalayan orogen, Zanskar, NW India. Geol Soc, Lond, Spec Publ 74:445–459

Platt J, England P (1994) Convective removal of lithosphere beneath mountain belts-Thermal and mechanical consequences. Am J Sci 294:307–336

Ramsay JG (1967). Folding and fracturing of rocks. McGraw-Hill Book Company, New York, p 568

Robinbson DM, Martin AJ (2014) Reconstructing the greater Indian margin: a balanced cross-section in Central Nepal focusing on the lesser Himalayan duplex. Tectonics 33:2143–2168

Robinson DM, DeCelles PG, Copeland P (2006) Tectonic evolution of the Himalayan thrust belt in Western Nepal: implications for channel flow models. Bull Geol Soc Am 118:865–885

Schelling D, Arita K (1991) Thrust tectonics, crustal shortening, and the structure of the far-eastern Nepal Himalaya. Tectonics 10:851–862

Schmid SM, Pfiffner OA, Froitzheim N, Schönborn G, Kissling E (1996) Geophysical-geological transect and tectonic evolution of the Swiss-Italian Alps. Tectonics 15:1036–1064

Searle MP, Law RD, Godin L, Larson KP, Streule MJ, Cottle JM, Jessup MJ (2008) The Himalayan main central thrust in Nepal. J Geol Soc Lond 165:523–534

Searle M, Windley B, Coward M, Cooper D, Rex A, Rex D, Tingdong L, Xuchang X, Jan M, Thakur V (1987) The closing of tethys and the tectonics of the Himalaya. Geol Soc Am Bull 98:678–701

Seeber L, Armbruster JG (1981) Great detachment Earthquake along the Himalayan arc and long-term forecasting. Earthq Predict: Int Rev Am Geophys Union 4:259–277

Sharma ML, Malik S (2006) Probabilistic seismic hazard analysis and estimation of spectral strong ground motion on bed rock in northeast India. In: Fourth international conference earthquake engineering, Taiwan, Paper No. 015, pp 1–10

Shrestha SB, Shrestha JN, Sharma SR (1987) Geological map of Mid-Western Nepal. Scale: 1:250,000. Department of Mines and Geology, Kathmandu

Stöcklin J (1980) Geology of Nepal and its regional frame. J Geol Soc Lond 137:1–34
Stöcklin J, Bhattarai KD, Singh-Chhetri V, Bhandari AN (1982) Photogeological map of part of Central Nepal. Scale: 1:100,000
Talalov VA (1972) Geology and ores of Nepal. UNDP report, Department of Mines and Geology, Kathmandu, 4 volumes (unpublished)
Upreti BN, Rai SM, Sakai H, Koirala DR, Takigami Y (2003) Early Proterozoic granite of the Taplejung window, far eastern lesser Nepal Himalaya. J Nepal Geol Soc 28:9–18
Upreti BN, Le Fort P (1999) Lesser Himalayan crystalline nappes of Nepal: problems of their origin. Geol Soc Am Spec Pap 328:225–238
Valdiya KS (1980) Geology of Kumaun lesser Himalaya. Wadia Institute of Himalayan Geology, Dehra Dun, p 291 (with a map in colours)
Webb AAG, Yin A, Harrison TM, Célérier J, Burgess P (2007) The leading edge of the greater Himalayan crystalline complex revealed in the NW Indian Himalaya: implications for the evolution of the Himalayan orogeny. Geology 35(10):955–958
West WD (1939) The structure of the Shali 'window' near Simla. Rec Geol Surv India 74(1):133–163 (with 4 plates and a geological map in colours; scale: 1 inch = 2 miles)
Yin A (2000) Mode of Cenozoic east-west extension in Tibet suggesting a common origin of rifts in Asia during the Indo-Asian collision. J Geophys Res: Solid Earth 105:21745–21759
Yin A, Kapp PA, Murphy MA, Manning CE, Mark Harrison T, Grove M, Lin D, Xi-Guang D, Cun-Ming W (1999) Significant late Neogene east-west extension in northern Tibet. Geology 27:787–790
Yoshida M, Igarashi Y, Arita K, Hayashi D, Sharma T (1984) Magnetostratigraphic and pollen analytic studies of the Takmar series, Nepal Himalayas. J Nepal Geol Soc 4:101–120. (Special Issue)

The manufacturer's authorised representative in the EU is Springer Nature Customer Service Centre GmbH, Europaplatz 3, 69115 Heidelberg, Germany. If you have any concerns regarding our products, please contact ProductSafety@springernature.com

Printed and bound by CPI Group (UK) Ltd, Croydon, CR0 4YY
30/09/2026
02225602-0006